INTRODUCTION TO SIMAN V AND CINEMA V

INTRODUCTION TO SIMAN V AND CINEMA V

JERRY BANKS
Georgia Institute of Technology

BARRY B. BURNETTE

HENRY KOZLOSKI

JAMES D. ROSE

JOHN WILEY & SONS, INC.
New York Chichester Brisbane Toronto Singapore

Acquisitions Editor	Charity Robey
Marketing Manager	Susan Elbe
Senior Production Editor	Nancy Prinz
Senior Designer	Kevin Murphy
Manufacturing Manager	Susan Stetzer
Cover Illustration	Steven Lyons

This book was set in 10/12 ITC Century Book by Publication Services and printed and bound by Courier Stoughton. The cover was printed by Phoenix Color Corp.

Library of Congress Cataloging–in–Publication Data

Introduction to SIMAN V and CINEMA V / Jerry Banks
... [et al.].
p. cm.
Includes index.
ISBN 0-471-30960-5 (paper)
1. Digital computer simulation. 2. SIMAN V (Computer program language) I. Banks, Jerry.
QA76.9.C65I59 1994
003′.35133–dc20 94-31556
CIP

Printed in the United States of America

10 9 8 7 6 5 4 3 2 1

To Rose and Bob Ward, may they have many more fun days together; and to Nancy Banks for her patience and understanding.

Jerry Banks

To Mr. and Mrs. H. Boyd Burnette, and to the rest of my family and friends for their support and encouragement in working on this text and throughout my academic career.

Barry B. Burnette

To Jo Spivey, who taught me the value of words and how to turn a phrase; and to Lorraine Allen for her support and patience during this project.

Jimmy D. Rose

To Henry and Teresa Kozloski, for providing a world full of hope and opportunity to me.

Henry Kozloski

Foreword

Simulation modeling is widely recognized as one of the most powerful techniques available for studying large and complex systems. It has become an essential tool for analyzing a diverse range of systems. It has been used to analyze both man-made process flow systems, such as manufacturing and service systems, as well as natural systems, such as ecological and biological systems.

Since its introduction in the early 1980s, SIMAN has gained widespread popularity as a simulation modeling tool for studying a wide range of complex systems. It is used extensively in industry, government, and universities throughout the world. Although its primary area of application has been manufacturing, it has also been used to study service systems, communication systems, transportation systems, health delivery systems, and so on.

The SIMAN language has changed significantly over the past decade. It has gone through several major revisions during this time that have greatly extended the functionality of the language and also simplified its use. The most recent version is SIMAN V, and it is a major upgrade of the language. This book is the first on the market to discuss fully modeling using the SIMAN V language.

This book has been reviewed by me, Customer Services, Deb Sadowski (vice president, Software Development), and Dave Sturrock (vice president, Applications). In my view, Banks et al. have succeeded in providing students with an up-to-date, in-depth coverage of all the important aspects of modeling with SIMAN V. They have presented the concepts in a clear, intuitively appealing way that addresses both the needs of the newcomer to simulation as well as the experienced modeler. They have included numerous examples throughout the book that illustrate important concepts in the language.

This book focuses on the practical aspects of building models with SIMAN V. It covers all of the essential features required to fully master the language, beginning with basic concepts and continuing through material handling and other advanced modeling features. In addition to SIMAN modeling concepts, the book also provides comprehensive coverage of the output processor and interactive run controller as well as Cinema V animation.

This book is not intended to fully address the science or theory related to simulation. This book is very much a "how to model with SIMAN" book, and in my view, it is an excellent companion book for use with a general simulation modeling textbook such as Banks and Carson or Law and Kelton. It is also an excellent standalone reference for people experienced with simulation to learn modeling with SIMAN V.

I am honored to have been asked by Professor Banks to write this foreword. Professor Banks has been involved with simulation for many years, and he is a highly respected scholar and writer in the field. He is experienced with using and teaching many different modeling tools and has a broad perspective on both the art and science of simulation modeling. I feel very fortunate that he has devoted his time and talent to writing this SIMAN textbook.

C. Dennis Pegden, President
Systems Modeling Corporation
September 1, 1994

Preface

This text provides an introduction to SIMAN V, a leading-edge simulation language, and Cinema, the animator for the language. The text is intended to accompany a one-quarter or one-semester course in simulation, a short course in SIMAN and Cinema, or as a self-study guide.

The text contains many examples. Sufficient explanation is provided to understand these examples and to complete the exercises that appear at the end of each chapter.

The text was written primarily for the person new to simulation and new to SIMAN. It was written for the person who wants to start simulating with SIMAN as quickly as possible. The basics of the language are emphasized to provide this quick start. The text will also be helpful to those who learned an earlier version of SIMAN, and want to learn about some of the powerful extensions that have been added to the latest version of the language. Cinema is described within ARENA, a graphical environment that uses CAD-like drawing features.

We recommend that the persons new to simulation study Chapter 1 to obtain a quick overview of simulation, and Chapters 2 through 8 to obtain a basic understanding of SIMAN, and that they work at least one exercise at the end of each chapter. The example given in Chapter 2 is solved using both SIMAN V at the command level and within the ARENA environment. After this short introduction to the ARENA environment, its next appearance is in Chapter 14 where animation in Cinema is introduced.

For those persons who need to understand material handling constructs, Chapters 9 and possibly 10 or 11, as required by the situation, should be added to the set. For those who are going to be responsible for complex modeling, Chapters 12 and 13 will give an introduction to the topic.

There are four appendices for quick reference purposes. These appendices contain a summary of the blocks, elements, attributes, variables, mathematical functions, and distributions.

Two diskettes accompany this text. They contain SIMAN V, the input and output processors. Their installation is explained in the README.DOC file. The diskettes also contain the examples presented in the text. Note that Section 2.5 and Chapter 14 require additional components of the ARENA environment. The README.DOC file also explains how these additional components may be obtained from Systems Modeling Corporation.

For those who want to learn about the basics of simulation, in addition to the programming aspects, we recommend a general simulation text, such as *Discrete-Event System Simulation*, by Banks and Carson (Prentice-Hall, Inc., Englewood Cliffs, N.J., 1984), or for a more advanced treatment of the subject, we recommend *Simulation Modeling and Analysis* by Law and Kelton (McGraw-Hill, New York, 1991).

This text is an introduction to the process-interaction approach of SIMAN and Cinema. It is not intended to be a complete reference to the software. For more depth on the topics, and for information on the event-scheduling and continuous approaches, we recommend the *SIMAN V Reference Guide,* the *Variables Guide*, the *ARENA User's Guide*, and the *ARENA Template Reference Guide*, all distributed by Systems Modeling Corporation.

Thanks are due to many people from Systems Modeling Corporation that have made this text possible. We appreciate the help of Randy Sadowski and Dennis Pegden who were receptive to the idea, and who provided encouragement and continued feedback. Our ongoing contact at Systems Modeling was Dave Sturrock, and he was extremely helpful. Deb Sadowski reviewed, commented, and provided numerous helpful suggestions on several versions of the text, and we greatly appreciate her insights. We also appreciate the insight and suggestions of S. Manivannan, then of Georgia Tech.

Jerry Banks
Barry Burnette
Henry Kozloski
James Rose
September 1, 1994

Contents

1

The Simulation Process

This book provides an introduction to SIMAN V, a powerful simulation language, and Cinema V, the animator for SIMAN V. Prior to embarking on the discovery of SIMAN V and Cinema V, it is useful to introduce some basic concepts of simulation. This chapter contains that introduction.

1.1 Simulation

Simulation is the imitation of the operation of a real-world process or system over time. Simulation involves the generation of an artificial history of the system, and the observation of that artificial history to draw inferences concerning the operating characteristics of the real system that is represented.

Simulation is an indispensable problem-solving methodology for the solution of many real-world problems that is used to: describe and analyze the behavior of a system, ask "what if" questions about the real system, and aid in the design of real systems. Both existing and conceptual systems are modeled with simulation.

EXAMPLE 1.1 Ad Hoc Simulation

Consider the operation of a tool crib where mechanics arrive for service between 1 and 10 minutes apart in time, integer values only, each value equally likely. The mechanics are served between 1 and 6 minutes, also integer valued and equally likely. Restricting the times to integer values is an abstraction of reality, since time is continuous, but this aids in presenting the example. The objective is to simulate the tool crib attendant, by hand, for 20 mechanics, and to compute measures of performance such as the percentage of idle time, the average waiting time per mechanic, and so on. Admittedly, 20 mechanics are far too few to draw conclusions about the operation of the system, but by following this example, the stage is set for further discussion in this chapter, and subsequent discussion about using the computer for performing simulation.

Since simulation is the emulation of reality, random interarrival and service times need to be generated. Assume that the interarrival times are generated using a spinner that has possibilities for the values 1 through 10. Further assume that the service times are generated using a die that has possibilities for the values 1 through 6.

Table 1.1 is called an *ad hoc simulation* table, i.e., this table's format is for the purpose of this problem, but does not pertain to all problems. Column 1 lists the 20 mechanics that arrive to the system. It is assumed that Mechanic 1 arrives at time zero, thus a dash is indicated in Row 1 of Column 2. Rows 2 through 20 of Column 2 were generated using the spinner. Column 3 shows the simulated arrival times. Since Mechanic 1 is assumed to arrive at time 0, and there is a 5 minute interarrival time, Mechanic 2 arrives at time 5. There is a 1 minute interarrival time for Mechanic 3, thus, the arrival occurs

Table 1.1 Ad Hoc Simulation

(1) Mechanic	(2) Time Between Arrivals	(3) Arrival Time	(4) Service Time	(5) Service Begins	(6) Time Service Ends	(7) Time in System	(8) Idle Time	(9) Time in Queue
1	-	0	2	0	2	2	0	0
2	5	5	2	5	7	2	3	0
3	1	6	6	7	13	7	0	1
4	10	16	5	16	21	5	3	0
5	6	22	6	22	28	6	1	0
6	2	24	4	28	32	8	0	4
7	9	33	3	33	36	3	1	0
8	1	34	4	36	40	6	0	2
9	10	44	1	44	45	1	4	0
10	3	47	3	47	50	3	2	0
11	5	52	1	52	53	1	2	0
12	2	54	2	54	56	2	1	0
13	3	57	3	57	60	3	1	0
14	5	62	6	62	68	6	2	0
15	10	72	2	72	74	2	4	0
16	5	77	2	77	79	2	3	0
17	3	80	4	80	84	4	1	0
18	4	84	5	84	89	5	0	0
19	7	91	3	91	94	3	2	0
20	7	98	1	98	99	1	4	0
						72	34	7

at time 6. This process of adding the interarrival time to the previous arrival time is called *bootstrapping*. By continuing this process, the arrival times of all 20 mechanics are determined. Column 4 contains the simulated service times for all 20 mechanics. These were generated by rolling the die.

Now, the simulation of the service process begins. At time 0, Mechanic 1 arrived and immediately began service. The service time was 2 minutes, so the service period ended at time 2. The total time in the system for Mechanic 1 was 2 minutes. The tool crib attendant was not idle, since the simulation began with the arrival of a mechanic. The mechanic did not have to wait for the tool crib attendant.

At time 5, Mechanic 2 arrived and immediately began service. The service time was 2 minutes so the service period ended at time 7. The tool crib attendant was idle from time 2 until time 5, so 3 minutes of idle time occurred. Mechanic 2 spent no time in the queue.

Mechanic 3 arrived at time 6, but service could not begin until time 7, as Mechanic 2 was being served until time 7. The service time was 6 minutes, so service was completed at time 13. Mechanic 3 was in the system from time 6 until time 13, or for 7 minutes as indicated in Column 7. Although there was no idle time, Mechanic 3 had to wait in the queue for 1 minute for service to begin.

This process continues for all 20 mechanics, and the totals shown in Columns 7, 8, and 9 are entered. Some performance measures can now be calculated as follows:

Average time in system = $72/20 = 3.6$ minutes

% idle time = $[34/99](100) = 34\%$

Average waiting time per mechanic = $7/20 = 0.35$ minutes

Fraction having to wait = $3/20 = 0.15$

Average waiting time of those that waited = $7/3 = 2.33$ minutes

This very limited simulation indicates that the system is functioning well. Only 15% of the mechanics had to wait. The tool crib attendant is idle about 1/3 of the time. Whether a slower tool crib attendant is to replace the current tool crib attendant depends on the cost of having to wait versus the savings from having a slower server.

This small simulation can be accomplished by hand, but there is a limit to the complexity of problems that can be solved in this manner. Also, the number of mechanics that must be simulated is much larger than 20, and the number of times that the simulation must be run for statistical purposes could be large. Hence, using the computer to solve real simulation problems is almost always appropriate.

Example 1.1 raises some issues that will be addressed in this chapter. The issues include the following:

1. How is the form of the input data determined?
2. What if the input data follows some other statistical distribution?
3. How does the user know that the simulation imitates reality?
4. What other kinds of problems can be solved by simulation?
5. How long does the simulation need to run?
6. How many different simulation runs should be conducted?
7. What statistical techniques should be used to analyze the outputs?

Each question raises a host of issues about which many textbooks and thousands of technical papers have been written. Whereas an introductory chapter cannot treat all of these questions in the greatest of detail, enough can be said to give the reader some insight that will be useful in understanding the framework of the remainder of the text.

1.2 Modeling Concepts

Several concepts underlie simulation; these include system and model, system state variables, entities and attributes, list processing, activities and delays, and finally the definition of discrete-event simulation. Additional information on these topics is available from Banks and Carson (1984) and Law and Kelton (1991). The discussion in this section follows Carson (1993).

1.2.1 System, Model, and Events

A model is a representation of an actual system. Immediately, there is a concern about the limits or boundaries of the model that supposedly represents the system. The model should be complex enough to answer the questions raised, but not too complex.

Consider an event as an occurrence that changes the state of the system. In Example 1.1, events include the arrival of a mechanic for service at the tool crib, the beginning of service for a mechanic, and the completion of a service.

There are both internal and external events, also called *endogenous* and *exogenous* events, respectively. For example, an endogenous event in Example 1.1 is the beginning of service of the mechanic, since that is within the system being simulated. An exogenous event is the arrival of a mechanic for service, since that occurrence is outside of the simulation. However, the arrival of a mechanic for service impinges on the system and must be taken into consideration.

This book considers *discrete-event* simulation models. These are contrasted with other types of models such as mathematical models, descriptive models, statistical models, and input–output models. A discrete-event model attempts to represent the components of a system and their interactions to such an extent that the objectives of the study are met. Most mathematical, statistical, and input–output models represent a system's inputs and outputs explicitly, but represent the internals of the model with mathematical or statistical relationships. An example is the mathematical model from physics,

$$\text{Force} = \text{Mass} \times \text{Acceleration}$$

based on theory. Discrete-event simulation models include a detailed representation of the actual internals.

Discrete-event models are dynamic, that is, the passage of time plays a crucial role. Most mathematical and statistical models are static in that they represent a system at a fixed point in time. Consider a spreadsheet model, a form of a statistical model, that represents the annual budget of a firm. Changes can be made in the budget and the spreadsheet can be recalculated, but the passage of time is usually not a critical issue. Further comments will be made about discrete-event models after several additional concepts are presented.

1.2.2 System State Variables

The *system state variables* are the collection of all information needed to define what is happening within the system to a sufficient level at a given point. The determination of system state variables is a function of the purposes of the investigation, so what may be the system state variables in one case may not be the same in another case even though the physical system is the same. Determining the system state variables is as much an art as a science. However, during the modeling process, any omissions will readily come to light.

Having defined system state variables, a contrast can be made between discrete-event models and *continuous* models based on the variables needed to track system state. The system state variables in a discrete-event model remain constant over intervals of time and change value only at certain well-defined points called *event times*. Continuous models have system state variables defined by differential or difference equations giving rise to variables that change continuously over time.

There are models that are both mixed discrete-event and continuous. There are also continuous models that are treated as discrete-event models after some reinterpretation of system state variables, and vice versa. SIMAN V has the capability to model continuous systems, but that capability is not treated in this text.

1.2.3 Entities and Attributes

An *entity* represents an object that requires explicit definition. An entity can be *dynamic* in that it "moves" through the system, or it can be *static* in that it serves other entities. In Example 1.1, the mechanic is a dynamic entity, whereas the tool crib attendant is a static entity.

An entity may have *attributes* that pertain to that entity alone. Thus, attributes should be considered as local values. However, many entities can have the same attributes. In Example 1.1, an attribute of the entity could be the time of arrival. Attributes of interest in one investigation may not be of interest in another investigation. Thus, if red parts and blue parts are being manufactured, the color could be an attribute. However, if the time in the system for all parts is of concern, the attribute of color may not be important.

1.2.4 Resources

A resource is a static entity that provides service to dynamic entities. The resource can serve one or more than one dynamic entity at the same time, that is, operate as a parallel server. A dynamic entity can request one or more units of a resource. If denied, the requesting entity joins a *queue,* or takes some other action (i.e., diverted to another resource, ejected from the system). (Other terms for queues include files, chains, buffers, and waiting lines.) If permitted to capture the resource, the entity remains for a time, then releases the resource.

There are many possible states of the resource. Minimally, these states are idle and busy, but other possibilities exist, such as failed, blocked, or starved, to mention a few.

1.2.5 List Processing

Entities are managed by allocating them to resources that provide service, by attaching them to event notices thereby suspending their activity into the future, or by placing them into an ordered list. Lists are used to represent queues. (Queues contain entities waiting for scarce resources or for system conditions to be met.)

Lists are usually processed according to FIFO (first-in-first-out), but there are many other possibilities. For example, the list could be processed by LIFO (last-in-first-out), according to the value of an attribute, or randomly, to mention a few. An example where the value of an attribute may be important is in SPT (shortest process time) scheduling. In this case, the processing time may be stored as an attribute of each entity, and the entities are in line according to the value of that attribute with the lowest value at the front of the line.

1.2.6 Activities and Delays

An *activity* is a duration of time whose span is known prior to commencement of the activity. Thus, when the duration begins, its end can be scheduled. The duration can be a constant, a random value from a statistical distribution, the result of an equation, it can come from an input file, or it can be computed based on the event state. For example, a service time may be a constant 10 minutes for each entity; it may be a random value from an exponential distribution with a mean of 10 minutes; it could be 0.9 times a constant value from time 0 to time 4 hours, and 1.1 times the standard value after time 4 hours; or it could be 10 minutes when the preceding queue contains less than or equal to four entities and 8 minutes when there are five or more in the preceding queue.

A *delay* is an indefinite duration that is caused by some combination of system conditions. When an entity joins a queue for a resource, the time that it will remain in the queue may be unknown, since that time may depend on other events that may occur. An example of another event would be the arrival of a rush order that preempts the resource. When the preempt occurs, the entity using the resource relinquishes its control instantaneously. Another example is a failure necessitating repair of the resource.

Discrete-event simulations contain activities that cause time to advance. Most discrete-event simulations also contain delays as entities wait. The beginning and ending of an activity or delay is an *event*.

1.2.7 Discrete-Event Simulation Model

Sufficient modeling concepts have been defined so that a *discrete-event simulation* model can be described as one in which the state variables change only at those points in time at which events occur. Events occur as a consequence of activity times and delays. Entities may compete for system resources, possibly joining queues while waiting for an available resource. Activity and delay times may "hold" entities for durations of time.

A discrete-event simulation model is conducted over time ("run") by a mechanism that moves simulated time forward. The system state is updated at each event along with capturing and freeing of resources that may or may not occur at that time.

1.3 World Views

There are four major views taken by the simulation community. They are: (1) the *process-interaction* method, (2) the *event-scheduling* method, (3) *activity scanning,* and (4) the *three-phase* method. The descriptions are rather concise; readers requiring greater explanation are referred to Balci (1988) or Pidd (1992).

1.3.1 The Process-Interaction Method

The simulation structure that has the greatest intuitive appeal is the *process-interaction* method. The basic idea behind the process-interaction approach is that the computer program should emulate the flow of an object through the system. The entity moves as far as possible in the system until it is delayed, enters an activity, or it exits from the system. When the entity's movement is halted, the clock advances to the time of the next movement of an entity.

This flow, or movement, describes in sequence all of the states that the object can attain in the system. For example, in a model of a self-service laundry a customer may enter the system, wait for a washing machine to become available, wash his or her clothes in the washing machine, wait for a basket to become available to then unload the washing machine, transport the clothes in the basket to a drier, wait for a drier to become available, unload the clothes into a drier, dry the clothes, and then leave the laundry. SIMAN V supports the process-interaction method.

1.3.2 The Event-Scheduling Method

The basic idea behind the *event-scheduling* method is to advance time to when something next happens. This usually releases a resource(s), that is, a scarce entity(ies), such as a machine(s) or transporter(s). The event then reallocates available objects or entities by scheduling activities where they can now participate. For example, in the self-service laundry, if a customer's washing is finished and there is a basket available, the basket could be allocated immediately to the customer and unloading of the washing machine could commence.

Time is advanced to the next scheduled event (usually an end of activity) and activities are examined to see if any can now start as a consequence. SIMAN V supports the event-scheduling method.

1.3.3 Activity Scanning

Activity scanning, also known as the two-phase approach, is similar to rule-based programming. (If a specified condition is met, then a rule is fired, meaning that an action is taken.) Activity scanning produces a simulation program composed of independent modules waiting to be executed. Scanning takes place at fixed time increments at which a determination is made concerning whether or not an event occurs at that time. If an event occurs, the system state is updated.

1.3.4 The Three-Phase Method

In the three-phase method, time is advanced until there is a state change in the system or until something next happens. The system is examined to determine all of the events that take place at this time, that is, all the activity completions that occur. Only when all resources that are due to be released at this time have been released is the reallocation of these resources into new activities started in the third phase of the simulation. In summary, the first phase is time advance; the second phase is the release of those resources scheduled to end their activities at this time; and the third phase is to start activities given the global picture about resource availability.

1.4 Advantages and Disadvantages of Simulation

Simulation is readily explained to a manager or customer. In addition, simulated behavior is comparable to what is happening in the real system or what is perceived for a system in design. Also, simulation is the only technique that can relate the various components of a complex system enabling analysis to be conducted. Lastly, simulation requires few simplifying assumptions. For these, and other reasons, simulation is frequently the technique of choice in problem solving.

In contrast to optimization models, simulation models are "run" rather than solved. In almost every instance, simulation is used for analysis purposes. Given a particular set of input and/or model characteristics, the model is run and the simulated behavior is observed. This process of changing inputs and/or model characteristics results in a set of scenarios that are evaluated. A good solution or the best among the simulated solutions is recommended for implementation.

Simulation has many advantages, and even some disadvantages, which are listed by Pegden, Shannon, and Sadowski (1990). The advantages include the following:

1. New policies, operating procedures, decision rules, information flows, organizational procedures, and so on can be explored without disrupting ongoing operations of the real system.
2. New hardware designs, physical layouts, transportation systems, and so on can be tested without committing resources for their acquisition.
3. Hypotheses about how or why certain phenomena occur can be tested for feasibility.
4. Time can be compressed or expanded allowing for an increase or a decrease in the speed of the phenomena under investigation. Thus, the simulation of the operation of a large computer for one nanosecond could take ten minutes, but the simulation of a factory for an 8-hour shift may only take five minutes.
5. Insight can be obtained about the interaction of variables.
6. Insight can be obtained about the importance of variables on the system's performance.
7. Bottleneck analysis can be performed indicating where work in process, information, materials, and so on are being excessively delayed.
8. A simulation study can help in understanding how the system operates rather than how individuals think the system operates.
9. "What if" questions can be answered. This is particularly useful in the design of new systems.

The disadvantages include the following:

1. Model building requires special training. It is an art that is learned over time and through experience. Furthermore, if two models are constructed by two competent individuals, they may have similarities, but it is highly unlikely that they will be the same.
2. Simulation results may be difficult to interpret. Since most simulation outputs are essentially random variables (they are usually based on random inputs), it may be hard to determine whether an observation is a result of system interrelationships or randomness.

3. Simulation modeling and analysis can be time consuming and expensive. Skimping on resources for modeling and analysis may result in a simulation model and/or analysis that is not sufficient to the task.
4. Simulation is used in some cases when an analytical solution is possible, or even preferable. This is particularly true in the simulation of some waiting lines where closed form queueing models are available.

In defense of simulation, these four disadvantages, respectively, can be offset as follows:

1. Vendors of simulation software have been actively developing packages that contain models that only need input data for their operation. Such models have the generic tag "simulators." Systems Modeling Corporation calls them "templates."
2. Some simulation software vendors have developed output analysis capabilities within their packages for performing very eloquent analysis. Systems Modeling Corporation conducts this type of analysis within the Output Processor, discussed in Chapter 6 of this text.
3. Simulation can be performed faster today than yesterday, and even faster tomorrow. This is attributable to the advances in hardware that permit rapid running of scenarios. It is also attributable to the advances in many simulation packages. For example, Systems Modeling Corporation software contains many constructs for modeling material handling using transporters such as fork lift trucks, conveyors, and automated guided vehicles.
4. Closed form models are not able to analyze most of the complex systems that are encountered in practice. In nearly 8 years of consulting practice of one of the authors, not one problem has been encountered that could have been solved by a closed form solution.

1.5 Areas of Application

The applications of simulation are vast. Recent presentations at the Winter Simulation Conference (WSC) can be divided into manufacturing, public systems, and service systems. WSC is an excellent way to learn more about the latest in simulation applications and theory. There are also numerous tutorials at both the beginning and advanced levels. WSC is sponsored by eight technical societies and the National

Institute of Standards and Technology (NIST). The technical societies are American Statistical Association (ASA), Association for Computing Machinery/Special Interest Group on Simulation (ACM/SIGSIM), Institute of Electrical and Electronics Engineers: Computer Society (IEEE/CS), Institute of Electrical and Electronics Engineers: Systems, Man and Cybernetics Society (IEEE/SMCS), Institute of Industrial Engineers (IIE), Operations Research Society of America (ORSA), The Institute of Management Sciences: College on Simulation (TIMS/CS), and The Society for Computer Simulation (SCS). The societies can provide information about upcoming WSCs, usually held Monday through Wednesday of the second full week of December.

1.5.1 Manufacturing Applications

Presentations included the following among many others:

Scheduling flexible manufacturing cells
Evaluating wafer fabrication cluster tools
Material handling systems
Power and free conveyor in an automotive paint shop
Design of a pharmaceutical manufacturing facility
Decision support in a continuous-flow manufacturing system
Smart card-based manufacturing
Evaluating tool delivery in a flexible manufacturing system

1.5.2 Public Systems Applications

Presentations included the following among many others:

Health systems
- Screening for abdominal aortic aneurysms
- Lymphocyte development in immune-compromised patients
- Asthma dynamics and medical amelioration
- Timing of liver transplants
- Diabetic retinopathy
- Evaluating nurse staffing and patient population scenarios
- Evaluation of automated equipment for a clinical processing laboratory
- Hospital surgical suite and critical care area

Military systems
- Air Force resource allocation
- Analysis of material handling equipment for prepositioning ships

Force allocation
Airland combat modeling
Theater airlift system productivity
C-141 depot maintenance
Air mobility command channel cargo system

Natural resources
Nonpoint source pollution
Weed scouting and weed control decision making
Surface water quality data

Public services
Emergency ambulance systems
Flow of civil lawsuits
Field offices within a government agency

1.5.3 Service System Applications

Presentations included the following among many others:

Transportation
Intelligent vehicle highway systems
Traffic control procedures at highway work zones
Taxi management and route control
Animation of a toll plaza
Port traffic planning model
Rapid transit modeling with automatic and manual controls

Computer systems performance
User transaction processing behavior
Database transaction management protocols
Evaluation of analytic models of memory queueing

Chemical process industries
Process plant design
Decision support for a specialty chemicals production plant

Air transportation
Human behavior in aircraft evacuations
Analysis of airport/airline operations
Combination carrier air cargo hub

Communications systems
Trunked radio network
Telephone service provisioning process
Picture archiving and communications systems
Modeling of broadband telecommunication networks
Virtual reality for telecommunication networks

1.6 Steps in a Simulation Study

Figure 1.1 shows a set of steps to guide a model builder in a thorough and sound simulation study. Similar figures and their interpretation can be found in other sources, such as Pegden, Shannon, and Sadowski

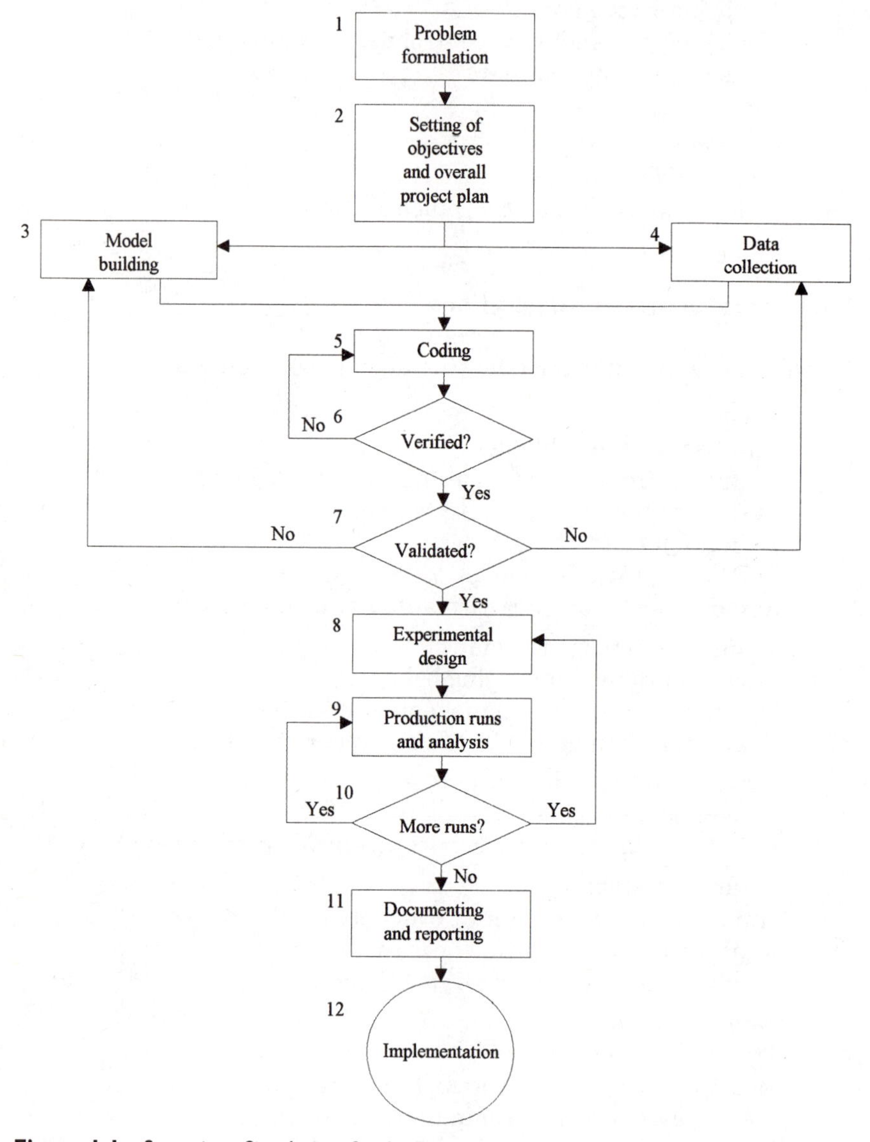

Figure 1.1 Steps in a Simulation Study. Reprinted with permission from J. Banks and J.S. Carson, *Discrete-Event System Simulation,* Prentice-Hall, Englewood, N.J., 1984, p. 12.

(1990) and Law and Kelton (1991). This presentation is built on that of Banks and Carson (1984).

1. **Problem formulation.** Every simulation study begins with a statement of the problem. If the statement is provided by those that have the problem (client), the simulation analyst must take extreme care to ensure that the problem is clearly understood. If a problem statement is prepared by the simulation analyst, it is important that the client understand and agree with the formulation. It is suggested that a set of assumptions be prepared by the simulation analyst and agreed to by the client. Even with all of these precautions, it is possible that the problem will need to be reformulated as the simulation study progresses.
2. **Setting of objectives and overall project plan.** Another way to state this step is "prepare a proposal." This step should be accomplished regardless of location of the analyst and client, viz., as an external or internal consultant. The objectives indicate the questions that are to be answered by the simulation study. The project plan should include a statement of the various scenarios that will be investigated. The plans for the study should be indicated in terms of time that will be required, personnel that will be used, hardware and software requirements if the client wants to run the model and conduct the analysis, stages in the investigation, output at each stage, and cost of the study and billing procedures, if any.
3. **Model building.** The real-world system under investigation is abstracted by a conceptual model, a series of mathematical and logical relationships concerning the components and the structure of the system. It is recommended that modeling begin simply and that the model grow until the complex model has been developed. For example, the basic model with the arrivals, queues, and servers is constructed. Then, add the failures and shift schedules. Next, add the material handling capabilities. Finally, add the special features. It is not necessary to construct an unduly complex model, which will add to the cost of the study and the time for its completion without increasing the quality of the output.

   The client should be involved throughout the model construction process. This will enhance the quality of the resulting model and increase the client's confidence in its use.

   There are many opportunities for the reader to construct models throughout this text. However, maturity in model building will come from experience with real problems.

4. **Data collection.** Shortly after the proposal is "accepted," a schedule of data requirements should be submitted to the client. In the best of circumstances, the client has been collecting the kind of data needed in the format required and can submit that data to the simulation analyst in electronic format. Oftentimes, the client indicates that the required data is indeed available. However, when the data is delivered, it is found to be quite different than anticipated. For example, in the simulation of an airline reservation system, the simulation analyst was told "we have every bit of data that you want over the last 5 years." When the study commenced, the data delivered was the average "talk time" of the reservationist for each of the years. Individual values, not summary measures, were needed.

   Model building and data collection are shown as contemporaneous in Figure 1.1. This indicates that the simulation analyst can readily construct the model, while the data collection is progressing.

5. **Coding.** The conceptual model constructed in Step 3 is coded into a computer recognizable form, an operational model. (In SIMAN V, the operational model consists of a model frame containing the system logic and an experiment frame containing data and other information for running the simulation as discussed in Chapter 2.)

6. **Verified?** Verification concerns the operational model. Is it performing properly? Even with small textbook-sized models, it is quite possible that they have verification difficulties. These models are one to two orders of magnitude smaller than real models. It is highly advisable that verification take place as a continuing process. It is ill advised for the simulation analyst to wait until the entire model is complete to begin the verification process. Also, use of the Interactive Run Controller (IRC), discussed in Chapter 5, is highly encouraged as an aid to the verification process. The IRC, or debugger, is an executable routine that assists the user in finding and correcting errors. Verification is extremely important and is discussed later in this chapter.

7. **Validated?** Validation is the determination that the conceptual model is an accurate representation of the real system. Can the model be substituted for the real system for the purposes of experimentation? If there is an existing system, call it the base system, then an ideal way to validate the model is to compare its output to that of the base system.

Unfortunately, there is not always a base system (such as in the design of a new system). There are many methods for performing validation, and some of these are discussed later in this chapter.

8. **Experimental design.** For each scenario that is to be simulated, decisions need to be made concerning the length of the simulation run, the number of runs (also called replications), and the manner of initialization, as required. This topic is discussed later in this chapter and again in concert with the Output Processor in Chapter 6.
9. **Production runs and analysis.** Production runs, and their subsequent analysis, are used to estimate measures of performance for the scenarios that are being simulated. This topic is discussed later in this chapter and additionally in concert with the Output Processor in Chapter 6.
10. **More runs?** Based on the analysis of runs that have been completed, the simulation analyst determines if additional runs are needed and if any additional scenarios need to be simulated.
11. **Documentation and reporting.** Documentation is necessary for numerous reasons. If the simulation model is going to be used again at a later date by the same or different analysts, it may be necessary to understand how the simulation model operates. This will enable confidence in the simulation model so that the client can make decisions based on the analysis. Also, if the model is to be modified by the same or a different simulation analyst, this can be greatly facilitated by adequate documentation. One experience with an inadequately documented model is usually enough to convince a simulation analyst of the necessity of this important step.

    The result of all the analysis should be reported clearly and concisely. This will enable the client to review the final formulation, the alternatives that were addressed, the criterion by which the alternative systems were compared, the results of the experiments, and analyst recommendations, if any.
12. **Implementation.** The simulation analyst acts as a reporter rather than an advocate. The report prepared in Step 11 stands on its merits and is just additional information that the client uses to make a decision. If the client is involved throughout the study period, and the simulation analyst follows all of the steps rigorously, then the likelihood of a successful implementation is increased.

1.7 Random Number and Random Variate Generation

Example 1.1 used input values that were generated by a spinner and a die. Almost all simulation models are constructed within the computer, so spinners and dies are not devices that will be used. Instead, the computer will generate independent random numbers that are distributed uniformly between 0 and 1, that is, U(0, 1). These random numbers can then be converted to the desired statistical distribution, or random variate, using one of several methods. Random variates are used to simulate interarrival times, batch sizes, processing times, repair times, and time until failurc, among others. Many researchers have written on the two topics in this section.

SIMAN V has a built-in random number generator (rng) that produces a sequence of acceptable random numbers. The generator is based on the linear congruential method that was documented by Knuth (1969). The rng in SIMAN V has been extensively tested, and no evidence has been found to reject the presumed uniformity and independence of its resulting random numbers.

The numbers generated by SIMAN's rng are actually pseudorandom. They are deterministic, since they can be reproduced. Knowing the starting value, the values that follow it can be predicted, totally determining the sequence. There is no reason for concern, since the length of the sequence prior to repeating itself is very, very long. On a 32-bit computer, this sequence can be longer than 2 billion. There are not many simulations that will require such a long sequence.

The importance of a good source of random numbers is that all procedures for generating nonuniformly distributed random variates involve a mathematical transformation of a uniform random number. For example, suppose R_i is the ith random number generated from a U(0, 1). Suppose further that the desired random variate is exponentially distributed with parameter λ. These values can be generated from

$$X_i = -\frac{1}{\lambda}\ln(1 - R_i) \qquad (1.1)$$

where X_i is the ith random variate generated (e.g., the time between the arrival of the ith and the ith+1 entities). Suppose $\lambda = 1/10$ arrivals per minute. Using Equation 1.1, (called the random variate generator, or rvg), if $R_1 = 0.3067$, then $X_1 = 3.66$ minutes. The rvg was developed using the inverse-transform technique. Other techniques include convolution, acceptance-rejection, and composition. These techniques are discussed by Law and Kelton (1991).

SIMAN V has built-in rvgs for the most widely used distributions and several that are not so widely utilized. SIMAN V also provides a facility for generating a sample from an empirical distribution (a distribution of the raw input data) that is either discrete or continuous. It is important that the simulation analyst know how to use rvgs, but it is not usually important to be concerned with their generation.

1.8 Input Data

For each element in a system being modeled, the simulation analyst must decide upon a way to represent the associated random variables. The presentation of the subject that follows is based on Banks, Carson, and Goldsman (1990).

The techniques used may vary depending on:

1. The amount of available data.
2. Whether the data are "hard" or someone's best guess.
3. Whether each variable is independent of other input random variables, or related in some way to other outputs.

In the case of a variable that is independent of other variables, the choices are as follows:

1. Assume that the variable is deterministic.
2. Fit a probability distribution to the data.
3. Use the empirical distribution of the data.

These three choices are discussed in the next three subsections.

1.8.1 Assuming Randomness Away

Some simulation analysts may be tempted to assume that a variable is deterministic, or constant. This value could have been obtained by averaging historic information. The value may even be a guess. If there is randomness in the model, this technique can surely invalidate the results.

Suppose that a machine manufactures parts in exactly 1.5 minutes. The machine requires a tool change according to an exponential distribution with a mean of 12 minutes between occurrences. The tool change time is also exponentially distributed with a mean of 3 minutes. An inappropriate simplification would be to assume that the machine operates in a constant time of 1.875 minutes and ignore the randomness. The consequences of these two interpretations are very large on such measures as the average number or time waiting before the machine. In Chapter 7, the demonstration of the effects are left as an exercise for the reader.

1.8.2 Fitting a Distribution to Data

If there are sufficient data points, say 100 or more, it may be appropriate to fit a probability distribution to the data using conventional methods. (Advanced methods for distribution fitting, such as that described by Wagner and Wilson (1993) are available to the interested reader.) When there is a small amount of data, the tests for goodness-of-fit offer little guidance in selecting one distribution form over another.

There are also underlying processes that give rise to distributions in a rather predictable manner. For example, if arrivals occur one at a time, are completely at random without rush or slack periods, and are completely independent of one another, a Poisson process occurs. In such a case it can be shown that the number of arrivals in a given time period follows a Poisson distribution and the time between arrivals follows an exponential distribution.

Systems Modeling Corporation provides built-in capability to perform input data analysis within its ARENA software. If a goodness-of-fit test is being conducted without the aid of input data analysis software, the following three-step procedure is recommended:

1. **Hypothesize a candidate distribution.** First, ascertain whether the underlying process is discrete or continuous. Discrete data arise from counting processes. Examples include the number of mechanics that arrive at a tool crib each hour, the number of tool changes in an 8-hour day, and so on. Continuous data arise from measurement (time, distance, weight, etc.). Examples include the time to produce each part and the time to failure of a machine, and so on.

   Discrete distributions frequently used in simulation include the Poisson, binomial, and geometric. Continuous distributions frequently used in simulation include the uniform, exponential, normal, triangular, lognormal, gamma, and Weibull. These distribution are described in virtually every engineering statistics text.

2. **Estimate the parameters of the hypothesized distribution.** For example, if the hypothesis is that the underlying data are normal, then the parameters to be estimated from the data are the sample mean and the sample variance.

3. **Perform a goodness-of-fit test, such as the *chi-squared* test.** The test measures the sum of squared differences between the expected and observed values, each of which squared term is divided by the expected value. The sum is compared to the appropriate tabulated value. If the test rejects the hypothesis, that is a strong indication that the hypothesis is not true. In

that case, return to Step 1, or use the empirical distribution of the data following the process described in Section 1.8.3.

The three-step procedure is described in virtually every engineering statistics text and in many simulation texts, such as Banks and Carson (1984) and Law and Kelton (1991). Even if software is being used to aid in the development of an underlying distribution, understanding the three-step procedure is recommended.

1.8.3 Empirical Distribution of the Data

When only a small amount of data are available, an attempt to fit a distribution is inappropriate as indicated previously. Also, when all possibilities have been exhausted for fitting a distribution using conventional techniques, then the empirical distribution can be used. The empirical distribution uses the data as generated.

An example will help to clarify the discussion. The times to repair a conveyor system after a failure, denoted by x, for the previous 100 occurrences are given as follows:

Interval(hours)	Frequency of Occurrence
$0 < x \leq 1.0$	27
$1.0 < x \leq 2.0$	13
$2.0 < x \leq 3.0$	31
$3.0 < x \leq 4.0$	18
$4.0 < x \leq 8.0$	11

No distribution could be fit to the data using conventional techniques. It was decided to use the data as generated for the simulation. That is, samples were drawn, at random, from the continuous distribution shown above. This required interpolation so that simulated values might be in the form 2.89 hours, 1.63 hours, and so on.

1.8.4 When No Data Are Available

There are many cases where no data are available. This is particularly true in the early stages of a study, when the data are missing, when the data are too expensive to gather, or when the system being modeled is not in existence.

One possibility in such a case is to obtain a subjective estimate, sometimes called a guesstimate, concerning the system. Thus, if the estimate that the time to repair a machine is between 3 and 8 minutes, a crude assumption is that the data follows a uniform distribution with a minimum value of 3 minutes and a maximum value of 8 minutes. The uniform distribution is referred to as the "distribution of maximum ignorance," since it assumes that every value is equally likely, and that is quite unlikely. A better "guess" occurs if the most likely value can also be estimated. This would take the form "the time to repair the machine is between 3 and 8 minutes with a most likely time of 5 minutes." Now, a triangular distribution can be used with a minimum of 3 minutes, a maximum of 8 minutes, and a most likely value (mode) of 5 minutes.

As indicated previously, there are processes that give rise to distributions. For example, if the time to failure follows the assumptions of the Poisson process indicated previously, and the machine operator says that the machine fails about once every 2 hours of operation, then an exponential distribution for time to failure could be assumed initially with a mean of 2 hours.

Estimates made on the basis of guesses and assumptions are strictly tentative. If, and when, data, or more data, become available, both the parameters and the distributional forms should be updated.

1.9 Verification and Validation

In the application of simulation, the real-world system under investigation is abstracted by a conceptual model. The conceptual model is then coded into the operational model (SIMAN V model and experiment frames, both recognized by the computer). Hopefully, the operational model is an accurate representation of the real-world system. However, more than hope is required to insure that the representation is accurate. There is a checking process that consists of two components:

1. **Verification.** A determination of whether the computer implementation of the conceptual model is correct. Does the operational model represent the conceptual model?
2. **Validation.** A determination of whether the conceptual model can be substituted for the real system for the purposes of experimentation.

The checking process is iterative. If there are discrepancies among the operational and conceptual models and the real-world system, then the relevant operational model must be examined for errors, or the conceptual model must be modified in order to better represent the real-world system (with subsequent changes in the operational model). The verification and validation process should then be repeated.

1.9.1 Verification

The verification process involves examination of the simulation program (SIMAN V model frame) to insure that the operational model accurately reflects the conceptual model. There are many common sense ways to perform verification.

1. **Follow the principles of structured programming.** The first principle is *top-down design,* that is, construct a detailed plan of the model frame before coding. The second principle is *program modularity,* that is, break the model frame into submodels. Write the model frame in a logical, well-ordered manner. It is highly advisable (we would say mandatory, if we could mandate such) to prepare a detailed flow chart indicating the macro activities that are to be accomplished. This is particularly true for real-world sized problems. It is quite possible to think through all of the computer code that is needed to solve the problems at chapter ends of this text. However, the computer code required is tiny compared to that of real-world problems.
2. **Make the operational model (both the SIMAN V model frame and experiment frame) as self-documenting as possible.** This requires comments on virtually every line and sometimes between lines. Every section of the model frame should be defined with the reason for its existence. All data in the experiment frame should be defined. Imagine that one of your colleagues is trying to understand the computer code that you have written, but that you are not available to offer any explanation.
3. **Have more than one person check the model frame.** There are several techniques that have been used for this purpose. One of these can be called *code inspection.* There are four parties as follows: (1) the moderator or leader of the inspection team, (2) the designer or individual that prepared the conceptual model, (3) the coder or person that prepared the operational model, and (4) the tester or the person given the verification responsibility. An inspection meeting is held at which time a narration of the design is provided and the operational model is discussed, line-by-line, along with the documentation. Detected errors are documented and classified. There is then a rework phase followed by another inspection. Alternatives to code inspection include the review except that the interest is not line-by-line, but in design deficiencies. Another alternative is the audit that verifies that the development of the

computer code is proceeding logically. It verifies that the stated requirements are being met.

4. **Check to see that the value of the input data are being used appropriately.** For example, if the time unit is minutes, then all of the data should be in terms of minutes, not hours or seconds.
5. **For a variety of input data values, insure that the outputs are reasonable.** Many simulation analysts are satisfied when they receive output. But that is not far enough. If there are 100 entities in a waiting line, when 10 would be rather high, there is probably something wrong. For example, the resource actually has a capacity of two, but was modeled with a capacity of one.
6. **Use the interactive run controller (IRC) to check that the program operates as intended.** Chapter 5 introduces the IRC. It is possible to skip this chapter with no loss in modeling complexity. Do not do that! The IRC is a very important verification tool that should be used for all real-system models. An example of one of the capabilities of the IRC is the trace that permits following the execution of the model frame step-by-step.
7. **Animation is a very useful verification tool.** Using animation, the simulation analyst can detect actions that are illogical. For example, it may be observed that a resource is supposed to fail as indicated by turning red on the screen. While watching the animation, the resource never turned red. This could signal a logical error. Animation is achieved within System Modeling Corporation's ARENA environment using Cinema V, as discussed in Chapter 14.

1.9.2 Validation

A variety of subjective and objective techniques can be used to validate the conceptual model. Sargent (1992) offers many suggestions for validation. Subjective techniques include the following:

1. **Face validation.** A conceptual model of a real-world system must appear reasonable "on its face" to those that are knowledgeable (the "experts") about the real-world system. For example, the experts can validate that the model assumptions are correct. Such a critique by experts would aid in identifying deficiencies or errors in the conceptual model. The credibility of the conceptual model would be enhanced as these deficiencies or errors are eliminated.

2. **Sensitivity analysis.** As model input is changed, the output should change in a predictable direction. For example, if the arrival rate increases, the time in queues should increase, subject to some exceptions. (An example of an exception is as follows: If a queue increases, it may be the case that resources are added within the model, negating the prediction.)

3. **Extreme condition tests.** Does the model behave properly when input data is at the extremes? If the arrival rate is set extremely high, does the output reflect this change with increased numbers in the queues, increased time in the system, and so on?

4. **Validation of conceptual model assumptions.** There are two types of conceptual model assumptions: *structural* assumptions (concerning the operation of the real-world system) and *data* assumptions. Structural assumptions can be validated by observing the real-world system and by discussing the system with the appropriate personnel. No one person knows everything about the entire system. Many people need to be consulted to validate conceptual model assumptions.

   Information from intermediaries should be questioned. A simulation consulting firm often works through other consulting firms. An extremely large model of a distant port operation (more than 10,000 miles away) was constructed. It was only after a visit by the simulation consulting firm to the port that it was discovered that one of the major model assumptions concerning how piles of iron ore are formed was in error.

   Assumptions about data should also be validated. Suppose that it is assumed that time between arrivals of mechanics to a tool crib during peak periods is independent and in accordance with an exponential distribution. In order to validate conceptual model assumptions, the following would be in order:

   a. Consult with appropriate personnel to determine when peak periods occur.
   b. Collect interarrival data from these periods.
   c. Conduct statistical tests to insure that the assumption of independence is reasonable.
   d. Estimate the parameter of the assumed exponential distribution.
   e. Conduct a goodness-of-fit test to insure that the exponential distribution is reasonable.

5. **Consistency checks.** Continue to examine the operational model over time. An example explains this validation procedure. A simulation model is used annually. Before using this

model, make sure that there are no changes in the real system that must be reflected in the structural model (the SIMAN V model frame). Similarly, the data should also be validated (in the SIMAN experiment frame). For example, a faster machine may have been installed in the interim period, but may not have been included in the information provided.

6. **Turing tests.** Persons knowledgeable about system behavior can be used to compare model output to system output. For example, suppose that five reports of actual system performance over five different days are prepared and five simulated outputs are generated. These 10 reports should be in the same format. The 10 reports are randomly shuffled and given to a person, say an engineer, that has seen this type of information. The engineer is asked to distinguish between the two kinds of reports, actual and simulated. If the engineer identifies a substantial number of simulated reports, then the model is inadequate. If the engineer cannot distinguish between the two, then there is less reason to doubt the adequacy of the model.

   Objective techniques include the following:

7. **Validating input–output transformations.** This technique's basic principle is the comparison of output from the operational model to data from the real system. Input–output validation requires that the real system currently exist. One method of comparison uses the familiar t-test, discussed in most engineering statistics texts. This procedure is shown by example.

EXAMPLE 1.2 t-test

Suppose that the average time mechanics spend in the queue before a tool crib attendant is the performance measure in a simulation study. The system is observed over a week, and the average is 3.0 minutes. The simulation is run five times, 1 day per run, with values 1.63, 2.22, 3.12, 1.09, and 2.11 minutes. The hypothesis to be tested is

$$H_0: \; E[X_i] = 3.0 \text{ minutes}$$

versus

$$H_1: \; E[X_i] \neq 3.0 \text{ minutes}$$

where X_i is the random variable corresponding to the average time in the queue during the ith simulation run. Define

$\mu_0 = 3.0 (= E[X_i]$ under H_0). The mean, variance, and standard deviations are given by the following equations:

$$\bar{X} = \sum_{i=1}^{n} \frac{X_i}{n} \tag{1.2}$$

$$S^2 = \frac{\sum_{i=1}^{n} X_i^2 - n\bar{X}^2}{n - 1} \tag{1.3}$$

$$S = \sqrt{S^2} \tag{1.4}$$

Using Equation 1.2, the mean of the five values is

$$\bar{X} = 2.07$$

Using Equation 1.3, the variance of the five values is

$$S^2 = 0.56$$

Finally, using Equation 1.4, the standard deviation of the five values is

$$S = 0.75$$

A t-test can be used to conduct the hypothesis test. The test statistic is

$$t_0 = \frac{\bar{X} - \mu_0}{S/\sqrt{n}} \tag{1.5}$$

Using Equation 1.5, t_0, the computed value of t, is given by

$$t_0 = \frac{2.03 - 3.0}{0.75/\sqrt{5}} = -2.89$$

Assume that $\alpha = 0.05$. This is a two-tailed test, as the performance measure could be too high or too low, thus α is divided by 2. The tabulated value of t with 4 degrees of freedom at $\alpha/2 = 0.025$, or $t_{n-1,1-\alpha/2}$, is given in statistics texts as

$$t_{4,0.975} = 2.78$$

Hence, reject the null hypothesis, H_0, at level 0.05, indicating that the operational model does not produce realistic times in the queue. Hence, changes in the conceptual or operational model are needed, with a revalidation.

8. **Validation using historical input data.** Instead of running the operational model with artificial input data, we could drive

the operational model with the actual historical record. It is reasonable to expect the simulation to yield output results very close to those observed from the real-world system. One method of comparison uses the familiar paired t-test, discussed in most engineering statistics texts. This procedure is shown by example.

EXAMPLE 1.3 Paired t-test

Suppose that we have collected interarrival and service time data from the tool crib given in Example 1.1 over a 5-day period. Let W_j denote the observed time in the queue for these 5 days, where $j = 1, 2, \ldots, 5$. For fixed j, we can drive the operational model with the actual interarrival and service times to obtain the simulated average times in the queue Y_j. We measure $D_j = W_j - Y_j$. If we are willing to assume that the D_j's are approximately normal random variables and that they are independent and identically distributed, then we proceed with the hypothesis that

$$H_0: \; E[D_j] = 0$$

versus

$$H_1: \; E[D_j] \neq 0$$

Define $\mu_0 = 0$ ($= E[D_j]$ under H_0). Assume that the W_j, Y_j and D_j values are given as shown in Table 1.2. Using variations on Equations 1.2, 1.3, and 1.4, it can be shown that the mean of the D_j values and their standard deviation are given by $\bar{D} = 0.044$, and $S = 0.57$, respectively.

The calculated value of t, or t_0, is determined by

$$t_0 = \frac{\bar{D} - 0}{S/\sqrt{n}} \tag{1.6}$$

Table 1.2 Data for Example 1.3

	Day				
Value	**1**	**2**	**3**	**4**	**5**
W_j	2.53	3.16	1.98	2.42	2.77
Y_j	2.89	2.78	1.45	3.18	2.34
D_j	−.36	.38	.53	−.76	.43

where n is the number of replications. Using Equation 1.6, the value of t_0 is given by

$$t_0 = \frac{0.044 - 0}{0.57/\sqrt{5}} = 0.17$$

Assume that $\alpha = 0.05$. Again, this is a two-tailed test, as the difference can be greater than or less than zero. As in Example 1.2, $t_{4,0.975} = 2.78$. Hence, we fail to reject H_0 at level 0.05.

1.10 Experimentation and Output Analysis

The analysis of simulation output begins with the selection of performance measures. Performance measures can be time persistent, based on counting of occurrences, or arise from the tabulation of expressions including means, variances, and so on. Examples include time in system, time in queue, number in queue, resource utilization, and throughput rate.

The simulation of a stochastic system results in performance measures that contain random variation. Proper analysis of the output is required to obtain sound statistical results from these replications. Specific questions that must be addressed when conducting output analysis are:

1. What is the appropriate run length of the simulation?
2. How do we interpret the simulated results?
3. How do we analyze the differences between replications?

The Output Processor in Chapter 6 can be used to answer these questions by:

1. Displaying data graphically.
2. Analyzing data statistically.
3. Interacting with data files.

In this chapter, some of the topics that are important to experimentation and output analysis are introduced.

1.10.1 Statistical Confidence and Run Length

A confidence interval for the performance measure being estimated by the simulation model is a basic component of output analysis. A confidence interval is a numerical range that has a probability of $1 - \alpha$ of including the true value of the performance measure, where $1 - \alpha$ is the confidence level for the interval. For example, let us say that the performance measure of interest is the average time in the queue, ϕ, and a $100(1 - \alpha)$ percent confidence interval for ϕ is desired. If many replications are performed and independent confidence intervals on ϕ are constructed from those replications, then $100(1 - \alpha)$ percent of those intervals will contain the true value of ϕ. Consider the following example:

EXAMPLE 1.4 Confidence Intervals

Given the data in Table 1.3. Both a 95% ($\alpha = 0.05$) and a 99% ($\alpha = 0.01$) confidence interval are desired. Assuming that the values for X are normally distributed, the half-width, h, will give a $1 - \alpha$ confidence interval for the true mean, μ, that is centered around X. The half-width, h, is computed as follows:

$$h = t_{n-1,1-\alpha/2} \frac{S}{\sqrt{n}} \tag{1.7}$$

where $t_{n-1,1-\alpha/2}$ is the tabulated value of t.

Since α reflects a confidence level of the entire confidence interval, in order to obtain a two-sided confidence interval, we use $\alpha/2$ to compute the half-width. Using Equations 1.2, 1.3, and 1.4, $\bar{X} = 67.4$ and $S = 3.57$. In addition,

$$t_{4,0.975} = 2.78 \text{ (95\% confidence)}$$
$$t_{4,0.995} = 4.60 \text{ (99\% confidence)}$$

resulting in:

$$h = 4.44 \text{ (95\% confidence)}$$
$$h = 7.34 \text{ (99\% confidence)}$$

Table 1.3 Data for Example 1.4

Replication Number	Average Time in the Queue
1	63.2
2	69.7
3	67.3
4	64.8
5	72.0

The confidence interval is given by

$$(\bar{X} - h, \bar{X} + h)$$

Therefore, the 95% confidence interval is (62.96, 71.84), whereas the 99% confidence interval is (60.06, 74.74).

As demonstrated in Example 1.4, the size of the interval depends on the confidence level desired. The higher level of confidence (99%) requires a larger interval compared with the lower confidence level (95%). In addition, the number of replications, n, and their standard deviation, S, are used in calculating the confidence interval. In simulation, each replication is considered one data point. The run length of a simulation corresponds to the number of replications conducted. Therefore, the three factors that influence the width of the confidence interval are:

1. Number of replications (n).
2. Level of confidence ($1 - \alpha$).
3. Variation within the performance measure (S^2).

The relationship between these factors and the confidence interval is:

1. As the number of replications increases, the width of the confidence interval decreases.
2. As the level of confidence increases, the width of the interval increases. In other words, a 99% confidence interval is larger than a 95% confidence interval.
3. As the variation increases, the width of the interval increases.

Since the variation within the model is determined by system parameters, we can only influence the number of replications and the level of confidence. Based upon the above statements, the relationship between the number of replications and the level of confidence can be described as follows:

1. If the confidence level is fixed, a larger number of replications will result in a smaller confidence interval.
2. If the size of the confidence interval is fixed, a larger number of replications will result in a higher degree of confidence.
3. If the number of replications is fixed, a higher level of confidence will produce a larger confidence interval.

1.10.2 Terminating Versus Nonterminating Systems

The procedure for output analysis differs based on whether the system is *terminating* or *nonterminating*. In a terminating system, the duration of the simulation is fixed. The duration can be fixed by specifying a finite length of time to simulate or by limiting the number of entities created or disposed. An example of a terminating system is a bank that opens at 9:00 A.M. and closes at 4:00 P.M. Some other examples of terminating systems include a check processing facility that operates from 8:00 P.M. until all checks are processed, a ticket booth that remains open until all the tickets are sold or the event begins, and a manufacturing facility that processes a fixed number of jobs each day and then shuts down.

By definition, a terminating system is one that has a fixed starting condition and an event definition that marks the end of the simulation. The system returns to the fixed initial condition, usually "empty and idle," before the system begins operation again. The objective of the simulation of terminating systems is to understand system behavior for a "typical" fixed duration. Since the initial starting conditions and the length of the simulation are fixed, the only controllable factor is the number of replications.

Therefore, the analysis procedure for terminating systems is to simulate a number of replications, compute the variance of a selected performance measure, and determine if the resulting confidence interval is within acceptable limits. For example, if the estimate of the performance measure is the average number of parts in the queue, the first step is to conduct a pilot run of n replications. Next, compute the confidence interval for the average number of parts in the queue using the observations recorded from each replication. Then, if the confidence interval is too large, determine the number of additional replications required to bring it within limits. Finally, conduct the additional replications and recompute the new confidence interval using all of the data. Iterate the last two steps until the confidence interval is of satisfactory size. This procedure can be completed using the Output Processor and is described by example in Chapter 6.

In a nonterminating system, the duration is not finite; the system is in perpetual operation. An example of a nonterminating system is an assembly line that operates 24 hours a day, 7 days a week. Another example of this type of system is the manufacture of glass fiber insulation for attics. If operation of the system is stopped, the molten glass will solidify in the furnace, having to be tediously chipped away before restarting the system. The objective in simulating a nonterminating system is to understand the steady-state behavior. In steady state, the effects of starting empty and idle are no longer apparent. To accurately study steady-state behavior, the effects of the initial conditions, or transient phase, must be removed from the simulation results. This can be accomplished by *swamping, preloading,* or *deletion.*

The first method, swamping, suppresses the initial condition effects by conducting a very long simulation run, so long that any initial conditions have only a minuscule effect on the long-run value of the performance measure. For example, if the initial conditions last for 100 hours, simulate for 10,000 hours. Two problems with the swamping technique are that computation time is often a scarce resource and the bias from starting empty and idle will always exist, even if it is small.

The second method, preloading, primes the system before the simulation starts by placing entities in processes and queues. In other words, make the initial conditions match steady-state conditions. This requires knowledge of how the system looks in steady state. Thus, if we are simulating a bank that has one line forming before three tellers, we need to observe the bank in operation in order to obtain information about the usual situation. For example, we may find that the three tellers are usually busy, and that there are about four people in line. This is how the simulation would begin when using the preloading technique. The bank is a very simple system to observe. However, for more complex systems, this initialization procedure becomes somewhat difficult, especially if the system is still in the design phase.

The third method, deletion, excludes the initial transient phase, which is influenced by the initial conditions. Data is collected from the simulation only after the transient (warm-up) phase has ended. This idea is demonstrated in Figure 1.2.

The difficulty with the deletion method is the determination of the length of the transient phase. Although elegant statistical techniques have been developed, a satisfactory method is to plot the performance measure of interest over time and visually observe when steady state is reached. Welch (1983) provides a formal description of this method. In this text, we will use the deletion method.

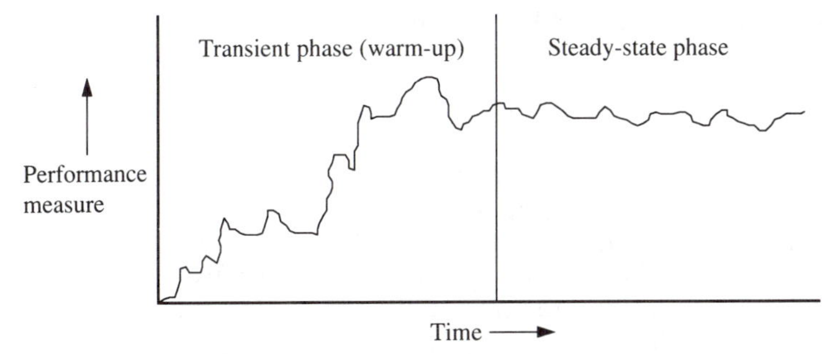

Figure 1.2 Deletion of Initial Conditions for a Nonterminating System.

The analysis procedure for nonterminating systems is first to conduct a few pilot replications. The length of the transient phase can be determined from these initial replications. The next step is to conduct *n* pilot replications, excluding the transient phase. Then, compute the confidence interval for the performance measure using the observations from the second set of pilot replications. If the confidence interval is too large, determine the number of additional replications required to bring it within limits. Finally, conduct the additional replications, excluding the transient phase, and recompute the new confidence interval using all the data. Iterate the last two steps until the confidence interval is of satisfactory size. This procedure can be completed using the Output Processor and is described by example in Chapter 6.

1.11 Summary

This chapter began with a definition of simulation, including an example. Underlying concepts were presented including system and model, system state variables, entities and attributes, list processing, activities and delays, and finally, the definition of discrete-event simulation. Next, four world views were discussed, including process interaction, event scheduling, activity scanning, and the three-phase method. The advantages and disadvantages of simulation were presented, with amelioration of the disadvantages. Next, areas of application from presentations at the Winter Simulation Conference were shown. The steps in a simulation study were given with a brief discussion of each. The manner in which random numbers and random variates are generated was presented next. Three ways that might be used for generating input data were described. However, the first method, assuming randomness away, is discouraged. The extremely important concepts of verification and validation were then discussed. The discussion of validation included two statistical procedures. The all important topic of experimentation and output analysis was introduced; it is indicated that this last topic is discussed further in Chapter 6.

1.12 References

Balci, O. (1988), "The Implementation of Four Conceptual Frameworks for Simulation Modeling in High-level Languages," in *Proceedings of the 1988 Winter Simulation Conference,* eds., M.A. Abrams, P.L. Haigh, and J.C. Comfort, Institute of Electrical and Electronics Engineers, Piscataway, N.J., pp. 287–95.

Banks, J. and J.S. Carson (1984), *Discrete-Event System Simulation,* Prentice-Hall, Englewood Cliffs, N.J.

Banks, J., J.S. Carson, and D. Goldsman (1990), "Computer Simulation," in *Handbook of Statistical Methods for Engineers and Scientists,* ed. H.M. Wadsworth, McGraw-Hill, New York, pp. 12.1–12.36.

Carson, J.S. (1993), "Modeling and Simulation World Views," in *Proceedings of the 1993 Winter Simulation Conference,* eds., G.W. Evans, M. Mollaghasemi, E.C. Russell, and W.E. Biles, Institute of Electrical and Electronics Engineers, Piscataway, N.J., pp. 18–23.

Knuth, D.W. (1969), *The Art of Computer Programming,* Vol. 2: *Semi-Numerical Algorithms,* Addison-Wesley, Reading, Mass.

Law, A.M. and W.D. Kelton (1991), *Simulation Modeling and Analysis,* 2nd ed., McGraw-Hill, New York.

Pegden, C.D., R.E. Shannon, and R.P. Sadowski (1990), *Introduction to Simulation Using SIMAN,* McGraw-Hill, New York.

Pidd, M. (1992), *Computer Modelling for Discrete Simulation,* John Wiley and Sons, Chichester, England.

Sargent, R.G. (1992), "Validation and Verification of Simulation Models," in *Proceedings of the 1992 Winter Simulation Conference,* eds., J.J. Swain, D. Goldsman, R.C. Crain, and J.R. Wilson, Institute of Electrical and Electronics Engineers, Piscataway, N.J., pp. 104–14.

Wagner, M.A.F., and J.R. Wilson (1993), "Using Univariate Bézier Distributions to Model Simulation Input Processes," in *Proceedings of the 1993 Winter Simulation Conference,* eds., G.W. Evans, M. Mollaghasemi, E.C. Russell, and W.E. Biles, Institute of Electrical and Electronics Engineers, Piscataway, N.J., pp. 365–373.

Welch, P.D. (1983), "The Statistical Analysis of Simulation Results," in *The Computer Performance Modeling Handbook,* ed. S. Lavenberg, Academic Press, Orlando, Fla.

1.13 EXERCISES

E1.1 Continue Example 1.1 for 20 more customers and compute the new performance measures.

E1.2 In Example 1.1, assume that the interarrival times are between 1 and 4 minutes apart, again all integer values and equally likely. Also, assume that there are two tool crib attendants, Alice and Bob, working independently. Mechanics prefer Alice, but if she is not available, they go to Bob. Simulate the tool crib for 30 customers and compute the new performance measures, differentiating the performance measures for Alice and Bob.

E1.3 In Exercise 1.2, Alice and Bob have decided to take turns. Perform the simulation and compute the performance measures.

E1.4 In Exercise 1.2, Alice works faster than Bob. Alice's time is between 1 and 5 minutes, whereas Bob's time is between 1 and 7 minutes. Perform the simulation and compute the performance measures.

E1.5 Suppose that the average time in the system is the performance measure of interest. This measure is observed over a week and found to be 32.83 minutes. A simulation has been developed, and the average times in the system for each of 5 days are as follows: 35.64, 28.51, 34.86, 32.81, and 33.98 minutes. Use the t-test to test the hypothesis that the mean of the simulated values is 32.83 minutes at the $\alpha = 0.05$ level of significance.

E1.6 Suppose that the average time in the system is the performance measure of interest. This measure is observed for 5 days with the following results: 30.57, 34.12, 29.65, 35.18, and 34.63 minutes. A simulation has been developed, and the average times in the system for each of 5 days are as follows: 35.64, 28.51, 34.86, 32.81, and 33.98 minutes. Use the paired t-test at the $\alpha = 0.05$ level of significance to determine if there is a difference between the actual and simulated values.

E1.7 Suppose that the average time in the system is the performance measure of interest. This measure is observed for 5 days with the following results: 30.57, 34.12, 29.65, 35.18, and 34.63 minutes. Determine the 95% confidence level for the performance measure.

2

Introduction and Overview

SIMAN stands for SImulation Modeling and ANalysis. The capabilities of the language include process interaction, event scheduling, and continuous simulation, or any mix of these approaches. This text concentrates on the first of those approaches. This chapter introduces the SIMAN modeling framework. The reader will also learn about the SIMAN model and experiment, how to enter these in computer code, and how to conduct the simulation using the computer. Instruction in modeling with SIMAN begins in earnest in Chapter 3.

2.1 The Model and Experiment Frames

Consider a part that goes through a drilling operation. The part is the entity that moves through the system. First the part or entity is CREATEd. Next, it moves into a QUEUE where it may or may not have to wait. Then, it SEIZEs the drill. A DELAY occurs while the drilling operation occurs. The drill is then RELEASEd and the part is DISPOSEd. The words in capital letters indicate blocks in SIMAN V. (Note: Blocks represent an action or event that can affect the moving or dynamic entity and other entities, both dynamic and static, as well.) The sequence of blocks for this entity is as follows:

CREATE
QUEUE
SEIZE
DELAY
RELEASE
DISPOSE

Entering these blocks and some additional information results in the building of the model frame.

In this model, the entities will be DISPOSEd when they complete the drilling process. If the entity is disposed, this is indicated as a modifier to the RELEASE block or as a separate block with the name DISPOSE. In some models, entities remain in the system from the time that they are created until the simulation ends.

Many entities can be in the system at the same time. A QUEUE block may have a limit, but can hold multiple entities. The DELAY block in the example may have a capacity for one or more entities.

The experiment frame provides the conditions for running the model. In the situation just described, the experimental conditions might include the distribution of time between arrivals and its parameter(s), the distribution of time and its parameter(s) for the drilling operation, the number of drills, the statistics and counters that we would like to generate, and the length of the simulation run. Most of the output in SIMAN is controlled by actions taken by the simulation analyst.

2.2 SIMAN Attributes and Variables

Attributes represent values associated with individual entities. Thus, they are local. Examples include job type, time job entered the system, process time for a job on each machine, and so on.

Global variables, called *variables,* represent values that describe the state of the system. These values are available to all entities. Examples include the current number of jobs in the system, the total number of jobs that have been in the system, the quantity of a particular resource that is currently available, the total value added by all entities that have been disposed, and so on.

General-purpose attributes and variables have predefined symbolic names given by A(1), A(2),... for attributes or V(1), V(2),... for variables. Alternately, the symbolic names may be user defined. Examples include *JobType, JobCount,...*. (Note: Italics are used to indicate user-supplied names or information in the model and experiment.) If *JobType* is the first attribute identified in the experiment frame, it can be represented also as A(1). Similarly, if *JobCount* is the fourth variable used, it can be represented also as V(4). General-purpose attributes and variables have meaning that is determined by the user.

Special-purpose attributes and variables are symbolic names that are reserved and have a predefined meaning. Examples are NQ and NR for variables. These represent the number of entities in a queue and the number of entities in a resource, respectively. Operationally, NQ(*DrillQ*) and NR(*Drill*) could refer to the number of entities in a queue before the resource *Drill* and the number of entities using the resource *Drill*. Special rules apply to the use of these attributes and variables.

2.3 Block Format

The general format for a block is as follows:

Label NAME Operands: MODIFIERS; Comment

The entries in the block are free format, and blank spaces are permitted. Delimiters, other than the ";" are used within the block as described in this and following chapters. In preparing block statements, liberal use of lines and comments help readability and documentation.

The Label, if used, is a unique alphanumeric block identifier. In model frames that direct entities based on probabilities or conditions, labels are necessary. For example, if 0.10 of the entities fail inspection and are sent to a location in the manufacturing process for rework, that location could be indicated by a Label in the model frame.

The NAME is the action, usually a verb, that is the function of the block. Examples are CREATE, QUEUE, SEIZE, DELAY, RELEASE, and DISPOSE.

Operands are constants, expressions, or alphanumeric names that provide information for the action. For example, an operand of the DELAY block indicates the length of the delay. As another example, an operand of the QUEUE block specifies the maximum length of the queue. Some operands are required, some are optional.

There are five block MODIFIERS; three that are used often in Chapters 3 and 4 are as follows:

MARK(*AttributeID*) marks the specified *AttributeID* with the arrival time to the block. Thus, MARK(*ArrivalTime*) would place the time of arrival at the current block in the attribute *ArrivalTime*.
DISPOSE removes the entity from the system. (DISPOSE can also be a block.)
NEXT(Label) sends the entity to the block with Label indicated. The usual flow of entities is from block to block. NEXT(Label) interrupts this flow.

These modifiers can be used on most blocks. However, there are exceptions, and these will be noted as the blocks are discussed.

EXAMPLE 2.1 The Single-Machine Problem

A single-machine job shop processes jobs that arrive according to an exponential distribution with a mean of 8 minutes. The jobs go into a queue with a capacity of two. The machine serving the jobs is a drill with a capacity of two jobs. The operation time on the drill is distributed uniformly between 6 and 9 minutes. After the drilling operation, the jobs are finished. Determine the average number of jobs waiting for the drill and the number of jobs completed in 8 hours.

The model listing is shown in Exhibit 2.1. The purpose of this model is to show how a problem statement is converted to block form, and how those blocks are written. Some explanation is necessary as given in the following paragraphs:

EXHIBIT 2.1 Model frame for Example 2.1.

```
!Model Frame for Example 2.1
!The Single Machine Problem
!
BEGIN;
CREATE:
   IAT;          Interarrival time
QUEUE,
   Buffer,       !Wait for Drill
   2;            Maximum capacity
SEIZE:
   Drill;        Capture the resource Drill
DELAY:
   DrillTime;    Use the Drill
RELEASE:
   Drill;        Free the resource Drill
COUNT:
   JobsDone;     Count the jobs completed
DISPOSE;         Destroy the job
```

Model frames may or may not start with a BEGIN statement. (Note: This text uses the BEGIN statement in each model to obtain the model listing on the default drive.) There are four symbols recognized by the parser as delimiters; the ",", ":", ";" and the "!". The first two are used within the block statement. The ";" is recognized as the end of a block. Thus, anything entered after the ";" is not processed. Many times, it is convenient to begin a line with a ";" or a "!" to use that line for documentation purposes or even to separate parts of the model. Note the use of the "!" in the first two lines of Exhibit 2.1. (When using the ";" to comment a line, the delimiter must be in the first column.) The "!" is for an in-line comment or in commenting an entire line. (When using the "!" to comment a line, the delimiter can be in the first or in any column.) In the QUEUE block of Exhibit 2.1, it is used prior to an indication of the queue capacity. The comments following the ";" and the "!" are not required, but are certainly good practice. Others reading the code are helped in understanding what was entered, and the person writing the code is also helped when returning to the model after a time lapse. Most real models are very long, thousands of blocks in many cases, and remembering what was done months in the past would be very difficult when modifications or extensions of the model are required.

The time between creations and the time for drilling are given by expressions in the experiment frame. This follows the practice of keeping data out of the model frame.

The experiment frame for Example 2.1 is shown in Exhibit 2.2. Again, note the use of BEGIN and the four delimiters "'", ":", ";" and "!". Also, note that free format is used. The experiment frame is readily related to the problem statement in Example 2.1. Although the PROJECT element does not appear to provide much information, its presence is necessary to obtain output from the simulation. The DSTATS element provides time-persistent information including the average and the maximum number of entities in the queue *Buffer* and the number of entities in the queue *Buffer* at the end of the simulation.

EXHIBIT 2.2 **Experiment frame for Example 2.1.**

```
!Experiment Frame for Example 2.1
!The Single Machine Problem
!
BEGIN;
PROJECT,        Single Machine,Team;
RESOURCES:      Drill;
QUEUES:         Buffer;
DSTATS:         NQ(Buffer);                 Queue Length
EXPRESSIONS:    1, IAT, EXPO(8, 1):         !Exponential interarrivals
                2, DrillTime, UNIF(6, 9, 1); Uniform Drill times
COUNTERS:       JobsDone;
REPLICATE,      1, 0, 480;                  One rep for 480 minutes
```

The EXPRESSIONS element references the interarrival time (*IAT*) as exponentially distributed with a mean of 5, and the *DrillTime* from a uniform distribution with a lower limit of 6 and an upper limit of 9. Random number stream 1 is being used in each case to generate the values.

The REPLICATE element indicates that there will be one run of this model starting at time zero and going to time 480. These times are in units defined by the user; in this case all times are in minutes.

2.4 Runtime Procedure

The user creates two text or ASCII files; *filename*.MOD and *filename*.EXP. These correspond to the model frame and the experiment frame, respectively. These two files are then compiled using the files MODEL.EXE and EXPMT.EXE. The results are *filename*.M and *filename*.E. These two files are then linked using the file LINKER.EXE with the result being *filename*.P. Using SIMAN.EXE, the simulation can now occur with the results found in *filename*.OUT.

This can be restated as follows:

1. User creates two text files:

   filename.MOD
   filename.EXP

2. Compile the model using the following procedure:

Command	**Result**
MODEL *filename filename*	*filename*.M

   Note: Extension .MOD is not required. Also note that including a BEGIN statement before any blocks in the model will result in a listing of the model on the default drive. This listing is identified as *filename*.MLS. If errors occur, they can be found in that file.

3. Compile the experiment using the following procedure:

Command	**Result**
EXPMT *filename filename*	*filename*.E

   Note: Extension .EXP is not required. Also note that including a BEGIN statement before any other statements in the experiment will result in a listing of the model on the default drive. This listing is identified as *filename*.ELS. If errors occur, they can be found in that file.

4. Link the compiled files using the following procedure:

Command	**Result**
LINKER *filename filename filename*	*filename*.P

Note: Extensions .M and .E are not required. Any errors that occur can be found in *filename*.LLS.

5. Conduct the simulation using the following procedure:

Command	**Result**
SIMAN *filename*	*filename*.OUT

Note: Extension .P is not required.

Some simulation analysts choose to write a "make" file that acts as a "shell" to execute MODEL.EXE, EXPMT.EXE, and LINKER.EXE resulting in a file with a .P extension.

Assume that the model and experiment frames for Example 2.1 are contained in ASCII text files EXAMP21.MOD and EXAMP21.EXP. Also assume that these files reside in the default directory C:\SIMAN. To compile the model frame, at the DOS prompt enter

C:\SIMAN>`MODEL EXAMP21 EXAMP21` [User enters that shown in Courier type]

The model processor will create a file named EXAMP21.M. To compile the experiment frame, enter

C:\SIMAN>`EXPMT EXAMP21 EXAMP21`

The experiment processor will create a file named EXAMP21.E. To link the two files, enter

C:\SIMAN>`LINKER EXAMP21 EXAMP21 EXAMP21`

The linker will create the simulation program file named EXAMP21.P. To run the simulation, enter

C:\SIMAN>`SIMAN EXAMP21`

The file EXAMP21.OUT is created when the simulation is completed. The output file is an ASCII text file that can be viewed with the same editor used to create the model and experiment frames. If you have a serial printer installed as "LPT1," a hard copy of the output may be generated by entering

C:\SIMAN>`COPY EXAMP21.OUT LPT1`

The output from the simulation is shown in Exhibit 2.3. Only the lines above the DISCRETE-CHANGE VARIABLES would be obtained if the DSTATS and COUNTERS elements were omitted.

***EXHIBIT 2.3* Summary Report for Example 2.1.**

Summary for Replication 1 of 1

Project: Single Machine Run execution date: 7/30/1993
Analyst: Team Model revision date: 7/30/1993
Replication ended at time: 480.0

Discrete-Change Variables

Identifier	Average	Variation	Minimum	Maximum	Final Value
NQ(Buffer)	.48160	1.3681	.00000	2.0000	1.0000

Counters

Identifier	Count	Limit
JobsDone	46	Infinite

The output file is listed on the screen in addition to appearing on the default drive. However, it is scrolled on the screen, so that only the last information is observed. In a model like Example 2.1, the output is so short that it will fit on one screen.

2.5 Solution Using ARENA

It is also possible to construct the model and experiment using the templates BLOCKS.TPO and ELEMENTS.TPO within the ARENA environment. This procedure is described briefly in this section. It is also possible to use the COMMON.TPO template and other features of ARENA for constructing programs as well as drawing layouts and animating models. These more advanced features are described in Chapter 14 and in the *Arena User's Guide* (1994). Reading Section 14.1 on the User Interface or the same-named chapter in the *Arena Users Guide* will explain the terminology (main menu bar, dialog box, etc.) used in this section.

Note that the capability to construct the model and experiment in the manner described in this section is not provided with the student version of the software distributed with this text. Consult the README.DOC file to determine how to obtain the software version with the added capability.

EXAMPLE 2.2 Using ARENA to Solve Example 2.1

Following is a step-by-step procedure explaining how to use the ARENA environment to solve the problem posed in Example 2.1:

Step 1. To the ARENA> (ARENA prompt) enter `ARENA`.(Enter what is shown in Courier print.)

Step 2. Click on **Model** in the main menu bar.

Step 3. Click on **New** in the pull-down menu.

Step 4. Click on **Panel** in the menu bar.

Step 5. Click on **Attach** in the pull-down menu.

Step 6. Blocks will be highlighted in the dialog box.
(If not highlighted, click on it.)
Click on **Attach** in the dialog box.

Step 7. Click on **BEGN** (i.e., BEGIN) in the blocks panel.

Step 8. Click on a location in the upper-left hand of the workspace region. Observe the placement of the BEGIN statement.

Step 9. Click on **CREA** (i.e., CREATE) in the blocks panel.

Step 10. Click on a location in the upper-left hand of the workspace region, but below and to the right of the BEGIN statement.
Observe the placement of the CREATE block.
Any object in the workspace region may be grabbed and moved by clicking on its handle and dragging it to the desired location.

Step 11. Double-click on the **CREATE** block causing the dialog box to open.

Step 12. Click on the **data box** to the right of *Interval* causing it to be highlighted.

Step 13. Enter `IAT`.

Step 14. Click on **Accept.** (Continue this process for all blocks in the model. Large and small arrowheads at the bottom of the blocks panel provide access to other blocks.)
For the SEIZE block, use the following procedure:

Double-click on the **SEIZE** block causing the dialog box to open.
Click on **Insert**
Enter `Drill` as the Resource ID in the highlighted area.
Click on **Accept** in the Resources dialog box.
Click on **Accept** in the SEIZE Block dialog box.

A similar Insert procedure is used on the RELEASE block and many of the elements.

Step 15. Click on **Panel** in the menu bar.

Step 16. Click on **Attach** in the pull-down menu.

Step 17. Click on **ELEMENTS.TPO** in the dialog box.

Step 18. Click on **Attach** in the dialog box.

Step 19. Click on **BEGIN** in the elements panel.

Step 20. Click on a location in the workspace region.

Step 21. Click on **PROJECT** in the elements panel.

Step 22. Click on a location in the workspace region.

Step 23. Double-click on the **PROJECT** element, causing the dialog box to open.

Step 24. Enter *Title* and *Analyst Name,* then click on Accept. (Continue this process for all elements in the experiment.)

Step 25. Click on the **Tilted A** in the leftmost side of the title bar.

Step 26. Click on **Save** in the drop-down menu.

Step 27. Enter a filename in the save file dialog box.

Step 28. Click on **Save.**

Step 29. Click on the **Run tab.**

Step 30. Click on **Go.**

Step 31. After the simulation is completed, click on **End** in the run panel.

Step 32. Click on **ARENA** in the main menu bar.

Step 33. Click on **View Text File.**

Step 34. Click on ***.TXT** to the right of *File Spec* in the dialog box.

Step 35. Delete *.TXT using the Delete key.

Step 36. Enter filename.OUT in *File Spec.*

Step 37. Click on **Load** in the dialog box.

Step 38. Click on the **MAXIMIZE button** to the right of the filename.

Step 39. Click on **Arena** in the leftmost side of the title bar.

Step 40. Click on **Quit.**

In Chapters 3 through 13, it is assumed that the command level of SIMAN is being used to construct the model and experiment. As such, files are made using a text editor.

2.6 Summary

This chapter introduces SIMAN V and the components of the model and experiment frames. The introduction includes an example that is solved at the command level and also within the Arena environment. This text concentrates on the use of SIMAN at the command level. However, there may be those who wish to learn SIMAN within the Arena environment.

2.7 REFERENCE

ARENA User's Guide (1994), Systems Modeling Corporation, Sewickley, Penn.

2.8 EXERCISES

E2.1 Enter the model and experiment of Example 2.1 into text files, then run the simulation and view the output.

E2.2 After completing E2.1, change the maximum capacity on the queue *Buffer* to three, and run the simulation. Compare the output to that of Example 2.1.

E2.3 After completing E2.1, change the maximum capacity on the queue *Buffer* to 20, and run the simulation. Compare the output to that of Example 2.1.

E2.4 After completing E2.1, conduct three replications of Example 2.1. Compare the different values of the average value of *NQ(Buffer)*.

E2.5 After completing E2.1, change the distribution of time in the drill to exponential with a mean of 7.5 minutes and run the simulation. Compare the output to that of Example 2.1.

E2.6 After completing E2.1, change the length of the simulation to 4800 minutes, and run the simulation. Compare the output to that of Example 2.1.

3

Basic Blocks and Elements

A SIMAN V simulation consists of a model frame and an experiment frame. The model frame is composed of blocks that describe or "model" the system to be simulated. The experiment frame contains the simulation data and establishes the statistics to be collected. The model and experiment frames are individually compiled and then linked to form the SIMAN V simulation program.

EXAMPLE 3.1 Problem Statement

Jobs are started at a production area according to an exponential distribution with a mean of 5 minutes. The production process consists of three operations; drilling, milling, and grinding. There are 2 drills, 3 mills, and 2 grinders. The drills and grinders can have a maximum of two jobs waiting in the queue to be processed, and the mills can have up to three.

Upon arrival to the drill area, a job is processed for 6 to 9 minutes, uniformly distributed. The job is then milled with a process time that is triangularly distributed with a minimum, mode, and maximum of 10, 14, and 18 minutes, respectively. Lastly, the job is processed in the grinder area according to the following discrete distribution: 25% require 6 minutes, 50% require 8 minutes, and 25% require 12 minutes. Figure 3.1 shows the layout for the system.

Jobs that cannot enter a queue due to capacity limitations are ejected from the system. Transportation times between resources are assumed to be negligible. Random number stream 1 is used for all processes and arrivals.

We wish to simulate the system for 40 hours and answer the following questions:

1. How many jobs are completed?
2. What is the utilization rate for each class of resource?

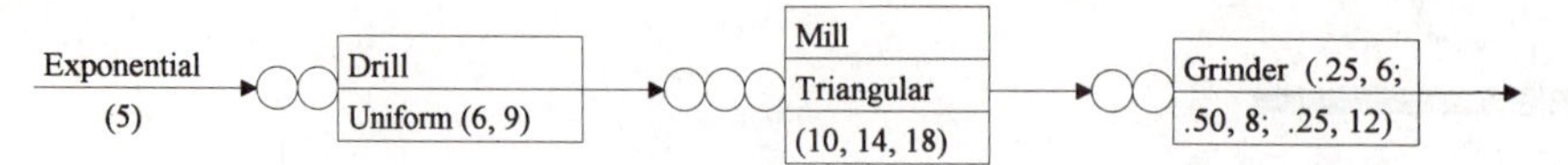

Figure 3.1 Processing systems for Example 3.1.

3. What is the total number of jobs ejected due to full queues?
4. What is the average number of jobs in each queue?

3.1 Blocks and Elements

The system described in Example 3.1 can be simulated using several basic SIMAN V blocks and elements. These blocks and elements are shown in Table 3.1 along with a brief description.

3.2 Basic Model Frame Concepts

SIMAN V model frame blocks are used to control the flow of entities through the system. In this sense, each block can be thought of as an operator. An entity enters the block, is subject to whatever operation the block defines, and is then sent to the next block. By default, the entity will proceed sequentially through the model frame blocks. In certain situations, a full queue for instance, a block may be used to redirect the entity to another location in the model frame.

Prior to introducing the basic SIMAN V modeling blocks, there are two simulation concepts that should be explained in more detail. The first of these is the random number stream. An introduction to random numbers and random variates was given in Chapter 1. The random number generator is responsible for creating real numbers, which are uniformly distributed between 0 and 1, that is, $U(0, 1)$. As stated in Chapter 1, these random numbers are really pseudorandom, since the same sequence can be repeated indefinitely. This is accomplished by passing the generator the same "seed" value each time it is restarted. In SIMAN V, there are ten different random number streams available. In a sense, this means that there are ten very large sets of random numbers from which random values may be retrieved. The default random number stream is 1. The second item of interest is the generation of nonuniform random numbers, or random variates. SIMAN V contains several random variate generators. These generators transform $U(0, 1)$ values into values that conform to other distributions. The most common distributions are given in the Table 3.2.

Table 3.1 Blocks/Elements Introduced in this Chapter

CREATE block	Introduces entities into the system
QUEUE block	Provides buffer space for entities to accumulate
SEIZE block	Gives control of a resource to an entity
DELAY block	Specifies the length of a time delay
RELEASE block	Releases a seized resource
COUNT block	Increments or decrements a counter value
DISPOSE block or modifier	Removes entities from the system
PROJECT element	Identifies a project, the analyst, and the date it was created
EXPRESSIONS element	Defines one- or two-dimensional arrays, distributions, and mathematical expressions
QUEUES element	Identifies and defines the queues referenced in the model frame
RESOURCES element	Identifies and defines the resources referenced in the model frame
COUNTERS element	Identifies and defines the counters referenced in the model frame
DSTATS element	Collects time-persistent statistics on resources, queues, and other user-defined values
REPLICATE element	Specifies the run length and number of replications

Table 3.2 Common Distributions

Distribution	Abbreviation	Parameters
Discrete	DISC	(CumP1, Val1, CumP2, Val2, . . . , Stream)
Exponential	EXPO	(Mean, Stream)
Normal	NORM	(Mean, StdDev, Stream)
Poisson	POIS	(Mean, Stream)
Triangular	TRIA	(Minimum, Mode, Maximum, Stream)
Uniform	UNIF	(Minimum, Maximum, Stream)

The last parameter for each distribution is used for specifying the random number stream. This parameter may be omitted if only the default random number stream is to be used.

The following sections are intended to introduce the most basic SIMAN V blocks and elements. In many instances, the blocks have an operand that may be specified in the form of an expression. The term "expression" refers to any SIMAN V expression that returns a numeric value. Expressions are discussed in detail in Section 3.3.3.

3.2.1 The BEGIN Statement

The BEGIN statement specifies whether or not a listing is written to a file when the model is compiled. If used, the BEGIN statement is entered as the first line of the model. The format for the BEGIN statement is as follows:

BEGIN, *Model Listing, Model Name*;

The operands are described as follows:

Operand	Description	Default
Model Listing	Option for generating a compile time listing [YES or NO]	YES
Model Name	Name associated with the model frame [*filename*.MOD]	Model frame filename

The first operand, *Model Listing,* is specified as YES or NO and determines whether or not a source code listing is to be written to a file as it is compiled. If *Model Listing* is omitted, then SIMAN V uses YES as the default. During initial development or major revision, it is advisable to accept the default by simply entering "BEGIN;." The listing generated by the model processor can be very useful in debugging a new or vastly revised model frame. Once the model frame is finalized, the line can be changed to "BEGIN, NO;." For very large models, this setting will significantly improve compilation time.

The second operand, *Model Name,* is specified when linking multiple model frames. The default for *Model Name* is the name of the file containing the model frame with extension ".mls". For example, if the model is saved in a file named *facility.mod* and *Model Name* has been defaulted, then the listing will be written to a file named *facility.mls*. The default for *Model Name* is used throughout this text.

If the BEGIN statement is omitted, then the *Model Listing* operand is defaulted to NO. However, any errors encountered during compilation will still be written to the default listing file.

3.2.2 The CREATE Block

The CREATE block introduces entities into the system. Generally, a CREATE block is the first block following the BEGIN statement. There is no limit on the number of CREATE blocks contained in any one model frame. The format of the CREATE block is as follows:

CREATE, *Batch Size, Offset Time*: *Interval, Max Batches*;

The operands are described as follows:

Operand	Description	Default
Batch Size	Number of entities generated [integer or expression truncated to an integer]	1
Offset Time	Time of the first arrival [real or expression]	0.0
Interval	Time between arrivals [real or expression]	Infinite
Max Batches	Maximum number of times entities will be generated [integer or expression truncated to an integer]	Infinite

The first operand, *Batch Size,* determines the number of entities that will be introduced into the system each time the CREATE block is activated. *Batch Size* may be entered as an integer or a mathematical expression that will be truncated to an integer. If *Batch Size* is omitted from the CREATE block, then a default value of 1 is used.

The second operand, *Offset Time,* specifies the delay between the start of the simulation and the arrival of the first group of *Batch Size* entities. *Offset Time* may be entered as a real number or a SIMAN V expression. The default for *Offset Time* is 0.0. Therefore, if the *Offset Time* parameter is omitted, then the first group of *Batch Size* entities will be generated at simulation startup.

The third operand, *Interval,* determines the arrival rate or amount of time between each creation of *Batch Size* entities. *Interval* can be entered as a real number or any valid SIMAN V expression. The default value for *Interval* is infinite. If the *Interval* operand is omitted, then the CREATE block will generate *Batch Size* entities one time.

The last operand, *Max Batches,* specifies the maximum number of times that the CREATE block can generate a group of *Batch Size* entities. *Max Batches* may be entered as an integer or a SIMAN V expression, which will be truncated to an integer. The default for *Max*

Batches is infinite. Note that specifying *Max Batches* is one of several ways to control simulation run time.

Following are some examples of the CREATE block:

CREATE;

This CREATE block uses the default values for all four parameters. This will cause a single entity to be generated at the start of the simulation.

CREATE: 15, 100;

This example accepts the default values for *Batch Size* and *Offset Time*. This will cause the CREATE block to generate one entity each 15 time units, with the first arriving at simulation startup. *Max Batches* limits the number of times an entity can be created to 100.

CREATE, 1, 0: 15, 100;

Since the *Batch Size* and *Offset Time* values in this example are equivalent to the defaults, this CREATE block is exactly the same as the previous example.

CREATE, 10: EXPO(3);

In this example, ten entities are generated at the start of the simulation. Additional entities are created according to the exponential distribution with a mean of 3, using the default random number stream. There is no limit to the number of times that a group of 10 entities can be created. When *Offset Time* is defaulted, the comma separating *Batch Size* and *Offset Time* is also omitted.

CREATE, , 12: ArrivalRate;

This CREATE block generates one entity 12 time units after the start of the simulation. Additional entities are created according to an expression named *ArrivalRate*. Expressions are discussed in Section 3.3.3. The format of the block (CREATE, ,) causes the SIMAN V model processor to use the default value for *Batch Size*.

3.2.3 The LABEL

By default, entities flow sequentially through the model. Sometimes it is necessary to send an entity to another location in the model. In order to do this, the entity is redirected to a labeled block. A label is a unique alphanumeric value on the same line and prior to the desired destination block. Model understanding is aided by using labels that indicate why an entity would be sent to a particular location. Following are three sets of labeled blocks:

```
Baked COUNT:
                Cakes Baked;
      DISPOSE;
Reject COUNT:
                Parts Rejected;
      DISPOSE;
Print COUNT:
                Boxes Printed;
      DISPOSE;
```

The *Baked, Reject,* and *Print* labels are descriptive of the state of an arriving entity and the action about to be taken. Combined with the respective blocks, each line is self-explanatory. Notice that each labeled block is followed by a DISPOSE block. Once the entity has been counted, it will be destroyed without any further processing. The DISPOSE block is discussed in Section 3.2.9.

Note: Since spacing does not matter to SIMAN, we will sometimes show blocks and elements with their operands on different lines, as above. Alternately, we will show blocks and elements all on one line as below. This is entirely a matter of style.

Now consider the following labeled blocks:

```
AS2J COUNT: Counter1;
     DISPOSE;
BKL3 COUNT: Counter2;
     DISPOSE;
```

The *AS2J* and *BKL3* labels do not reveal any information concerning the state of an arriving entity. We do not know why an entity would be sent to these locations and the generic names add to the confusion. Obfuscated code is hard to trace and even harder to modify. Simple, understandable naming conventions will make the life of a modeler much easier.

3.2.4 The QUEUE Block

The QUEUE block provides buffer space for entities awaiting further processing. The format of the QUEUE block is as follows:

QUEUE, *Queue ID, Capacity, Balk Label:* repeats;

The QUEUE block has two additional operands, which are discussed in Chapter 4. The operands shown here are described as follows:

Operand	Description	Default
Queue ID	Queue number [integer or expression truncated to an integer] or name	—
Capacity	Queue capacity [integer or expression truncated to an integer] or FULLWHEN(expression)	Infinite
Balk Label	LABEL of the destination block for balking entities	Dispose of entity

The first operand, *Queue ID,* is a required operand that refers to the name or sequential number of the queue as defined in the experiment frame QUEUES element. If a specific *Queue ID* is used in more than one model frame QUEUE block, then the associated queue in the experiment frame QUEUES element must be defined with the keyword SHARED. In other words, the queue *Queue ID* must be defined as a SHARED queue in the experiment frame if an entity can enter the queue *Queue ID* from more than one location in the model frame.

The second operand, *Capacity,* is an integer or expression that specifies the maximum number of entities that may be waiting in the queue *Queue ID* at any time. Alternately, FULLWHEN(*expression*) may be used to determine whether or not the queue is full when the entity arrives. If the logical expression is true, then the queue is considered full and the entity is denied entry. Otherwise, the queue is not yet full and the entity is allowed to enter. Negative values for *Capacity* are interpreted by SIMAN V as infinite, which is also the default. If *Capacity* is specified by the value of an expression, then the value of the expression only affects the arriving entity. If the queue currently contains more entities than the result of the expression, then the additional entities are left in the queue and the arriving entity is sent to the block specified by *Balk Label* or destroyed if *Balk Label* is omitted.

The third operand, *Balk Label,* is the LABEL that identifies the destination block for entities denied entry into the queue due to capacity limitations. If no *Balk Label* is specified, then entities that encounter a full queue are destroyed without notification.

Following are some examples of the QUEUE block:

```
QUEUE, 1;
```

In this example, the arriving entity enters the first queue specified in the QUEUES element of the experiment frame. The omitted *Capacity* operand causes the block to utilize the default value of infinity. Since there is no limitation on the number of entities residing in the queue, a *Balk Label* would be pointless.

```
QUEUE, DrillQ;
```

In this example, the entity enters a queue named *DrillQ*. If *DrillQ* is the first queue specified in the experiment frame QUEUES element, then this QUEUE block is identical to the first example.

```
QUEUE, Buffer, 0;
```

In this example, an entity enters the queue named *Buffer* with a *Capacity* of 0. If the next block is not immediately available, the entity is destroyed. Otherwise, the entity is released to the next sequential block and incurs no delay in the queue.

```
QUEUE, Shelf1, FULLWHEN(Pallets > 4), Store;
```

When an entity arrives to this QUEUE block, SIMAN V checks to see if the value of *Pallets* is greater than 4. If so, then the entity is sent to the block labeled *Store.* If *Pallets* is 4 or less, then the entity is added to the queue *Shelf1.*

```
             :
Print  QUEUE, PrinterQ, 0, Spool;
             :
       other blocks
             :
Spool  QUEUE, SpoolQ, 4, Reject;
             :
```

This last example contains two different QUEUE blocks from the same model frame. The first QUEUE block in this example is labeled *Print,* and the second is labeled *Spool.* An arriving entity could reach the first QUEUE block through sequential processing or it may have been redirected to *Print* from another location in the model frame. Note that the *PrinterQ Capacity* operand is specified as 0. If the following block is not immediately available, then the entity is sent to the block labeled *Spool.* When the entity arrives at *Spool*, it will attempt to enter the queue named *SpoolQ*. The *Capacity* of *SpoolQ* is specified as 4. If there are already four entities waiting in the queue, then the arriving entity will be sent to the block labeled *Reject.*

3.2.5 The SEIZE Block

The SEIZE block is used to allocate one or more resources to the arriving entity. If an entity enters a SEIZE block and attempts to gain control of more units of the resource than are currently available, then the entity is returned to the preceding queue to wait until the required number of units of the resource become available. If a QUEUE block does not immediately precede the SEIZE block, or an alternative method of controlling buffer space is not used, then the entity is stored in an internal queue until the resource becomes available. Use of internal queues is discouraged since queue statistics (waiting time, number in queue) are not available for entities residing in internal queues. In this text, QUEUE blocks will always precede SEIZE blocks. The format for the SEIZE block is as follows:

SEIZE, *Priority: Resource ID, Number of Units*: repeats;

The operands are described as follows:

Operand	Description	Default
Priority	Priority value [integer, real, or expression]	1.0
Resource ID	Number [integer or expression truncated to an integer] or name of the resource to be seized	—
Number of Units	Number of units of the resource to seize [integer or expression truncated to an integer]	1

Additional SEIZE block concepts are discussed in Section 7.4.

In a model frame that has more than one SEIZE block trying to allocate the same resource, the *Priority* operand can be used to determine which seizing entity will capture the resource. A lower *Priority* value has priority over a higher value. A negative *Priority* value is interpreted by SIMAN V to be 0.0.

The second operand, *Resource ID,* is the alphanumeric name, number, or expression truncated to an integer that identifies a resource as defined in the RESOURCES element of the experiment frame.

The last operand, *Number of Units,* is an integer or expression truncated to an integer that indicates the number of units of the resource *Resource ID* to allocate to the seizing entity. The default value of *Number of Units* is 1.

Following are some examples of the SEIZE block:

SEIZE, 1.0: 1;

In this example, *Priority* is specified as 1.0, *Number of Units* is defaulted to 1, and *Resource ID* is specified as one. An arriving entity will attempt to seize 1 unit of the first resource listed in the RESOURCES element of the experiment frame. If all the units of resource 1 are busy, then the entity will be returned to the preceding queue until one of the units becomes available.

SEIZE: Drill;

In this SEIZE block, the arriving entity attempts to gain control of 1 unit of the resource *Drill*. The entity is returned to the preceding queue if it cannot capture a unit of the resource. If *Drill* is the first resource listed in the experiment frame RESOURCES element, then this example is identical to the previous one since the default *Priority* of 1.0 is accepted.

SEIZE, 3.0: Workers, 2: Hoist: Tool3A;

In this third example, an arriving entity attempts to seize 2 *Workers,* 1 *Hoist,* and 1 *Tool3A*. The *Priority* value of 3.0 applies to all three resources. The entity will wait in the preceding queue until all requested resources are available.

```
        :
SEIZE, 4.0: Machine, 2;
        :
other blocks
        :
SEIZE, PrVal: Machine, 2;
        :
```

This last example illustrates the use of two SEIZE blocks that reference the same resource. The first block attempts to seize 2 units of resource *Machine* with a priority of 4.0 and the second with a priority determined by the entity attribute *PrVal*.

Attributes are discussed in detail in Chapter 4. For now, we will define an attribute to be a value assigned to an entity. For example, arriving entities could have the attribute *PrVal* that is assigned the value 1, 2, or 3. A similar concept, also discussed fully in Chapter 4, is that of a variable, a value available to all entities.

When there is an entity waiting at each of the SEIZE blocks, the decision as to which SEIZE block is allocated the first 2 free units is made based on priority. If *PrVal* is less than 4.0, then the second block seizes the 2 units of *Machine*. If *PrVal* is greater than 4.0, then the first block is allocated the two units of *Machine*. If *PrVal* is equal to 4.0, then SIMAN V allocates the two machines based on the FIFO selection rule. In other words, the entity that has been waiting the longest is allocated the 2 units of *Machine*.

3.2.6 The DELAY Block

The DELAY block delays an entity for a specified amount of time. For example, when an entity that has successfully seized a resource is allocated to an entity, the next logical step is for the entity to be processed by the resource. The DELAY block simulates this processing time. When the specified amount of time has elapsed, the entity leaves the DELAY block and continues through the model frame. The format of the DELAY block is as follows:

DELAY: *Duration*;

The operand is described as follows:

Operand	Description	Default
Duration	Length of the delay [real number or expression]	0.0

Following are some examples of the DELAY block:

DELAY: 8.5;

This DELAY block will delay an arriving entity for 8.5 time units.

DELAY: EXPO(5, 2);

This DELAY block will hold the entity for an amount of time specified by a sample value from the exponential distribution with a mean of 5 using random number stream 2.

DELAY: OrderSize/ShiftSize;

In the third example, the length of the delay is determined as a function of variables *OrderSize* and *ShiftSize*. For a given *ShiftSize*, the length of the delay will increase with the *OrderSize*. Conversely, for a given *OrderSize*, the length of the delay will decrease as *ShiftSize* increases.

```
Wash    DELAY:     WashTime;
        COUNT:     Washed;
        DISPOSE;
Rinse   DELAY:     RinseTime;
        COUNT:     Rinsed;
        DISPOSE;
```

In this example, an arriving entity may arrive to the first DELAY block through sequential processing or it may have been redirected to this block, which is labeled *Wash*. When the entity enters the first DELAY block, it is delayed for an amount of time specified by a sample from a distribution named *WashTime* in the experiment frame EXPRESSIONS element. When the entity exits the DELAY block, a counter named *Washed* is incremented and the entity is disposed. Notice that no entities can enter the following DELAY block through sequential processing. Any entity arriving to the DELAY block labeled *Rinse* will be redirected to this location in the model frame. The length of the DELAY is determined by a distribution named *RinseTime* in the experiment frame EXPRESSIONS element.

3.2.7 The RELEASE Block

The RELEASE block releases at least 1 unit of a specified resource. As seen in Section 3.2.5, the SEIZE block allocates *Number of Units* of a resource to an arriving entity. Similarly, when the resource completes processing an entity, the captured units must be released before another entity can seize the resource. The format for the RELEASE block is as follows:

RELEASE: *Resource ID, Quantity to Release*: repeats;

The operands are described as follows:

Operand	Description	Default
Resource ID	Number [integer or expression truncated to an integer] or name of the resource to release	—
Quantity to Release	Number of units of the resource to release [integer or expression truncated to an integer]	1

Following are some examples of the RELEASE block:

RELEASE: 1;

This block releases 1 unit of the first resource defined in the experiment frame RESOURCES element.

RELEASE: Drill;

This example releases 1 unit of the resource named *Drill* in the experiment frame RESOURCES element. If *Drill* is the first entry in the RESOURCES element, then the first two examples are identical.

:

RELEASE: Workers: Hoist: Tool3A;

:

other blocks

:

RELEASE: Workers;

:

When an entity enters the first RELEASE block of this example, 1 unit of each resource *Workers, Hoist,* and *Tool3A* is released. The entity leaves the RELEASE block and continues processing until it reaches the next RELEASE block where another unit of the resource *Workers* is released.

RELEASE: Machine, 2;

This example releases 2 units of the resource named *Machine*.

3.2.8 The COUNT Block

The COUNT block is used to increment or decrement the value of a counter by a specified amount. The format for the COUNT block is as follows:

COUNT: *Counter ID, Counter Increment*;

The operands are described as follows:

Operand	Description	Default
Counter ID	Number [integer or expression truncated to an integer] or name of the counter	—
Counter Increment	Amount to be added to the counter [integer or expression truncated to an integer]	1

The *Counter ID* is the alphanumeric name, number, or expression truncated to an integer that identifies a sequentially numbered counter in the COUNTERS element of the experiment frame.

The value for *Counter Increment* may be positive or negative. The default value for *Counter Increment* is one. A *Counter Increment* value less than zero serves to decrement the counter.

Following are some examples of the COUNT block:

```
COUNT: 1;
```

When an entity enters this COUNT block, the first counter defined in the experiment frame COUNTERS element is incremented by the default value of one.

```
COUNT: Parts Made, 1;
```

This COUNT block increases the counter named *Parts Made* by one. If *Parts Made* is the first counter in the experiment frame COUNTERS element, then this example is identical to the previous example COUNT block.

```
Absent  COUNT: EmployeeID + 2;
```

The third example is a COUNT block labeled *Absent*. The value of *EmployeeID* + 2 specifies an integer *Counter ID* value. The *EmployeeID* + 2 sequentially numbered entry in the experiment frame COUNTERS element is incremented by one.

```
Audit    COUNT: Profit, Revenue - Cost;
```

The fourth example is a COUNT block labeled *Audit*. When an entity enters the COUNT block, the counter named *Profit* in the experiment frame COUNTERS element is changed by the amount *Revenue* – *Cost*, where *Revenue* and *Cost* are attributes or variables. If *Revenue* is greater than *Cost*, then *Profit* is increased. If *Cost* is greater than *Revenue*, then *Profit* is decreased. Otherwise, the counter *Profit* is left unchanged.

3.2.9 The DISPOSE Block/Modifier

The DISPOSE block, as the name suggests, disposes of entities. There is no limit on the number of DISPOSE blocks that may be used in the model frame. DISPOSE may be used as a standalone block or as a modifier of another block. The DISPOSE block has no operands and appears as

```
DISPOSE;
```

An entity may be redirected to the DISPOSE block or may arrive sequentially.

Following are examples of the DISPOSE block and modifier:

```
                :
          QUEUE, ReworkQ, 5, Trash;
                :
          other blocks
                :
    Trash DISPOSE;
                :
```

This example uses the QUEUE block *Balk Label* operand to redirect the entity to a DISPOSE block when the queue *ReworkQ* is full. Upon arrival, if there are already five entities residing in the queue *ReworkQ*, the entity is sent to the DISPOSE block labeled *Trash*.

```
RELEASE:    Worker;
COUNT:      Jobs Completed:
            DISPOSE;
```

In this example, DISPOSE is used as a modifier to the COUNT block. An entity reaches the COUNT block after releasing the resource *Worker*. The counter named *Jobs Completed* is incremented and the entity is ejected.

3.2.10 Model Frame for Example 3.1

In Example 3.1, the process starts when a job arrives to the drills and ends when the final product leaves the grinders. For this system, each job is represented by one entity, each type of machine is defined as a multiple capacity resource, and the buffer space preceding each class of resource is represented as a finite capacity queue. In addition to the required blocks and elements, the model frame includes a BEGIN statement. The model frame source code for Example 3.1 is as follows:

EXHIBIT 3.1 Model frame for Example 3.1.

```
! Example 3.1 Model Frame: The Three Machine Problem
!
BEGIN, NO;
        CREATE:
                ArrivalRate;     !Create jobs according to the EXPRESSION
                                 that defines the arrival rate
        QUEUE,
                DrillQ,          !Enter the drill queue and wait for a drill
                2,               !Maximum capacity of the drill queue is 2
                Balk;            If queue is full, go to the block labeled Balk
        SEIZE:
                Drill;           Capture an available drill
        DELAY:                   !Process according to the EXPRESSION
                DrillTime;       that defines the drilling process time
        RELEASE:
                Drill;           Free the drill when finished
        QUEUE,
                MillQ,           !Enter the mill queue and wait for a mill
                3,               !Maximum capacity of the mill queue is 3
                Balk;            If queue is full, go to the block labeled Balk
        SEIZE:
                Mill;            Capture an available mill
        DELAY:                   !Process according to the EXPRESSION
                MillTime;        that defines the milling process time
        RELEASE:
                Mill;            Free the mill when finished
        QUEUE,
                GrindQ,          !Enter grinder queue and wait for grinder
                2,               !Maximum capacity of grinder queue is 2
                Balk;            If queue is full, go to the block labeled Balk
        SEIZE:
                Grinder;         Capture an available grinder
        DELAY:                   !Process according to the EXPRESSION
                GrindTime;       that defines the grinding process time
        RELEASE:
                Grinder;         Free the grinder when finished
        COUNT:
                Jobs Completed: !Count completed jobs
                DISPOSE;         and destroy the entity
Balk    COUNT:
                Number Balked: !Count jobs that balk
                DISPOSE;         and destroy the entity
```

Inspection of Exhibit 3.1 reveals a logical flow that parallels the system being simulated. An entity "flows" through the model frame similar to the way that a job flows through the machine shop.

Notice that each of the DELAY blocks has a descriptive alphanumeric name for the *Duration* operand. These names identify a particular distribution defined in the experiment frame EXPRESSIONS element.

3.3 Basic Experiment Elements

The experiment frame generates output, defines the simulation boundaries, and provides data required by the model frame. The experiment frame is contained in a separate file from the model frame. The two files are compiled independently and linked together to form the program file.

Having two separate frames may seem like extra work at first, but as the size and complexity of the simulation grows, the benefit of this structure is quickly realized. Consider a simulation model that is composed of several thousand lines of code and 20 or more different instances of various distributions. Your intention is to run the simulation after changing several types of distributions and using updated parameters. Having all the distributions and their parameters grouped in one location will definitely simplify this task. Additionally, the model frame does not have to be recompiled if only the experiment frame is altered. The following sections contain detailed descriptions of the basic experiment frame statements and elements.

3.3.1 The BEGIN Statement

Similar to the model frame, the experiment frame may also include a BEGIN statement. If included, the BEGIN statement appears on the first nonblank line of the experiment frame. The format for the BEGIN statement is as follows:

BEGIN, *Listing, Run Controller;*

The operands are described as follows:

Operand	Description	Default
Listing	Determines if a compile time listing is generated [YES or NO]	YES
Run Controller	Turns the Interactive Run Controller on or off [YES or NO]	NO

The first operand, *Listing,* determines if a source code listing is written to a file during compilation. During initial development or major revision, the listing should be generated, as this will simplify the task of debugging.

The second operand, *Run Controller,* turns the Interactive Run Controller (IRC) on or off at compilation time. If *Run Controller* is specified as YES, the IRC is invoked at the start of the simulation. By the completion of Chapter 4, the complexity of the simulations will be such that the benefits of the IRC can be realized. Therefore, Chapter 5 is devoted entirely to the IRC.

All errors encountered during compilation are written to a listing file regardless of the specification for the *Listing* operand. The listing file will have the same name as the experiment frame file with the extension *.els.*

3.3.2 The PROJECT Element

The PROJECT element labels the summary report with a title, the modeler identification, and the date. If the PROJECT element is omitted from the experiment frame, then the summary report will not be generated. The format of the PROJECT element is given by

PROJECT, *Title, Analyst, Date, Summary Report*;

The operands are described as follows:

Operand	Description	Default
Title	Project title	Blank
Analyst	Modeler name	Blank
Date	Date in mm/dd/yyyy format	System dependent
Summary Report	Generate a summary report [YES or NO]	YES

The first operand, *Title,* is an alphanumeric name assigned to the simulation. There is no limit on the length of the *Title,* but only the first 24 characters will be written to the summary report. Slashes or dashes ("/", "\", "-") may not be used in the *Title.*

The second operand, *Analyst,* is also an alphanumeric value and serves to identify the modeler. There is no limit on the length of *Analyst,* but only the first 24 characters will be written to the summary report. Slashes or dashes may not be used in *Analyst* either.

The *Date* operand is entered in the mm/dd/yyyy format. If no date is specified, then SIMAN V will use the compilation date, if provided by the system.

The *Summary Report* operand specifies whether or not a report will be generated at the completion of the simulation. Specifying NO for *Summary Report* or omitting the PROJECT element will suppress report generation. The default value for *Summary Report* is YES.

Following are two examples of the PROJECT element:

PROJECT;

This PROJECT element will cause the summary report to be generated. Since *Title, Analyst,* and *Date* have been defaulted, the report will not have a title, the author of the model will not be identified, and the date generated will be the system date at compilation time, if available.

PROJECT, New Assembly Line, Team, 04/23/1996;

This PROJECT element will generate a summary report with a *Title* of *New Assembly Line,* the *Analyst* is identified as *Team,* and the *Date* as 04/23/1996.

3.3.3 The EXPRESSIONS Element

The EXPRESSIONS element defines expressions and arrays that may be referenced by name in the model frame or with the ED variable. The format for the EXPRESSIONS element is

EXPRESSIONS: *Number, Name(1-D Array Index, 2-D Array Index), Expressions*: repeats;

The operands are described as follows:

Operand	Description	Default
Number	Expression number [integer]	Sequential integers
Name	Expression name [alphanumeric identifier]	Blank
1-D Array Index	Integer index for a named array	No array
2-D Array Index	Second index for a two-dimensional named array	No array
Expressions	SIMAN V expression	—

The first operand, *Number,* is an integer expression identifier. If used, *Number* operands must be entered in ascending order and the expression may be referenced in the model frame with ED(*k*) where *k* evaluates to *Number.* The *Number* operand and the trailing comma may be omitted.

The second operand, *Name,* is an alphanumeric expression identifier. If *Number* is omitted, *Name* must be specified. The experiment processor will assign sequential integer values to each expression *Name.* The expression can then be referenced in the model frame with *Name* or ED(*k*), where *k* is the default expression *Number.* If *Number* is specified, then *Name* may be omitted, but the trailing comma must be included. However, the *Name* operand may not be omitted when defining one- or two-dimensional arrays.

The third and fourth operands, *1-D Array Index, 2-D Array Index,* specify the sizes of a one- or two-dimensional named array. *1-D Array Index* and *2-D Array Index* may be one of the following forms:

Name(N)	One-dimensional array with N values
Name(Low .. High)	One-dimensional array with (High - Low + 1) values and the first value is at Low
Name(N1, N2)	Two-dimensional array with N1 rows and N2 columns
Name(L1 .. H1, L2 .. H2)	Two-dimensional array with (H1 - L1 +1) rows and (H2 - L2 + 1) columns

Named array values are separated by commas and entered one column at a time. Array values that are not specified will be assigned the value of the last entry in the array.

The last operand, *Expressions,* may be any valid SIMAN V expression. *Expressions* may reference other *Expressions,* but circular references are not allowed. Appendix D lists all the distributions that may be used with the EXPRESSIONS element.

Below is an example EXPRESSIONS element:

```
EXPRESSIONS:
        1, , EXPO(12, 3):
        Process, NORM(7.5, 1.2, 8):
        3, Package, UNIF(10, 15, 4):
        4, , DISC(0.6, 0, 1.0, 1, 7) * 9.5;
```

The first entry defines an exponential distribution with a mean of 12 and random number stream 3. *Number* is specified as 1 and *Name* is left blank. To obtain a sample value from this distribution, the modeler would specify ED(k), where k evaluates to 1. The second entry defines a normal distribution with a mean and standard deviation of 7.5 and 1.2, respectively. Random number stream 8 is used. The *Name* operand is specified as *Process* and the *Number* is defaulted. In order to obtain a sample value, the model frame would contain a block that references *Process* or ED(k), where k evaluates to 2.

The third entry defines a uniform distribution with a minimum of 10, a maximum of 15, and uses the fourth random number stream. *Name* is specified as *Package,* and *Number* is 3. To retrieve a sample value from this distribution, the model frame would contain a reference to *Package* or ED(k), where k has the value of 3. Note that if *Number* had been defaulted, ED(k) would still return a sample from this distribution, if k is equal to 3. The last line of the example EXPRESSIONS element generates a sample that has value 0 60% of the time and 9.5 40% of the time. The discrete distribution (DISC) generates a 0 or a 1 with probabilities 0.6 and 0.4, respectively, and utilizes the seventh random number stream. The resultant 0 or 1 is multiplied by 9.5 to give the sample value for ED(k), where k has the value of 4. The purpose of this example is to illustrate the use of a mathematical function for the *Expressions* operand. An alternative method to achieve the same results would be DISC(0.6, 0, 1.0, 9.5, 7).

Following is an EXPRESSIONS element that defines two one-dimensional arrays and a matrix:

```
EXPRESSIONS:
        TypeFact(3), 2.2, 1.3, 0.9:
        MachTime(2), 4.7, 3.5:
        Finish(3, 2), TypeFact(1) * MachTime(1),
                      TypeFact(2) * MachTime(1),
                      TypeFact(3) * MachTime(1),
                      TypeFact(1) * MachTime(2),
                      TypeFact(2) * MachTime(2);
```

In this example, *TypeFact* is a one-dimensional array that contains process time factors. *MachTime* is a one-dimensional array that contains machine dependent standard process times. For example, *TypeFact(1)* represents a process time factor for a type one job and has a value of 2.2. *MachTime(2)* represents the standard process time of machine two and has a value of 3.5 minutes. This means that a type one job on machine two would require 7.7 ($2.2 * 3.5$) minutes to complete processing. The *Finish* array has three rows and two columns. Note that the *Finish* array expressions reference preceding expressions and that the array element *Finish*(3, 2) is defaulted to the last specified entry in the array

(*Finish*(2, 2)), which has the value 4.55 (1.3 * 3.5). Each entry in the *Finish* array is the product of two elements of the first two arrays.

An example showing how an array could be used in the model frame follows:

```
QUEUE,     1;
SEIZE,     Machine;
DELAY:     Finish(JobType, MachID);
```

In this code fragment, entities are placed in queue number 1 until a unit of the resource *Machine* becomes available. Once the resource is seized, the length of the delay is returned from *Finish* based on the values of *JobType* and *MachID*. If *JobType* is 1 and *MachID* is 2, then the delay will be 7.7 minutes.

The elements of the *Finish* array are entered one column at time. The order of the array is illustrated in Table 3.3.

Table 3.3 Elements of the *Finish* Array

Reference	Expression	Value
Finish(1,1)	TypeFact(1) * MachTime(1)	2.2 * 4.7
Finish(2,1)	TypeFact(2) * MachTime(1)	1.3 * 4.7
Finish(3,1)	TypeFact(3) * MachTime(1)	0.9 * 4.7
Finish(1,2)	TypeFact(1) * MachTime(2)	2.2 * 3.5
Finish(2,2)	TypeFact(2) * MachTime(2)	1.3 * 3.5

3.3.4 The QUEUES Element

The QUEUES element defines the queues referenced by the associated model frame. The format for the QUEUES element is as follows:

QUEUES: *Number, Name, Ranking Criterion, Block Label*:
repeats;

The operands are described as follows:

Operand	Description	Default
Number	Queue number [optional]	Consecutive integers beginning with 1
Name	Unique queue name	Blank

Operand	Description	Default
Ranking Criterion	Rule for ordering entities in the queue	FIFO
Block Label	LABEL of block in the model frame associated with this queuc	Determined during simulation run

The first operand, *Number,* is an optional integer value that can be used to identify individual queues or specify the total number of queues. If *Number* is defaulted, the comma following *Number* is omitted and the *Name* must be specified. In this case, the experiment processor will assign sequential values to *Number* for each *Name.*

The second operand, *Name,* is the alphanumeric identifier of the queue. If defaulted, the trailing comma must be included and *Number* must be specified.

The third operand, *Ranking Criterion,* determines which method is used to order the arriving entities in the queue. The four general ranking rules available in SIMAN V are as follows:

FIFO	First In First Out
LIFO	Last In First Out
LVF(Expression)	Low Value First based on *Expression*
HVF(Expression)	High Value First based on *Expression*

FIFO is the default ordering rule used by SIMAN V. The entities leave the queue in the same order that they arrived.

LIFO inserts entities in the queue based on their arrival time. The newest arrival is placed in the front of the queue and leaves before entities already waiting.

LVF positions entities in descending order based on the value of *Expression.* Typically, *Expression* will be an entity attribute or a function whose value depends on an attribute. Entities that encounter a LVF(*JobType*) queue will be ordered according to respective *JobType* values. Ties are broken on a FIFO basis.

HVF is the opposite of LVF. Entities with higher values for *Expression* are placed in front of the queue, and low values are placed in the rear.

The last operand, *Block Label,* can be the label of a QUEUE block in the model frame or the key word SHARED. A QUEUE block in the model frame of the form

Label QUEUE, *Expression, Capacity, Balk Label*;

can be linked to an entry in the QUEUES element by entering *Label* for the *Block Label* operand. This is most advantageous when *Expression* is a mathematical function that evaluates to a *Number* operand instead of specifying the *Name* operand. If a specific queue may be entered from multiple locations in the model frame, the key word SHARED must be used for *Block Label*. Following is an example experiment frame QUEUES element with several different types of queues:

```
QUEUES:
        ReceivingQ:
        ProcessingQ, HVF(JobType):
        PackagingQ, LVF(ArrivalTime):
        4, , , SHARED:
        10;
```

This QUEUES element defines 10 queues that may be used in the associated model frame.

The first entry uses the FIFO default ranking rule and may be referenced in the model frame as

```
QUEUE, 1;
```

or

```
QUEUE, ReceivingQ;
```

or

```
QUEUE, Expression;
```

where *Expression* is a function that evaluates to 1. The operands, *Capacity* and *Balk Label,* have been omitted here as they will be in the remainder of examples in this section. This implies that the example queues have an infinite capacity and will therefore incur no balking.

The second entry in the QUEUES element uses the HVF ranking rule based on the value of an attribute named *JobType.* Entity attributes are discussed further in Chapter 4. This queue may be referenced in the model frame as

```
QUEUE, 2;
```

or

```
QUEUE, ProcessingQ;
```

or

QUEUE, Expression;

where *Expression* evaluates to 2. When an entity enters the QUEUE block, the value of its *JobType* attribute will determine its initial position in the queue. The arriving entity is placed in front of all entities with a lower *JobType* and behind entities with a higher *JobType.* SIMAN V uses FIFO to break any ties.

The third entry in the QUEUES element uses the LVF ranking rule and may be referenced in the model frame as

QUEUE, 3;

or

QUEUE, PackagingQ;

or

QUEUE, Expression;

where *Expression* evaluates to 3. An arriving entity is placed in front of all entities with a higher *ArrivalTime* and behind entities with a lower *ArrivalTime.*

The fourth entry in the QUEUES element defines a shared queue with no name that uses the default FIFO ranking rule. The queue may be referenced in the model frame as

QUEUE, 4;

or

QUEUE, Expression;

where *Expression* evaluates to 4. Unlike the other examples, this queue may be referenced in multiple locations in the model frame.

The last entry in the QUEUES element is the integer 10. This means that there is a total of ten queues, the first four of which have been individually defined. The remaining six queues have no *Name* and use the default FIFO ranking rule. These six queues may be referenced in the model frame as

QUEUE, Number;

or

QUEUE, Expression;

where *Number* is an integer in the range 6 to 10 and *Expression* evaluates to an integer in the same range.

3.3.5 The RESOURCES Element

The RESOURCES element defines resources referenced by the associated model frame. The format for the RESOURCES element is

RESOURCES: *Number, Name,* CAPACITY*(Expression)*:
repeats;

Additional operands available for the RESOURCES element are discussed in Section 7.2. The basic operands for the RESOURCES element are described as follows:

Operand	Description	Default
Number	Resource number [optional]	Consecutive integers beginning with 1
Name	Unique alphanumeric resource name	—
CAPACITY	Keyword to indicate integer capacity	CAPACITY
Expression	Initial resource capacity	1

The first operand, *Number,* is an optional integer value that numerically identifies the resource. If defaulted, the trailing comma is also omitted and the experiment processor will assign a sequential integer for *Number.* If used, *Number* values have to be in ascending order.

The second operand, *Name,* is the unique alphanumeric identifier assigned to the resource. Unlike the QUEUES element, the RESOURCES element *Name* operand requires an entry.

The third operand, CAPACITY, is a SIMAN V keyword that indicates that the resource capacity is an integer value or expression.

The last operand, *Expression,* specifies the number of units of resource *Name* that are available. The value for *Expression* may be an integer or expression truncated to an integer. Since CAPACITY is the default for the CAPACITY operand, CAPACITY*(Expression)* may be replaced with just *Expression.*

Following is an example experiment frame RESOURCES element:

```
RESOURCES:
        1, Drill:
        2, Machine, CAPACITY(5):
        Worker, 2:
        Hoist:
        5, Tool3A;
```

This RESOURCES element defines 1 *Drill,* 5 units of *Machine,* 2 units of *Worker,* 1 *Hoist,* and 1 *Tool3A.* Notice that the capacity of the resource *Worker* is specified as 2, which is equivalent to CAPACITY*(2).* Refer to Section 3.2.5 to see how these resources are referenced in the model frame.

3.3.6 The COUNTERS Element

The COUNTERS element defines counters used by the model frame to count occurrences during the simulation. The format for the COUNTERS element is

COUNTERS: *Number, Name, Limit, Initialize Option,*
Output File: repeats;

The operands are described as follows:

Operand	Description	Default
Number	Counter number [optional]	Consecutive integers beginning with 1
Name	Unique alphanumeric counter name	Counter *Number*
Limit	Maximum counter value [positive integer]	Infinite
Initialize Option	Reset the counter for each replication [YES, NO, or Replicate]	Replicate
Output File	Output file for counter observations	No save

The first operand, *Number,* is an optional integer value that can be used to identify the counter. If used, *Number* values must be entered in ascending order. If defaulted, the trailing comma is also omitted and a *Name* must be assigned.

The second operand, *Name,* is an alphanumeric identifier of the counter. If *Name* is left blank, then the trailing comma is still required and a *Number* must be assigned.

The third operand, *Limit,* specifies the maximum value for the counter. If a counter reaches the specified *Limit,* then the simulation is terminated. If it is omitted, then SIMAN V assigns *Limit* to a default value of infinity.

The fourth operand, *Initialize Option,* determines if the counter is reset to zero between replications. The default value for *Initialize Option* is *Replicate.* Defaulting or specifying *Replicate* for *Initialize Option* causes the counter to be reset as specified by the REPLICATE element. If specified as NO, then the value of the counter at the end of one replication is the initial value at the start of the next replication. If specified as YES, the counter value is set to zero at the beginning of each replication.

The last operand, *Output File,* is the name of an additional output file dedicated to a counter. If used, the *Output File* contains detailed counter information that is not included in the summary report. *Output File* may be an alphanumeric filename or an integer. If an integer is used, then the file *FileName*.NNN will be created, where *FileName* is the same as the simulation program file name and NNN is the integer specified by the modeler. The *Output File* is used as input to the output processor. If used, the value entered for *Output File* must be unique.

Following is an example experiment frame COUNTERS element:

```
COUNTERS:
        5:
        6, Rejected Parts:
        Good Parts, 100, , "Products.dat";
```

The first entry defines five unlimited counters with no name that can be referenced in the model frame as

```
COUNT: Expression;
```

where *Expression* is an integer or an expression truncated to an integer in the range 1 to 5. The summary report will not contain entries for unnamed counters whose value does not change during the simulation. If one or more of the first five counters change during the simulation, the summary report will contain entries named Counter 1, Counter 2, and so on, depending on the counter number.

The second entry defines an unlimited counter named *Rejected Parts.* This counter may be referenced in a model frame COUNT block with the number 6, the name *Rejected Parts,* or an *Expression* that evaluates to 6. The summary report entry for this counter will be titled *Rejected Parts,* illustrating the benefit of specifying a *Name.*

The last entry defines a counter named *Good Parts* with a limit of 100 and *Output File* called *Products.dat.* The simulation will terminate if the counter reaches 100. Detailed counter data for *Good Parts* will be written to an output file named *Products.dat.*

3.3.7 The DSTATS Element

The DSTATS element collects time-averaged statistics for discrete events. Simulation components such as queues and resources change during the simulation. The number of entities in a queue can decrease, a resource can be seized, a transporter can be released, and so on. Each of these occurrences is a discrete event, or an event that changes the state of the system. The DSTATS element collects statistics on these events as they change through time. Statistics may be collected on user or system variables or expressions containing these variables. The summary report will contain the mean, variance, minimum, maximum, and final value for each entry in the DSTATS element. The format for the DSTATS element is as follows:

DSTATS: *Number, Expression, Name, Output File*: repeats;

The operands are described as follows:

Operand	**Description**	**Default**
Number	Statistic number [optional]	Consecutive integers beginning with 1
Expression	SIMAN V expression for which time-dependent statistics are to be collected	—
Name	Unique alphanumeric name of the DSTAT that will also be used as a title in the summary report	Expression
Output File	Output file for DSTAT observations	No save

The first operand, *Number,* is an optional integer statistic identifier. The *Number* and the trailing comma may be omitted. If *Number* is specified, then it must be entered in ascending order.

The second operand, *Expression,* is the SIMAN V or user-defined expression for which statistics are collected throughout the simulation.

The third operand, *Name,* is the summary report title associated with statistics collected on this *Expression.* If *Name* is defaulted, then SIMAN V will use the value of *Expression* as the name of the associated statistics in the summary report.

The last operand, *Output File,* is the name of an additional output file dedicated to a DSTAT *Expression.* If it is used, then the *Output File* will contain detailed information on the values of *Expression* over time. The use of *Output File* is identical to that of the COUNTERS *Output File* operand discussed in Section 3.3.6.

The following example makes use of two SIMAN V special purpose variables. The first, NR(*Resource Id*), returns the number of busy units of the resource *Resource ID.* The second, NQ(*Queue ID*), returns the number of units waiting in queue *Queue ID.*

```
DSTATS:
        NR(Machine)/3,      Machine Utilization:
        NR(4)/10,           Worker Utilization:
        NQ(3),              Service Line, "Service.dat";
```

The first entry in the DSTATS element collects statistics on the resource called *Machine.* The special purpose variable returns the number of units of *Machine* that are currently in use. In this case, *Machine* has capacity 3. If 1 unit of *Machine* is currently in use, *Expression* evaluates to 1/3 or 0.333. If *Machine* is operating at capacity, the result of *Expression* will be 1.000. The *Expression* is updated throughout the simulation and the time-persistent results are written to the summary report.

The second entry collects statistics on resource 4. From the entry, it is apparent that resource 4 has 10 units and is probably called *Worker.* The summary report will contain an entry titled *Worker Utilization* and all statistics collected for this *Expression.*

The third entry collects statistics on the number of entities in queue number 3. The summary report will contain an entry titled *Service Line* and the associated statistical values. Detailed DSTAT information collected on queue number 3 is also written to a file named *Service.dat.*

3.3.8 The REPLICATE Element

The REPLICATE element determines the number of simulation replications, the time for the first replication to begin, the length of each replication, initialization techniques between multiple replications, and the warm-up time. The format of the REPLICATE element is given by

REPLICATE, *Number of Reps, Beginning Time, Replication Length, Initialize System, Initialize Statistics, Warm Up Period*;

The operands are described as follows:

Operand	Description	Default
Number of Replications	Number of replications [integer]	1
Beginning Time	Start time for the first replication [constant]	0.0
Replication Length	Maximum runtime for each replication[constant]	Infinite
Initialize System	Reset system to empty and idle [YES or NO]	YES
Initialize Statistics	Reset statistics between replications [YES or NO]	YES
Warm Up Period	Warm-up time to reach steady state [constant]	0.0

The first operand, *Number of Replications,* is an integer value that specifies the number of times the simulation will be run. A summary report is generated for each replication.

The second operand, *Beginning Time,* determines the starting time of the first replication. The third operand, *Replication Length,* indicates the maximum run time for each replication. If the length of the simulation is only limited by *Replication Length,* then the first replication will run *Beginning Time* + *Replication Length* time units.

The fourth operand, *Initialize System,* specifies whether or not the system is cleared and reset between replications. If *Initialize System* is set to YES, then any entities residing in the model at termination are destroyed prior to the next replication. Otherwise, a subsequent replication will start with entities from the previous replication still in the system. The fifth operand, *Initialize Statistics,* specifies whether or not collected statistics are cleared between replications.

The last operand, *Warm Up Period*, determines the amount of time to run the simulation before collecting statistics. The *Warm Up Period* operand is generally used to allow a nonterminating system to reach steady state prior to collecting statistics.

If the REPLICATE element is not included in the experiment frame, then one replication will continue to run until some other mechanism such as a COUNTERS limit terminates the simulation.

Some examples of the REPLICATE element are as follows:

```
REPLICATE, 1, 0, 2400;
```

This REPLICATE element will result in one replication lasting 2400 time units.

```
REPLICATE, 3, 0, 500, , , 100;
```

This example specifies three replications of 500 time units each. The system will warm up for 100 time units at which time the statistics are cleared.

```
REPLICATE, 10, 0, 100, NO, YES;
```

This example specifies ten replications of 100 time units each. The system state at the end of each replication is carried over to the next replication and the statistics are reset between replications.

```
REPLICATE, 4, 0, 200, YES, NO, 15;
```

This REPLICATE element will result in four replications of 200 time units each. The system is set to empty and idle at the end of each replication. The statistics are cleared after a 15-time-unit warm-up period.

3.3.9 Experiment Frame for Example 3.1

For Example 3.1, the experiment frame will be used to define resources and queues, supply sample values for four distributions, collect statistics and counter values, and determine the length of the simulation. Exhibit 3.2 illustrates the complete experiment frame for Example 3.1.

EXHIBIT 3.2 **Experiment frame for Example 3.1.**

```
!Example Problem 3.1: The Three Machine Problem
!
BEGIN;
PROJECT,          Three Machine, Team;
```

EXHIBIT 3.2 *continued.*

```
EXPRESSIONS:  ArrivalRate,          EXPO(5,1):
              DrillTime,            UNIF(6,9,1):
              MillTime,             TRIA(10,14,18,1):
              GrindTime,            DISC(0.25, 6, 0.75, 8, 1.0, 12,1);
QUEUES:       1, DrillQ:
              2, MillQ:
              3, GrindQ;
RESOURCES:    1, Drill, 2:
              2, Mill, 3:
              3, Grinder, 2;
COUNTERS:     1, Jobs Completed:
              2, Number Balked;
DSTATS:       1, NQ(DrillQ),        Jobs in Drill Queue:
              2, NQ(MillQ),         Jobs in Mill Queue:
              3, NQ(GrindQ),        Jobs in Grinder Queue:
              4, NR(Drill)/2,       Drill Utilization:
              5, NR(Mill)/3,        Mill Utilization:
              6, NR(Grinder)/2,     Grinder Utilization;
REPLICATE, 1, 0, 2400;
```

3.3.10 Example 3.1: Solution

The summary report for Example 3.1 is given in Exhibit 3.3. At the beginning of this chapter, a problem was stated and the following four questions were asked:

EXHIBIT 3.3 **Summary report for Example 3.1.**

Summary for Replication 1 of 1

Project: Three Machine   Run execution date: 8/24/1994
Analyst: Team   Model revision date: 8/24/1994
Replication ended at time: 2400.0

DISCRETE-CHANGE VARIABLES

Identifier	Average	Variation	Minimum	Maximum	Final Value
Jobs in Drill Queue	.35486	1.7921	.00000	2.0000	.00000
Jobs in Mill Queue	.76285	1.1640	.00000	3.0000	.00000
Jobs in Grinder Queue	.23274	2.0067	.00000	2.0000	1.0000
Drill Utilization	.70734	.52211	.00000	1.0000	.00000
Mill Utilization	.86108	.29546	.00000	1.0000	1.0000
Grinder Utilization	.77446	.42441	.00000	1.0000	1.0000

EXHIBIT 3.3 *continued.*

COUNTERS

Identifier	Count	Limit
Jobs Completed	435	Infinite
Number Balked	46	Infinite

Simulation run complete.

1. How many jobs are completed?
2. What is the utilization rate for each class of resource?
3. What is the total number of jobs ejected due to full queues?
4. What is the average number of jobs in each queue?

Answers to the first and third questions are contained in the COUNTERS section of the summary report. There were 435 completed jobs and 46 jobs that balked, or encountered a full queue.

The utilization rates for each class of resource are contained in the Average column of the DISCRETE-CHANGE VARIABLES section. For example, the average Mill Utilization rate is about 86.1%. If the average number of mills in use is needed, then multiply the utilization rate by the number of mills.

The average number of jobs in each queue is also contained in the Average column of the DISCRETE-CHANGE VARIABLES section. For example, the average number of jobs waiting in the drill queue was approximately 0.355.

3.4 SUMMARY

This chapter introduces basic SIMAN V blocks and elements. Most of these concepts will be used in every simulation model and are the foundation for simulating more complex systems. This chapter also discussed the motivation behind the dual-frame structure of SIMAN V. The model frame contains the actual "simulation" code, whereas the experiment frame provides data, collects statistics, and generates reports, among other functions.

The QUEUE, SEIZE, DELAY, and RELEASE model frame block sequence is very common and represents the standard processing of an entity by a resource. The CREATE block is used to introduce entities into the model. The COUNT block increments or decrements a counter.

The value of the counter will be contained in the simulation output. DISPOSE may be used as a block or modifier and removes entities from the system.

The experiment frame identifies the simulation in the PROJECT element. Queues and resources referenced in the model frame are defined with the experiment frame QUEUES and RESOURCES elements. The COUNTERS element identifies the various counters used by the model frame and provides the counter title to appear in the simulation output. The DSTATS element collects time-averaged statistics for discrete events such as the number of entities in queue or number of resources in use. The REPLICATE element specifies the length of the simulation, number of replications, warm-up period, and system state after each replication.

3.5 EXERCISES

E3.1 Consider a system that manufactures electronic parts. There are three stations through which the parts must go: assembly, soldering, and inspection. Time between orders for parts is exponentially distributed with a mean of 15 minutes. The distributions of time spent (in minutes) at each workstation are given as follows:

Assembly	NORM(14, 2)
Soldering	EXPO(22)
Inspection	UNIF(13, 17)

Assembly and inspection are performed on one piece at a time. The soldering operation can process two jobs at a time. The shop closes after 8 hours. If parts are still being processed when the shop closes, they will be completed the following day. Simulate this system for 5 days. Reset the statistics after each run, but do not reset the system. Use random number stream 1 for all distributions. Collect and print statistics on the utilization at each station, their associated queues, and total number of parts produced during each 8-hour shift.

E3.2 Consider Exercise 3.1 with the following enhancements: When a new part is started, it is placed on a special tray. The part will remain on the tray until soldering is complete. There are only four of these trays available. Orders that arrive when all the trays are in use are sent to an alternate assembly area.

a. What is the utilization of each type of resource?

b. How many trays are needed to ensure no orders are sent to the alternate assembly area?

E3.3 Consider Example 3.1 with the following enhancements: Due to increased demand, the manufacturer has added another production area.

The new production area is designated as the Overflow Production Facility (OPF) and contains 1 drill, 1 mill, and 1 grinder. Processing on the OPF drill is distributed uniformly between 4.5 and 8 minutes. The OPF mill has an exponentially distributed process time with a mean of 5 minutes. The OPF grinder process times are normally distributed with a 7.5-minute mean and standard deviation of 1.2 minutes. Orders now arrive at the original production area according to an exponential distribution with a mean of 3.3 minutes. Orders that cannot enter a queue due to capacity limitations are sent to the same location in the OPF to continue processing. For example, if an order arrives to the original milling area when the queue is full, then that order is transferred to the OPF mill. When milling has been completed, the order will stay in the OPF and proceed to the OPF grinder. Orders that encounter a full queue in the OPF are ejected from the system. Each of the new machines may have up to two orders waiting to be processed. Transportation times are assumed to be negligible and random number stream 3 is to be used for all new processes. Simulate the system for 40 hours and answer the following questions:

- **a.** What is the total number of orders completed?
- **b.** Of these, how many were completed in the original facility?
- **c.** What is the utilization rate of the OPF grinder?
- **d.** What is the average number of orders in the original mill queue?
- **e.** What is the average number of orders in each queue in the OPF?
- **f.** What is the utilization rate of the original facility and the OPF?
- **g.** What is the average number of orders in the system?
- **h.** How many orders are ejected from each queue in the OPF?
- **i.** How many orders fail to complete processing?

E3.4 Three types of castings arrive at a workstation, each with exponentially distributed interarrival times. First, the castings are degreased by a single worker in uniformly distributed time of 45 ± 15 seconds. Then the worker takes 5 seconds to position the casting on an automated machine. Machine utilization is the highest priority. In other words, if the machine is idle with no castings waiting to be positioned, or the machine is busy when a new casting arrives, then the worker can start degreasing the new casting. When degreasing is finished, the worker will attempt to place a casting on the machine before starting to degrease another casting. The time to machine each type of casting and send it to the next station (an automated process) is normally distributed. The mean arrival times and processing means and standard deviations for each type of casting are as follows:

Casting Type	Interarrival Times Mean (sec)	Process Times Mean (sec)	Process Times Std. Dev. (sec)
1	198	32	5
2	167	43	7
3	204	36	6

There is no limit to the number of castings that may be waiting to be degreased or waiting to be placed on the machine. Use random number stream 7 for all distributions. Simulate for one 8-hour day and answer the following questions:

a. What is the worker utilization rate?

b. What is the average number of castings waiting to be placed on the machine?

c. What is the utilization of the machine?

d. How many total castings are completed?

e. Summarize the state of the system at the end of the simulation. For example:

What type of casting is on the machine?
How many of each type of casting are waiting for the machine?
What type of casting is being degreased?
How many of each type of part are waiting to be degreased?

E3.5 Perform each of the following independently. (Return to the base system of E3.4 after each modification.)

a. Type one castings now arrive two at a time, type two castings arrive five at a time, and type three castings arrive four at a time. There are now four workers and four machines. Include the average number of castings waiting for a worker in the summary report.

b. The machine has been replaced. The new machine is 6 seconds faster than the old one. Run this simulation for 5 days and generate five independent reports, one for each day. Do not initialize the system between each run, but do clear collected statistics between each replication. Answer the questions from E3.4 for the first of the five runs.

c. The buffer before the worker will hold only two castings. Castings that cannot enter the buffer are discarded. How many are discarded?

d. Workers, but not machines, take a 30-minute break at 10:00 AM and 2:00 PM. The break does not interrupt in-process work. Workers begin at 8:00 AM and go home at 4:00 PM. Answer the questions from E3.4. Comment about the utilization rate of the worker.

E3.6 A short-term parking lot holds 50 cars. Cars arrive according to an exponential distribution with a mean of 3 minutes. Cars stay in the lot according to a triangular distribution with a minimum of 30 minutes, maximum of 6 hours, and a mode of 2 hours. The simulation stops when the lot reaches capacity.

a. Beginning empty, how long is it before the lot fills?

b. What is the total number of cars entering the lot?

c. Comment on the utilization of the lot.

E3.7 A process consists of cleaning, welding, and inspecting. The cleaning process is broken into two phases. The first phase is 1.5 minutes and is done automatically. The second phase is 1.5 minutes and requires the operator. The operator accompanies the part from this point until the start of inspection. Welding requires a time that is uniformly distributed between 15 and 45 seconds. Inspection is accomplished in a time that is exponentially distributed with a mean of 2 minutes. There is only one operator. There is unlimited waiting space prior to the cleaning process. However, no parts are allowed to wait prior to welding and only one part can be waiting for inspection. Arrivals occur according to an exponential distribution with a mean of 5 minutes. Simulate the system until 500 parts have been completed.

a. Collect statistics on operator utilization.

b. Collect statistics on parts waiting to be cleaned.

c. Determine if any parts were rejected due to limited queue capacity.

4

Intermediate Concepts

In Chapter 3 you learned the basic modeling blocks and elements needed to create a workable SIMAN model. In this chapter we will expand upon these basic blocks to enhance the model and extend the simulation output.

EXAMPLE 4.1 Problem Statement

Consider the problem statement for Example 3.1. The same basic production process will remain intact. However, now the system produces jobs of two types, 70% of Type I and 30% of Type II. Type I jobs will follow a production path from the drill to the mill to the grinder, and finally to the inspector as shown in Figure 4.1. Any jobs that attempt to enter a queue at full capacity are ejected from the system. The maximum queue capacities for each resource are drill, 2, mill, 3, plane, 3, grinder, 2, and inspector, 4. Type II jobs will be processed using the drill, plane, grinder, and inspector. The delay time for the plane is based on a triangular distribution with parameters (20, 26, 32). The plane has a capacity of 3. The inspection time is normal with a mean of 3.6 and a standard deviation of 0.6. There is only one inspector. Processing time at the grinder is based on a discrete distribution: For Type I jobs, 10% take 6 minutes, 65% take 7 minutes, and 25% take 8 minutes; for Type II jobs, 10% take 6 minutes, 25% take 7 minutes, 30% take 8 minutes, 25% take 9 minutes, and 10% take 10 minutes. From historical information it is known that 90% of all jobs pass inspection, 5% are sent for rework, and 5% are scrapped. Reworked jobs are sent back to the drill. At the grinder, Type I jobs take priority over Type II jobs. The following information is desired:

1. Number of jobs completed by type.
2. Number of jobs ejected from the queues.

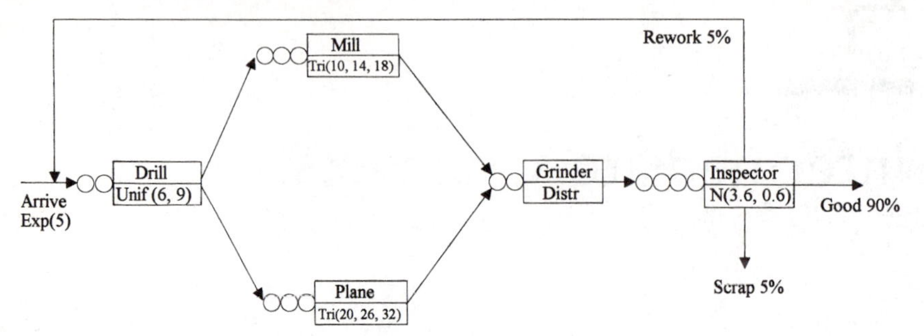

Figure 4.1 Processing Systems for Example 4.1

3. Number of failed jobs.
4. Number of reworked jobs.
5. Utilization of each resource.
6. Average number of entities in each queue over time.
7. Average flowtime (time in system) of a job.
8. Time between job exits.

The new blocks and element introduced in this chapter are shown in Table 4.1. These blocks and elements are essential to virtually all SIMAN simulations.

4.1 Intermediate Model Frame Concepts

A very important aspect of SIMAN that makes your models more flexible and allows for expanded output is the inclusion of general purpose attributes and variables. A SIMAN variable is global, meaning that a change in its value is registered throughout the entire model. A SIMAN attribute is a value that is identified locally. This attribute is part of the information carried by a specific entity. Thus, when an entity's attribute is changed, it is changed for that entity only. An attribute may refer to the color or size of an entity, while a variable may refer to revenue for each unit produced. The term "general purpose" is used because SIMAN reserves some attributes and variables for special purposes. For example, the variable *TNOW* is always used to hold the current simulation time. The modeler cannot use *TNOW* as a general-purpose variable. SIMAN attributes and variables may be used as operands in some cases and may be used as part of a valid SIMAN expression.

Table 4.1 Blocks/Elements Introduced in this Chapter

New Block/Element	Purpose
ASSIGN block	Used to fill a SIMAN variable or attribute with a value
TALLY block	Used to gather time-averaged date
BRANCH block	Used to create decision points within the model
ATTRIBUTES element	Defines the symbolic names and initial values of general-purpose attributes used in the model
VARIABLES element	Defines the symbolic names and initial values of general-purpose variables used in the model
PARAMETERS element	Defines the parameters used in experimental distributions (e.g., mean, standard deviation, minimum, maximum)
TALLIES element	Used in the experiment frame to define TALLY block output variables in the summary report

4.1.1 The ASSIGN Block

The ASSIGN block allows values to be given to attributes and variables. The format of the ASSIGN block is as follows:

ASSIGN: *Variable* or *Attribute* = *Value*: *repeats*;

Operand	Description	Default
Variable or Attribute	SIMAN variable or attribute	—
Value	Value to be assigned [expression].	—

Examples of the ASSIGN block are as follows:

```
ASSIGN:  JobType = ED(3);
```

This ASSIGN block gives the attribute *JobType* a value expressed by an experimental distribution.

```
ASSIGN:  Color = JobType + 7:
         Return = 34.5;
```

The value of the attribute *Color* is determined by adding 7 to the value of attribute *JobType*. In addition, the variable *Return* is set to a constant value of 34.5.

4.1.2 The BRANCH Block

Oftentimes, entities follow varying paths. Perhaps jobs of a different type are sent to various machining locations, perhaps we know that the probability is 0.05 that all completed jobs will need rework. SIMAN allows these determinations to be made using the BRANCH block. The BRANCH block has the following format:

```
BRANCH,     Max Number of Branches, Random Number Stream:
            WITH, Probability, Label, Primary:
            IF, Condition, Label, Primary:
            ELSE, Label, Primary:
            ALWAYS, Label, Primary;
```

Operand	**Description**	**Default**
Maximum Number of Branches	Maximum number of branches that may be taken	Infinite
Random Number Stream	Random number stream to use with the WITH rule [integer]	10 or a number specified in SETUP
Probability	Probability of selecting the branch [expression between 0 and 1]	—
Label	Label where a copy of the entity is routed	—
Condition	Branch condition [expression]	—
Primary	Whether the primary entity may select this branch	Yes

An example using the BRANCH block is as follows:

```
BRANCH, 1, 3:
        WITH, .8, Exit:
        WITH, .05, First:
        WITH, .15, Last;
```

This BRANCH block sends an entity to the label *Exit* with probability 0.8, to *First* with probability 0.05, and to *Last* with probability 0.15. The probabilities must sum to 1. *Exit, First,* and *Last* must all be valid labels within the model. A branch using the WITH operand is called a *probabilistic* branch. The *Random Number Stream* operand determines which random number stream will be used to calculate the probabilities. Random number stream 10 is the default value. This operand is only necessary when a probabilistic branch is used in the BRANCH block.

As another example of the BRANCH block consider the following:

```
BRANCH, 1:
        IF, JobType == 1, Type1:
        ELSE, Type2;
```

This BRANCH block sends an entity to label *Type1* if its *JobType* is 1. The logical operand, "==", indicates equality, whereas "=" indicates assignment of a value to a variable or attribute. Refer to Table 4.2 for a list of logical operands. Otherwise, the entity is sent to label *Type2*. This is called a *conditional* branch, and any valid logical SIMAN expression may be used as the *Condition*. If, for some reason, the entity cannot take any branch, it is disposed.

Operands may be used in combination, and that brings us to the *Max Number of Branches* operand. *Max Number of Branches* defines the number of branches an entity may choose. If the value is set to 1, then the first branch chosen is the only one taken. The default value for *Max Number of Branches* is the total number of available branches.

Consider the following BRANCH block:

```
BRANCH, 2:
        WITH, .8, Exit:
        WITH, .2, Redo:
        ALWAYS, Check;
```

Table 4.2 Conditional Branch Logical Operands

Logical Operand	Style 1 (FORTRAN)	Style 2
Exactly equal to	.EQ.	==
Not equal to	.NE.	<>
Less than	.LT.	<
Less than or equal to	.LE.	<=
Greater than	.GT.	>
Greater than or equal to	.GE.	>=
Logical AND	.AND.	
Logical OR	.OR.	

In this BRANCH block, the entity passing through it is essentially cloned. One entity will be directed to *Exit* or *Redo* with respective probabilities of 0.8 or 0.2. In addition, the cloned entity will always be sent to label *Check*. If *Max Number of Branches* branches have been chosen and an ALWAYS is encountered, a runtime error will occur. Branches using ALWAYS or ELSE are called *deterministic* branches.

The result of having two identical entities in the model at the same time may be quite useful. For example, the original entity may be used in the simulated manufacturing process, while the copy represents information. In a hospital, there exists a patient and the patient's chart; in a job shop, there may be component parts that are assembled into the final product. In the previous example, the label *Check* may be attached to a COUNT block, which immediately DISPOSEs the entity after it is counted. Thus, the secondary entity sent to *Check* performs its function and is removed from the model. Its parent, the primary entity, remains intact, unaffected by the fate of its clone.

4.1.3 The TALLY Block

The TALLY block records statistics on the mean, variance, minimum and maximum values, and the number of observations for display in the output file or another external file. The format for the TALLY block is as follows:

TALLY: *Tally ID, Value*: *repeats*;

Operand	Description	Default
TallyID	Tally location to record statistics [expression truncated to an integer, tally number, or a tally symbol name]	—
Value	Value to be recorded [expression, INTERVAL(AttributeID), BETWEEN(VariableID), or BETWEEN]	—

For Example 4.1, the TALLY block is used in the following manner:

```
Pass TALLY:
            Flowtime, INTERVAL(ArrTime);
     TALLY:
            Exit Period, BETWEEN;
```

Table 4.3 INTERVAL and BETWEEN Calculations

Created at	Tallied at	INTERVAL Time	BETWEEN Time
1734	1840	106	—
1743	1853	110	13
1750	1870	120	17

The block labeled *Pass* records statistics concerning the interval between *ArrTime,* marked at the CREATE block, and this TALLY block in the output variable Flowtime. The next TALLY block records statistics on the time between executions of the block.

More precisely, the *INTERVAL* operand records statistics on the difference between TNOW (the current time of the simulation clock) and the value of *AttributeID.* (INT can be used as an abbreviation for INTERVAL.)

BETWEEN stores the time between entity executions of the block. (BET can be used as an abbreviation for BETWEEN.) Adding a *VariableID* in parentheses stores those statistics in a global variable. If the BETWEEN option references the same variable at each execution, then the value is reset. The *VariableID* may be an indexed global variable if separate values must be stored.

Table 4.3 shows a possible numerical example for the above TALLY blocks. In Table 4.3, the first entity is created at time 1734 and executes the TALLY block at time 1840. The length of time between 1734 and 1840, the interval, is 106 time units. The next entity is created at 1743 and executes the TALLY block at time 1853. The intervening value is 110 time units. The time between the two executions of the TALLY block, the BETWEEN time, is 13 time units. The third entity, created at 1750, executes the TALLY block 120 time units later, 17 time units after the last entity.

4.1.4 Modifiers

The NEXT(*Label*) modifier sends the executing entity to the indicated label. It is one way to create a loop within the model. In Example 4.1 all jobs, regardless of type, begin with processing at the drill. Type I jobs then move on to the mill, and Type II jobs then move on to the plane. After their respective processing, Type I and II jobs move to the grinder. We can use the NEXT modifier in the following way to send Type I and Type II jobs to the grinder:

```
          :
          RELEASE:
                    Mill:
                    NEXT(Grinding);
          :
          RELEASE:
                    PLANE:
                    NEXT(Grinding);
Grinding  QUEUE,
                    GrinderQ, 2, Eject;
          SEIZE:
                    Grind;
          :
```

Although the modifier NEXT(*Grinding*) in the second RELEASE block is not needed, many modelers choose to include it in case intervening blocks are added later.

The JUMP(*Expression*) modifier performs a similar function to the NEXT modifier except that the executing entity is sent to the block determined by *Expression*. No time elapses during the execution of a NEXT or JUMP modifier, and no blocks between the current block and the Labeled block or block sequence number, respectively, are executed.

The MARK(*Attribute ID*) block assigns the value of TNOW to the attribute identified by *Attribute ID*. TNOW is an internal SIMAN variable that tracks the simulation time. *Attribute ID* is an integer attribute number or a symbol name.

In Example 4.1 we are interested in determining the flowtime of the jobs. We have used the TALLY block with the attribute *ArrTime* (see Section 4.1.3). As entities are created in the model, we need to mark their arrival time in the attribute *ArrTime* for use by the TALLY block as shown below:

```
CREATE:
        EX(1, 1):
        MARK(ArrTime);
```

The MARK modifier may be used with other modifiers by separating them with a colon as shown in the following example:

```
CREATE:
        EX(1, 1):
        MARK(ArrTime):
        NEXT(Start);
```

Here both the MARK and NEXT modifiers are used with the CREATE block. Otherwise, only one modifier can be used per block. Additionally, not all blocks can make use of every modifier. These exceptions will be discussed as they arise.

4.2 Intermediate Experiment Frame Concepts

The following SIMAN elements relate to the model frame blocks discussed in the preceding section.

4.2.1 The ATTRIBUTES and the VARIABLES Elements

When using general purpose attributes and variables within the model frame, they must first be defined within the experiment frame. This is accomplished using the ATTRIBUTES and VARIABLES elements, whose formats are as follows:

ATTRIBUTES: *Number, Name(Index), Value, . . .* : repeats;
VARIABLES: *Number, Name(Index), Value, . . .* : repeats;

Operand	**Description**	**Default**
Number	Sequential numerical identifier	Sequential
Name	Symbolic name for the ATTRIBUTE or VARIABLE	—
Index	Number if part of an array	—
Value	Information to be stored in the ATTRIBUTE or VARIABLE	—

As in many other experiment frame elements, the *Number* operand and its trailing comma may be omitted in order to invoke SIMAN's internal numbering system. You may explicitly specify a *Number* for each. If you choose to do this, then each attribute or variable must have a unique *Number,* and they must appear in ascending order.

Name(*Index*) refers to the unique name of the attribute or variable. The *Index* operand refers to an attribute or variable that is an array. You also may specify initial constant values for attributes and variables using the *Index* operand. If omitted, each attribute and variable will default to 0 as an initial value.

Examples are as follows:

```
ATTRIBUTES:  JobType, 1:
             Color;
```

JobType is the first attribute and has an initial value of 1. *Color* is the second attribute and has an initial value of zero. All entities will have a *JobType* of 1 and a *Color* of 0 until a new value is assigned.

```
VARIABLES:  Revenue:
            Stock, 100;
```

The variable *Revenue* has an initial value of zero while the variable *Stock* begins the simulation at a level of 100.

Arrays may be specified by inserting the subscript range in parenthesis. If a single number N is used, SIMAN assumes that the subscript range is 1 to N.

Consider the following examples:

```
VARIABLES:  1, Bin(15), 1:
            16-20, Shelf(5);
```

In this example, the variable *Bin* is allowed to reference subscripts between 1 and 15; all subscripts are initially set to a value of 1. The variable *Shelf* will allow subscripts from 1 to 5 and are set initially to 0. Notice that *Shelf(1)* is the 16th variable in the SIMAN model. Each subscript must have a unique number. Therefore, the *Shelf* series begins numbering at 16.

In order to avoid confusion concerning this, it is advisable to use the internal numbering feature of SIMAN in the following way:

```
VARIABLES:  Bin(15),1:
            Shelf(5);
```

Consider the following example:

```
ATTRIBUTES:  Temps(-2..2), 12, 20, 25, 20, 12:
             Param(2, 2), 3, 2, 5.6, 3.1;
```

Two attributes, *Temps* and *Param,* are defined in this example. *Temps* is a one-dimensional variable like *Bin* and *Shelf,* with subscripts ranging from -2 to 2 including zero. The numbers shown are the initial values of the five attributes *Temps(-2)* through *Temps(2)*. *Param* is a two-dimensional variable (notice the comma instead of the two dots separating the numbers). Initial values are also provided: *Param(1, 1)* = 3, *Param(1, 2)* = 5.6, *Param(2, 1)* = 2, and *Param(2, 2)* = 3.1. The values may be placed all on one line in the experiment frame. They are displayed here as a matrix to clarify how SIMAN will assign the values.

3	2
5.6	3.1

4.2.2 The PARAMETERS Element

The two-letter random variable functions used in Example 4.1 rely on the PARAMETERS element to define the parameters of the function. By defining parameters in this way, only the experiment frame needs to be changed for running the simulation under varying interarrival and service time data. The format is as follows:

PARAMETERS: *Number, Name, Pl, ..., Pu*: repeats;

The operands are described as follows:

Operand	Description	Default
Number	Parameter set number [integer]	Sequential
Name	Parameter set name [symbol name]	Blank
Pl, . . . , Pu	Lower through upper parameter values [constant]	—

Parameter sets are referenced by their *Number* and/or *Name*. Either the *Number* or *Name* (and their trailing commas) may be omitted. If numbers are used, then they must be entered in ascending sequential order. The values for *Pl, . . . , Pu* depend upon which random number function is being used.

In Example 4.1 the PARAMETERS Element is used in the following way:

```
PARAMETERS:  1, 5:
             2, .7, 1, 1.0, 2:
             3, 6, 9:
             4, 12, 16, 20:
             5, 20, 26, 32:
             6, .1, 6, .75, 7, 1.0, 8:
             7, .1, 6, .35, 7, .65, 8, .9, 9, 1.0, 10:
             8, 3.6, .6;
```

Eight parameter sets are defined for use with eight different distributions. For example, the first distribution occurs as:

```
CREATE:
        EX(1, 1):
        MARK(ArrTime);
```

The operand *EX(1, 1)* indicates that an exponential distribution is to be used with its parameters located in set 1 of the PARAMETERS element. The second number indicates that random number stream 1 should be used. Thus *EX(1, 2)* uses the second random number stream. The first parameter set is:

```
PARAMETERS:  1, 5:
                :
```

indicating that the value 5 will be used as the mean of the exponential distribution.

4.2.3 The TALLIES Element

This element, like the COUNTERS element for the COUNT block, defines valid TALLY variables for the output file. The format is as follows:

TALLIES: *Number, Name, Output File*: repeats;

Operand	Description	Default
Number	Tally number [integer]	Sequential
Name	Tally name and identifier for labeling the statistics in the summary report [symbol name]	*"Tally Number"*
Output File	External file name or unit for storing observations	No save

The summary report uses the symbol name of the TALLY variable. If no *Name* is given, then it appears in the report as "Tally *number*," where *number* is a sequential number. The current value of any TALLY variable may be examined using the variable TAVG(*identifier*) where the *identifier* is the TALLY *Number* or *Name*.

TALLIES recorded by SIMAN are the arithmetic average of all observations, independent of time. Time averages should be measured using the DSTATS element.

For Example 4.1, the TALLIES element is used in the following manner:

```
TALLIES:  Flowtime:
          Exit Period;
```

The two tallies are named *Flowtime* and *Exit Period.*

EXAMPLE 4.1: Solution

In the following sections the model and experiment frames for Example 4.1 are shown in their entirety as Exhibits 4.1 and 4.2. The summary report is shown in Exhibit 4.3. The answers to the questions asked in the Problem Statement follow Exhibit 4.3.

The Model Frame

Refer to the model frame in Exhibit 4.1 as the logic of the model is presented. An entity is created according to an exponential distribution with mean given by the first PARAMETERS set and using random number stream 1. The arrival time of the entity is then recorded in the attribute *ArrTime.* A number of attributes are then assigned to the entity including the *JobType* and all the processing times at each resource.

The entity proceeds through the drilling operation and then reaches a BRANCH block. Based on *JobType,* the entity will then proceed to the block labeled *Milling* or to the block labeled *Planing.* After the milling and planing operations, the entity is sent to the block labeled *Grinding.*

After the inspection process, the entity reaches another BRANCH block. The entity may take at most one branch, and any probabilistic branching will use random number stream 10. Based on the probabilities given, the entity will be sent to the block labeled *Rework,* the block labeled *Scrap*, or to the block labeled *Done.*

At block *Done* the entity's INTERVAL and BETWEEN times are calculated for tallies *Flowtime* and *Exit Period,* the entity is counted as a completed job and then destroyed. An entity entering block *Rework* is counted and then sent back to the drilling operation. Jobs sent to the *Scrap* or *Eject* blocks are counted and then destroyed.

EXHIBIT 4.1 Model frame for Example 4.1.

```
!Model Frame for Example 4.1
!
BEGIN,NO;
          CREATE:                   !Create arriving jobs according to
            EX(1, 1):               !Exponential distribution and save
            MARK(ArrTime);          Arrival times in attribute ArrTime
          ASSIGN:                   !Store job type and delay times
            JobType = DP(2, 1):     !Type I or Type II job
            DrillTime = UN(3, 1):   !Uniform drill time
            MillTime = TR(4, 1):    !Triangular mill time
            PlaneTime = TR(5, 1):   !Triangular plane time
            GrindTime = DP(JobType+5, 1):
                                    !Grind time is dependent on job type
            InspectTime = RN(8, 1);Normally distributed inspection time
Drilling  QUEUE,                    !Queue for drill with capacity 2
            DrillQ, 2, Eject;       Send overflow to Eject
          SEIZE:
            Drill;                  !Capture resource Drill
          DELAY:
            DrillTime;              !Process for DrillTime
          RELEASE:
            Drill;                  !Free the resource Drill
          BRANCH,1:
            IF, JobType == 1, Milling:
                                    !Send Type I to Milling
            ELSE, Planing;          Send Type II to Planing
Milling   QUEUE,                    !Queue for mill with capacity 3
            MillQ, 3, Eject;        Send overflow to Eject
          SEIZE:
            Mill;                   Capture resource Mill
          DELAY:
            MillTime;               Process for MillTime
          RELEASE:
            Mill:                   !Free the resource Mill
            NEXT(Grinding);
Planing   QUEUE                     !Queue for plane with capacity 3
            PlaneQ, 3, Eject;       Send overflow to Eject
          SEIZE:
            Plane;                  Capture resource Plane
          DELAY:
            PlaneTime;              Process for PlaneTime
          RELEASE:
            Plane:                  !Free the resource Plane
            NEXT(Grinding);
Grinding QUEUE,                     !Queue for grinder with capacity 2
            GrinderQ, 2, Eject;     Send overflow to Eject
```

EXHIBIT 4.1 *continued.*

```
        SEIZE:
          Grinder;               Capture resource Grinder
        DELAY:
          GrindTime;             Process for GrindTime
        RELEASE:
          Grinder;               Free the resource Grinder
Inspect QUEUE,                   !Queue for inspector with capacity 4
          InspectorQ, 4, Eject;  Send overflow to Eject
        SEIZE:
          Inspector;             Capture resource Inspector
        DELAY:
          InspectTime;           Process for InspectTime
        RELEASE:
          Inspector;             Free the resource Inspector
        BRANCH, 1, 10:           !Take only one branch
          WITH, 0.05, Rework:    !Rework 5% of the jobs
          WITH, 0.05, Scrap:     !Scrap 5% of the jobs
          ELSE, Done;            Remaining 90% are complete
Done    TALLY:                   !Tally the average flowtime
          Flowtime, INT(ArrTime);
        TALLY:                   !Tally the average time between exits
          Exit Period, BET;
        COUNT:
          JobType, 1:            !Count number of jobs by type and
          DISPOSE;               Destroy the entity
Rework  ASSIGN:                  !Reassign process times
          JobType = DP(2, 1):
          DrillTime = UN(3, 1):
          MillTime = TR(4, 1):
          PlaneTime = TR(5, 1):
          GrindTime = DP(JobType+5, 1):
          InspectTime = RN(8, 1);
        COUNT:
          Reworks, 1:            !Count the number of jobs reworked
          NEXT(Drilling);        Send reworks back to drill
Scrap   COUNT:
          Fails, 1:              !Count number failed jobs and
          DISPOSE;               Destroy the entity
Eject   COUNT:
          Ejections:             !Count the number of jobs ejected
          DISPOSE;               Destroy the entity
```

The Experiment Frame

Exhibit 4.2 shows the experiment frame for Example 4.2. The new elements ATTRIBUTES, PARAMETERS, and TALLIES are used.

EXHIBIT 4.2 **Experiment frame for Example 4.1.**

```
!Experiment Frame for Example 4.1
!
BEGIN,NO;
PROJECT, Example 4.1, Team;
ATTRIBUTES:   ArrTime:
              JobType:
              DrillTime:
              MillTime:
              PlaneTime:
              GrindTime:
              InspectTime;
RESOURCES:    Drill, 2:
              Mill, 3:
              Plane, 3:
              Grinder, 2:
              Inspector;
QUEUES:       DrillQ:
              MillQ:
              PlaneQ:
              GrinderQ, LVF(JobType):
              InspectorQ;
COUNTERS:     Type I Job Count:
              Type II Job Count:
              Fails:
              Reworks:
              Ejections;
PARAMETERS:   1, 5:
              2, .7, 1, 1.0, 2:
              3, 6, 9:
              4, 10, 14, 18:
              5, 20, 26, 32:
              6, .1, 6, .75, 7, 1.0, 8:
              7, .1, 6, .35, 7, .65, 8, .9, 9, 1.0, 10:
              8, 3.6, .6;
TALLIES:      Flowtime:
              Exit Period;
DSTATS:       NQ(DrillQ), Drill Queue:
              NQ(MillQ), Mill Queue:
              NQ(PlaneQ), Plane Queue:
              NQ(GrinderQ), Grinder Queue:
              NQ(InspectorQ), Inspect Queue:
              NR(1)/2, Drill Utilization:
              NR(2)/3, Mill Utilization:
              NR(3)/3, Plane Utilization:
              NR(4)/2, Grinder Utilization:
              NR(5), Inspector Util.;
REPLICATE, 1, 0, 2400;
```

Discussion of Output

In the problem statement for Example 4.1 eight items of information were requested. The summary report is given in Exhibit 4.3. The answers to the questions are as follows:

EXHIBIT 4.3 **Summary report for Example 4.1**

Summary for Replication 1 of 1

Project: Example 4.1 Run execution date : 8/24/1994
Analyst: Team Model revision date: 8/24/1994
Replication ended at time: 2400.0

TALLY VARIABLES

Identifier	Average	Variation	Minimum	Maximum	Observations
Flowtime	43.232	.30072	28.553	139.21	415
Exit Period	5.7232	.68849	2.2358	32.859	414

DISCRETE-CHANGE VARIABLES

Identifier	Average	Variation	Minimum	Maximum	Final Value
Drill Queue	.38334	1.6657	.00000	2.0000	1.0000
Mill Queue	.12960	3.0280	.00000	3.0000	.00000
Plane Queue	.08354	3.9159	.00000	2.0000	.00000
Grinder Queue	.24682	1.9683	.00000	2.0000	.00000
Inspect Queue	.15952	2.3134	.00000	2.0000	.00000
Drill Utilization	.74256	.48184	.00000	1.0000	1.0000
Mill Utilization	.63343	.50967	.00000	1.0000	1.0000
Plane Utilization	.53347	.62882	.00000	1.0000	.33333
Grinder Utilization	.71857	.48518	.00000	1.0000	.50000
Inspector Util.	.69698	.65937	.00000	1.0000	1.0000

COUNTERS

Identifier	Count	Limit
Type I Job Count	288	Infinite
Type II Job Count	127	Infinite
Fails	29	Infinite
Reworks	19	Infinite
Ejections	40	Infinite

Simulation run complete

1. Number of jobs completed by type jobs — 288 Type I jobs; 127 Type II jobs
2. Number of jobs ejected from the queues — 40
3. Number of failed jobs — 29
4. Number of reworked jobs — 19

These four items of information were determined by COUNT blocks within the model frame. The results of each COUNT block are found under the COUNTERS heading in the output file.

5. Utilization of each resource

Drill utilization	0.74256
Mill utilization	0.63343
Plane utilization	0.53347
Grinder utilization	0.71857
Inspector utilization	0.69698

These values are found under the DISCRETE-CHANGE VARIABLES heading of the output file. They were gathered by the NR(*ResourceID*) expressions in the experiment frame.

6. Average number of entities in each queue over time

Drill queue	0.38334
Mill queue	0.12960
Plane queue	0.08354
Grinder queue	0.24682
Inspect queue	0.15952

These values are found under the DISCRETE-CHANGE VARIABLES heading of the output file. They were gathered by the NQ(*Queue ID*) expressions in the experiment frame.

7. Average flowtime (time in system) of a job — 43.232 minutes
8. Time between job exits — 5.7232 minutes

This information was gathered by TALLY blocks in the model frame. They appear under the TALLY VARIABLES heading in the output file. Notice that, as in the DISCRETE-CHANGE VARIABLES section, the average value along with the variation, maximum, minimum, and final values are given.

4.3 SUMMARY

This chapter increases the complexity of problems that SIMAN can solve. The ASSIGN block and the ATTRIBUTES and VARIABLES elements allow for thorough description of entities. The TALLY block and the TALLIES element provide time-dependent data in the summary report. The BRANCH block offers the first tool for decision making within the model.

4.4 EXERCISES

E4.1 A production process consists of four operations; milling, planing, drilling, and inspection. There are four mills, two planers, three drills, and one automated inspection machine. The data for arrivals and processing is as follows:

Arrivals	Exponential	mean = 2 minutes
Milling	Triangular	(8, 12, 14)
Planing	Uniform	(6, 10)
Drilling	Normal	mean = 5, standard deviation = 1
Inspection	Constant	2 minutes

60% of the jobs are Type I, the remainder are Type II. Type I jobs are served at milling, move to drilling, and finally to inspection. Type II jobs go to planing then drilling then inspection. Jobs fail inspection with probability 0.10. Fifty percent of the failed jobs are returned to the beginning of the system, and the remainder are scrapped.

There are finite buffers in front of these machines. The buffer sizes are 8 in front of mills, 4 in front of planes, 6 in front of drills, and 3 in front of inspection. When jobs cannot join the queue because it is full, they will be ejected from the system. Simulate the system for 40 hours of operation.

- **a.** How many good jobs are made of each type?
- **b.** How many reworks occur?
- **c.** How many jobs are scrapped?
- **d.** What is the flowtime for each type of job?
- **e.** What is the utilization of each resource?
- **f.** How many jobs are ejected from the system because queues are full?

Use random number stream 10 for any BRANCH blocks. Use random number stream 1 for all other purposes.

E4.2 Reconsider Exercise 4.1. Change the mean interarrival time to 4.5 minutes. Based on a simulation run of 80 hours, what is the minimum number of buffer spaces in front of each resource such that no jobs are ejected?

E4.3 Customers enter a fast-food restaurant according to an exponential interarrival time with a mean of 1 minute. Customers have a choice of ordering one of three kinds of meals: (1) a soft drink, (2) fries, or (3) soft drink, fries, and a burger. Upon entering, the customer enters a single queue, awaits the availability of a cashier, gives the order to the cashier, then a cook prepares the order, the customer pays the cashier, and then the customer exits. A cashier may not take any additional orders until the current customer has paid. In this system, there are two cooks and two cashiers. One queue with a capacity of ten customers serves both cashiers. The time to order and pay is represented by a triangular distribution with parameters (0.4, 0.8, 1.2) minutes and (0.2, 0.4, 0.6) minutes, respectively. Use the information in the following table:

Type	Customers	Cooking Time
I	30%	Uniform(0.3, 0.8)
II	15%	Uniform(0.8, 1.1)
III	55%	Uniform(1.0, 1.4)

Model this system for 40 hours of operation.

a. How many of each order type were placed?

b. What is the utilization of the cashiers and cooks?

c. What is the average flowtime for a customer?

d. What is the average time in the queue?

e. How many customers leave the system because the wait is too long?

E4.4 Regular jobs arrive according to an exponential distribution with a mean of 6 minutes. These jobs are processed on Machine A followed by Machine B. Each machine has a processing time that follows an exponential distribution with a mean of 4.8 minutes. There is an unlimited queue in front of each machine. Rush jobs arrive according to an exponential distribution with a mean of 60 minutes and require the same processing times on Machine A and B as regular jobs. Simulate the operation for 100 hours with an interest in flowtime for regular and rush jobs, using the following queue disciplines:

a. FIFO for all jobs, or

b. Priority given to rush jobs over regular jobs, FIFO otherwise.

For this comparison, dedicate separate random number streams to the arrival, Machine A, and Machine B. Assign Machine A and B processing times to attributes immediately after each job is created. Also compare the number of regular and rush jobs that arrive and the number completed along with the queue statistics for both queue disciplines.

E4.5 A mail sorter receives mail pieces from a zip code sorting machine. The mail pieces arrive in batches uniformly distributed between 1 and 100 pieces (the capacity of the machine) according to an exponential distribution with mean 12 minutes. This sorter serves five zip codes, all equally likely on any given piece of mail. The sorter must look at the piece of mail and place it in the proper bin, a process represented by an exponential distribution with a mean of 6 seconds. In any given batch, 3% of the mail pieces will be incorrectly sorted. Simulate the system for 8 hours and determine the total number of pieces correctly sorted and the number of pieces incorrectly sorted. If the sorter increases the mean of the sorting process to 8 seconds, the percentage of errors decreases to 1.5%. What are the changes in the total number sorted and the total number of errors?

5

Using the Interactive Run Controller

The *Interactive Run Controller* (IRC) or debugger is an essential component of successful simulation model building. Even the best of simulation analysts makes mistakes or commits logical errors when building a model. The IRC assists in finding and correcting those errors in the following ways:

1. The simulation can be monitored as it progresses. This can be accomplished by advancing the simulation until a desired time has elapsed, then displaying model information at that time. Another possibility is to advance the simulation until a particular condition is in effect and then display information.
2. Attention can be focused on a particular block, group of blocks, or a particular entity. For instance, every time an entity enters a specified block, the simulation will pause so that information can be gathered. As another example, every time that a specified entity becomes active, the simulation will pause.
3. Values of selected model components can be observed. When the simulation has paused, the current value or status of variables, attributes, queues, resources, counters, and so on, can be observed.
4. The simulation can be temporarily suspended, or paused, not only to view information, but to reassign values or redirect entities.

This chapter presents many of the IRC's capabilities and shows by example how these are used. Rather than read the chapter only, it is advisable to follow the examples as they develop. All of the capabilities of the IRC cannot be shown in this chapter, since there are SIMAN concepts that have not been presented. (See the *VARIABLES Guide,* published by Systems Modeling, for additional capabilities.) However, it is extremely important that beginning modelers rely on the IRC for help rather than their colleagues, professors, or the technical support staff.

It is quite possible to become frustrated by mistakes. But we learn from our mistakes, and the IRC helps us in this learning process. Using the IRC will elevate your modeling skills.

5.1 Entering the IRC

There are two ways to enter the IRC. The first of these is "planned" in that the modeler knows before the simulation begins that an IRC session will be started. The second method is "unplanned" and occurs when it seems something is wrong with the model. For example, a simulation usually takes 30 seconds, but 60 seconds have already elapsed. The modeler can interrupt the simulation to see what is happening.

Method 1. Modify the BEGIN element in the Experiment frame as follows:

```
BEGIN,,YES;
```

The IRC will be invoked at time 0 and await commands from the keyboard. The IRC prompt is a ">."

Method 2. Press the **Escape** key during execution of the program. The IRC will be invoked at the current simulation time and await commands from the keyboard. This method can also be used to invoke the IRC at time 0 by pressing the Escape key just prior to the commencement of the simulation. For example, immediately after the Systems Modeling Corporation "razzle" appears on the screen, press the Escape key.

5.2 Exiting the IRC

There are two methods to leave the IRC, as follows:

Method 1. To end the simulation and generate a summary report, use the following:

`>END` (The command that you enter is shown in Courier print)

Method 2. To end the simulation without a summary report use the following:

```
>QUIT
```

5.3 Principal Features

To start using the IRC, we only need to understand the **GO, SHOW, VIEW,** and **STEP** commands, which are briefly defined as follows:

> **GO** means to execute until a set time is reached, a prescribed condition has been met, or the simulation has finished.
>
> **SHOW** is used to indicate the current values of variables, expressions, and attributes of active entities.
>
> **VIEW** is used to indicate IRC command settings, the model listing with block numbers assigned by the MODEL frame compiler, statistical information, and other information about model components.
>
> **STEP** moves the simulation along one or more blocks and suspends the simulation at that point.

The IRC commands are case insensitive. However, we are showing them in all capital letters.

EXAMPLE 5.1 The First IRC Session

Consider Example 3.1, the three-machine problem. We have augmented the model frame by using an ASSIGN block to provide the delay values for the three machines. Changes made to the experiment frame include a *Run Controller* response of *Yes* in the BEGIN statement, inclusion of an ATTRIBUTES element, and slight modification of the EXPRESSIONS element. The revised model and experiment are shown in Exhibits 5.1 and 5.2. Important changes are shown in italics.

EXHIBIT 5.1 **Revised model frame.**

```
!Example 5.1 Model Frame:  The Three Machine Problem
!
BEGIN;
        CREATE:               !Create jobs according to the EXPRESSION
              ArrivalRate;    that defines the arrival rate
        ASSIGN:               !Assign delay times to Attributes 1, 2, and 3
              DrillTime=ED(2):
              MillTime=ED(3):
              GrindTime=ED(4);
        QUEUE,
              DrillQ,         !Enter the drill queue and wait for a drill
              2,              !Maximum capacity of the drill queue is 2
              Balk;           If the queue is full, go to the block labeled Balk
```

EXHIBIT 5.1 *continued.*

```
        SEIZE:
                Drill;              Capture an available drill
        DELAY:                      !Process according to the EXPRESSION
                DrillTime;          that defines the drilling process time
        RELEASE:
                Drill;              Free the drill when finished
        QUEUE,
                MillQ,              !Enter the mill queue and wait for a mill
                3,                  !Maximum capacity of the mill queue is 3
                Balk;               If the queue is full, go to Balk
        SEIZE:
                Mill;               Capture an available mill
        DELAY:                      !Process according to the EXPRESSION
                MillTime;           that defines the milling process time
        RELEASE:
                Mill;               Free the mill when finished
        QUEUE,
                GrindQ,             !Enter grinder queue and wait for grinder
                2,                  !Maximum capacity of grinder queue is 2
                Balk;               If the queue is full, go to Balk
        SEIZE:
                Grinder;            Capture an available grinder
        DELAY:                      !Process according to the EXPRESSION
                GrindTime;          that defines the grinding process time
        RELEASE:
                Grinder;            Free the grinder when finished
        COUNT:
                Jobs Completed:     !Count completed jobs
                DISPOSE;            and destroy the entity
Balk    COUNT:
                Number Balked:      !Count jobs that balk
                DISPOSE;            and destroy the entity
```

EXHIBIT 5.2 Revised experiment frame.

```
!Example Problem 5.1:   The Three Machine Problem
!
BEGIN,                  ,Yes;
PROJECT,                Example 5.1, Team;
!
```

EXHIBIT 5.2 *continued.*

```
EXPRESSIONS:    1, ArrivalRate, EXPO(5, 1):
                2, , UNIF(6, 9, 1):
                3, , TRIA(10, 14, 18, 1):
                4, , DISC(0.25, 6, 0.75, 8, 1.0, 12, 1);
ATTRIBUTES:     1, DrillTime:
                2, MillTime:
                3, GrindTime;
QUEUES:         1, DrillQ:
                2, MillQ:
                3, GrindQ;
RESOURCES:      1, Drill, 2:
                2, Mill, 3:
                3, Grinder, 2;
COUNTERS:       1, Jobs Completed:
                2, Number Balked;
DSTATS:         1, NQ(DrillQ), Jobs in Drill Queue:
                2, NQ(MillQ), Jobs in Mill Queue:
                3, NQ(GrindQ), Jobs in Grinder Queue:
                4, NR(Drill)/2, Drill Utilization:
                5, NR(Mill)/3, Mill Utilization:
                6, NR(Grinder)/2, Grinder Utilization;
REPLICATE, 1, 0, 2400;
```

The IRC has been invoked so that it begins at simulation time 0 using the first of the two methods indicated previously. Give the command (the IRC is case insensitive):

```
0.0>STEP 1
```

On the screen, you see that we are at Block 1, the CREATE block, and that no time has elapsed. Now give the command:

```
0.0>STEP 10
```

On the screen you see that we are at the ASSIGN block and that the time is 10.8595. Perhaps you are wondering why we are not at the tenth block in the MODEL frame. The answer is that the ASSIGN block is where the tenth block execution of the simulation occurred. The entities move from block-to-block in this model, but the tenth event, or step, takes place at the ASSIGN block. If you want to prove this to yourself, you can give repeated Step 1 commands and see that the block executions do not occur necessarily in block-to-block sequence, and furthermore, the tenth time that you do this you will be at the ASSIGN block.

Oftentimes, we use the STEP command very crudely, such as

```
10.8595>STEP 100
```

On the screen you see that we are at time 68.0805 and we are at Block number 3, a QUEUE block.

The GO command continues the simulation until an IRC event is encountered. It also initiates or resumes the simulation. An example is

```
68.0805>GO UNTIL 240
```

On the screen you see that the simulation did pause at time 240.0.

The SHOW command indicates the current values of variables, expressions, and attributes of the active entity. Let us first see the value of *DrillTime*, an attribute of the active entity, by entering

```
240.0>SHOW DrillTime
```

On the screen, you see that this value is 8.717. You can find out how many entities are using the *Mill*, a variable, at time 240.0 by asking

```
240.0>SHOW NR(Mill)
```

The answer shown on the screen is 3, so the *Mill* is fully utilized at the moment.

We can also form expressions and show them as in the following:

```
240.0>SHOW NR(Drill)+NR(Mill)+NR(Grinder)
```

We find that there are six entities using the resources at time 240. The same result would be obtained from

```
240.0>SHOW NR(1)+NR(2)+NR(3)
```

The VIEW command indicates the source model listing, statistical information, and other information concerning model components. It also indicates IRC command settings. For example, we can indicate

```
240.0>VIEW SOURCE 5..10
```

and see those six blocks, with block numbers, from the source code. If we omitted the block numbers, all of the blocks would scroll, but for lengthy programs, this is not very helpful.

Other possibilities for viewing in this example include COUNTERS, DSTATS, ENTITY, SUMMARY, and QUEUE. For example,

```
240.0>VIEW DSTATS
```

shows data for all six of the discrete-change variables.

```
240.0>VIEW COUNTERS
```

shows that 34 parts had been finished at time 240.0 and none had been balked. To display information about an entity, use the command:

```
240.0>VIEW ENTITY
```

If we want to learn about the status of a queue, either its name or number can be used. Thus,

```
240.0>VIEW QUEUE 1
```

is the same as

```
240.0>VIEW QUEUE DrillQ
```

Additional display commands will be described later in this chapter.

5.4 The TRACE Command

Suppose that it is desired to follow a specific entity as it moves from block to block. This can help us determine if the logic is correct. The TRACE command can accomplish this result. Suppose that only a specific block or set of blocks are to be watched so that whenever an entity is in the block or blocks, a message will appear. The TRACE command can do this. We can also set trace conditions, for example, whenever a resource is full. Then, every block that is entered while the condition holds will be indicated on the screen. The TRACE command is extremely valuable in debugging, but its use must be judicious because an overabundance of information can result. To capture the trace information on your printer, press the Ctrl and Print Screen keys simultaneously. The same procedure is used to cancel the listing.

EXAMPLE 5.2 The Second IRC Session

The reader can follow along using the model in Example 5.1. A full trace showing all activity could be accomplished with the following set of commands:

```
>SET TRACE
>GO
```

As mentioned previously, this would not be very useful because too much information would be generated. A selective trace is much more helpful. An example is given by:

```
>SET TRACE BLOCKS 8,12
>GO
```

that shows information about entities as they pass through blocks 8 and 12 only. The trace can be stopped by pressing the Escape key. When the trace is stopped, the listing of all blocks for which tracing is activated can be shown by

```
>VIEW TRACE BLOCKS
```

The trace blocks can be cancelled by using

```
>CANCEL TRACE BLOCKS 8,12
```

Another example is as follows:

```
>SET TRACE ENTITY 5
>GO
```

This example shows the progress of entity 5 through the simulation, provided that entity 5 is currently active. The trace condition can be cancelled when it is no longer needed, that is,

```
>CANCEL TRACE ENTITY 5
```

A trace can be set on a condition as in

```
>SET TRACE CONDITION NQ(DrillQ).EQ.2
>GO
```

This trace condition shows information about the entities whenever the buffer before the drill is full. Again, when the trace condition is no longer needed, it can be cancelled by

```
>CANCEL TRACE CONDITION NQ(DrillQ).EQ.2
```

5.5 Intercepting Entities

An intercept temporarily suspends execution when the active entity completes a time delay or leaves a BRANCH block. An example will show how this IRC feature operates.

EXAMPLE 5.3 The Third IRC Session

Using the model in Examples 5.1 and 5.2, at TNOW = 0.0, we give the command:

```
0.0>GO UNTIL 240
```

Then we intercept the current entity when the entity executes an event as follows:

```
240.0>SHOW IDENT
```

We see that no entity is currently active. However, if we follow the sequence:

```
240.0>STEP 1
240.658>SHOW IDENT
```

we see that entity 2 is active. (You may wonder why the entity number is 2 when 34 parts have been finished. It seems that the entity number should be 35 or higher. However, SIMAN uses the entity numbers repeatedly. As soon as an entity has been disposed, its number is available for another entity.) To see its attributes, we issue the command:

```
240.658>VIEW ENTITY
```

and observe the following:

$$\begin{aligned} \text{DrillTime} &= 6.15548 \\ \text{MillTime} &= 11.0982 \\ \text{GrindTime} &= 8.0 \end{aligned}$$

Next, we activate an intercept for entity 2 as follows:

```
240.658>SET INTERCEPT 2
240.658>GO
```

Every time we give the GO command, an intercept occurs. (If we continue this after the entity is disposed, and there is nothing else to stop the simulation, we get a summary report.)

When we want to stop the intercept on entity z at some time xxx.xx we give the command:

```
xxx.xx>CANCEL INTERCEPT "z"
```

We can cancel all intercepts at once using a wild card indicator as follows:

```
xxx.xx>CANCEL INTERCEPT *
```

5.6 Setting Breakpoints

A *breakpoint* is a block at which execution is temporarily suspended whenever an entity enters. Breakpoints can be set and cancelled as well as viewed. An example will show how breakpoints are used.

EXAMPLE 5.4 The Fourth IRC Session

Using the same model as in Examples 5.1 and 5.2, at TNOW = 0, give the following command:

```
0.0>VIEW SOURCE 8-12
```

This shows the source code for the five blocks. Suppose that we want to set a breakpoint every time an entity enters the DELAY block for the *Mill*. The command is:

```
0.0>SET BREAK 9
```

Then we start the simulation with the following command:

```
0.0>GO
```

The first entity reaches Block 9, the DELAY block, at TNOW = 6.9652. If we want to continue this breakpoint, we give the command:

```
6.9652>GO
```

and see the next entity enter the breakpoint at 17.6537. We can have multiple breakpoints. Suppose we also desire a breakpoint at the block whose label is *Balk*. Then we give the following command:

```
17.6537>SET BREAK Balk
```

After the command

```
17.6537>GO
```

we have a break at time 19.4094 at Block 9. We can see the current breakpoints with the command:

```
19.4094>VIEW BREAK
```

We can cancel any or all breakpoints. The breakpoint on Block 9 can be cancelled with the command:

```
19.4094>CANCEL BREAK 9
```

To make sure that this happened, we issue the command:

```
19.4094>VIEW BREAK
```

Now issue the command:

```
19.4094>GO
```

and see that a break occurs at the block with label *Balk* at TNOW = 390.614.

To cancel all breakpoints at once, use the wild card indicator:

```
390.614>CANCEL BREAK *
```

If the command

```
390.614>GO
```

is now given, the end of the simulation will be reached and a summary report will be written.

5.7 Setting Watch Expressions

If we want to know when an attribute or variable changes its value, a watch expression can be used. There are two types of watch expressions. The first type temporarily suspends the simulation. The second type provides a message to the screen.

EXAMPLE 5.5 The Fifth IRC Session

Using the model of the previous examples in this chapter, we will now see how watch expressions can be used. Suppose that we would like to know whenever the number of entities using any resource changes value after TNOW = 240. We issue the following commands:

```
0.0>GO UNTIL 240
240.0>SET WATCH NR(Drill)
240.0>SET WATCH NR(Mill)
240.0>SET WATCH NR(Grinder)
240.0>GO
```

At time 240.658, the simulation is temporarily halted as the number in the *Mill* changes from 3 to 2. The watch on the *Mill* can be cancelled by

```
240.658>CANCEL WATCH NR(Mill)
```

We can now see if any watches remain with

```
240.658>VIEW WATCH
```

The remaining watches can be cancelled with the wild card indicator

```
240.658>CANCEL WATCH *
```

Suppose that we would like to know every time the queue before the *Grinder* is full, but we do not want to suspend the simulation; we just want the information to appear on the screen. We issue the following commands:

```
240.658>SET WATCH /nostop NQ(GrindQ)>1
240.658>GO
```

and we see that the condition occurs from TNOW = 458.8 to TNOW = 458.947, from TNOW = 540.315 to TNOW = 540.918, ..., and from TNOW = 2315.53 to TNOW = 2316.97.

5.8 The Calendar

The calendar contains events scheduled in the future. The calendar can be examined using the VIEW command.

EXAMPLE 5.6 The Sixth IRC Session

Using the model of the previous examples in this chapter, we will now examine the calendar of upcoming events in the simulation. At time 0.0, issue the command

```
0.0>VIEW CALENDAR
```

and see that there are two entities scheduled to arrive. Entity 2 is scheduled to arrive at time 0.0 at the CREATE block. The stopping entity, or entity 1, is scheduled to arrive at time 2400, the end of the simulation.

Now, advance the clock to time 1000 and look at the calendar using the following commands:

```
0.0>GO UNTIL 1000
1000.0>VIEW CALENDAR
```

and see that there are nine events scheduled in the future. The first of these events concerns Entity 15. It is scheduled to RELEASE the Drill at 1001.08. Other information is provided concerning this entity, such as its attribute values. The session can be terminated by issuing the command:

```
1000.0>QUIT
```

5.9 SUMMARY

Understanding the IRC and using it during model building is an essential part of simulation using SIMAN. Even experienced simulation analysts make mistakes that are difficult to spot, but much less difficult when using the IRC. Some beginning modelers become frustrated with their errors and turn to their fellow students or instructors for help with their programs. Much of their frustration can be eliminated, and they can be much more effective by becoming proficient with the IRC. In addition, when any of us are under the pressure of time, it can be very costly to spend minutes and even hours searching for someone to help spot an error.

Bc bold in using the IRC. No changes in the code can occur. The worst that can happen is that the simulation will have to be restarted from *TNOW* = 0.

Two more pieces of advice in debugging are as follows: First, if you find that your frustration level is rising, and you have been using the IRC for a long time, take a break. Second, some old-fashioned "desk checking" is very useful when nothing else is working. We take out the pencil and a hard copy of the program and follow entities through blocks, and so on, without using the computer.

5.10 REFERENCE

Variables Guide (1994), Systems Modeling Corporation, Sewickley, Penn.

5.11 EXERCISES

E5.1 Reconsider Exercise 3.1. For the first of the five replications only:

- **a.** At time 300, how many parts are in the Assembly queue?
- **b.** At time 300, what is the total number of parts in these three operations: assembly, soldering, and inspection?
- **c.** At time 360, how many parts have been completed?
- **d.** When does the next part complete processing after time 360?
- **e.** At time 420, what entities are scheduled to arrive?
- **f.** From time 420 to time 480, when is the Assembly idle?

E5.2 Consider the model and experiment used in the examples of this chapter.

- **a.** After 200 steps, let the simulation clock run to the next value of the clock that is a multiple of 100. How many entities are waiting in the DrillQ at that time?
- **b.** At time 200, where is (are) the closest entity (entities) to completion?
- **c.** At time 400, set a watch on the number of jobs using the Mill. What is the next change after that time?

d. At time 600, set a watch that will stop the simulation the next time that there are a total of three jobs in the queues. When does that happen?

e. At time 1800, view the DSTATS. Execute one step. Which DSTAT changed the most, absolutely?

f. At time 2200, how many jobs have been completed?

g. When is the next job completed after time 2200?

h. At what time does the last job in the simulation enter the Grinder?

E5.3 Reconsider Exercise E4.1. Using the IRC, answer the following questions:

a. When does the first Type I job complete its processing?

b. At what time does the first ejection occur from the Plane?

c. When does the utilization of the Inspector first reach 0.60?

d. At time 1800, how many Drills are being used?

e. At what time is the 40th job scrapped?

f. What is the total number of jobs completed at time 2200?

g. How many jobs are being processed on the Mill when the 600th Type II job completes processing?

E5.4 Reconsider Exercise E4.2. Using the IRC, answer the following questions:

a. At what time is the maximum queue first reached for the Mill?

b. If this maximum occurs at other times during the simulation, when does it next recur?

c. When the maximum Mill queue occurs for the second time, what is the total number of jobs in all queues?

d. There is a penalty of $4 for each unit in the Mill queue, $6 for each unit in the Plane queue, $8 for each unit in the Drill queue, and $10 for each unit in the Inspection queue at time 3600. What is the total penalty?

6

Output Analysis

Output analysis was introduced in Section 1.10, and it is appropriate here to review the earlier material prior to studying the material in this chapter.

6.1 Output Processor Fundamentals

The Output Processor is used to display data graphically, conduct statistical analyses, and manipulate the data captured in the data files. These data files are created by specifying the *Output File* operand in the SIMAN elements COUNTERS, DSTATS, OUTPUTS, FREQUENCIES, and TALLIES in the experiment frame.

The Output Processor is accessed by typing `output` at the SIMAN> prompt. Next click on Output. Select New under the Output pop-up menu. This will open the New Data Group window shown in Figure 6.1. The next step is to select the data files that will be evaluated. This is accomplished using the Add option under the Files menu. After selecting the Add option, the Add Data Files window will open, as shown in Figure 6.2.

Once opened, the next step is to specify the directory in which the files are stored. The default directory is the one in which the software has been installed (C:\SIMAN in this case). The default directory can be altered by clicking the mouse on the *Directory* field, entering the new directory path, and pressing the Enter key.

The file specification determines which files will be listed from the specified directory. The *File Spec* is defaulted as *.DAT, which will show all files ending with the file extension .DAT. If an alternate file specification is desired, click on the *File Spec* field, enter the new specification, and press the Enter key. The specification *.* lists all of the files in the directory.

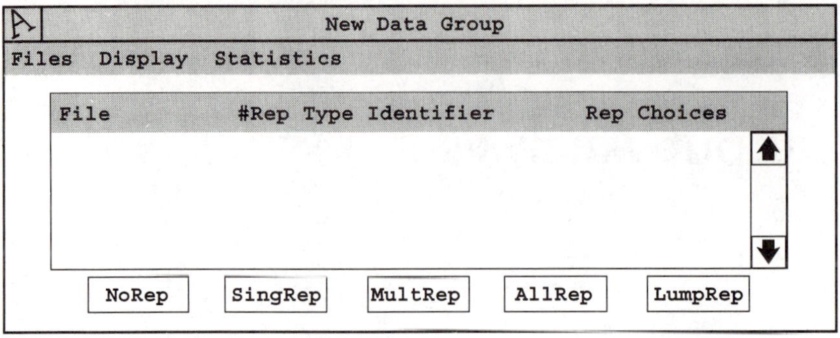

Figure 6.1 The New Data Group window.

After the *Directory* and *File Spec* have been entered, a list of files will appear. Scan through this list until the desired file is located and then highlight the file by selecting it with the mouse. Next, click on the **Add** button to add this file to the list of data files to be evaluated. Repeat the above process until all desired data files have been added. Finally, select the **Done** button to return to the New Data Group window.

Now, for each file, specify the replication treatment. This is accomplished by highlighting each file individually and then selecting from the five treatments. The five treatments are listed in Table 6.1.

Once all the files have been specified and the proper replication treatments have been selected, the Output Processor options can be executed. Now, let us become familiar with some of the Output Processor options.

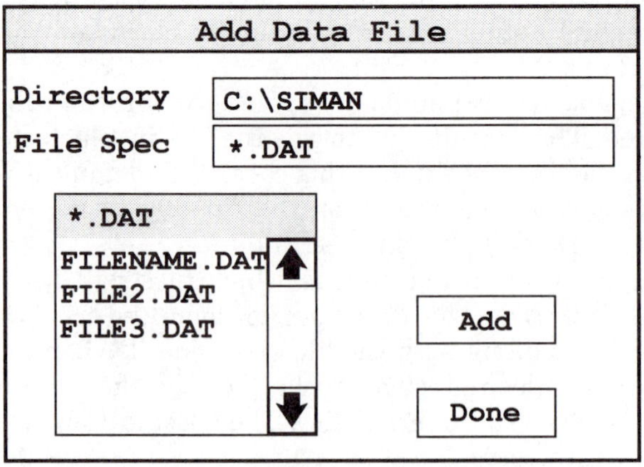

Figure 6.2 The Add Data File Option dialog box.

Table 6.1 Replication Treatments

Treatment	Action
NoRep	Selects no replications from the highlighted file(N)
SingRep	Selects a single replication from the file(S)
MultRep	Selects multiple replications from the file(M)
AllRep	Selects all of the replications in the file(A)
Lumped	Treats all replications as one group of data and performs the calculation on all simultaneously(L)

6.2 Output Processor Options

The Output Processor contains three types of options: File, Display, and Statistical. File options manipulate the data files. Display options display the data in various formats. Statistical options perform statistical operations on the data. The options are used to answer questions such as:

1. What is the appropriate number of replications?
2. How do we interpret the simulated results?
3. How do we analyze the differences between replications?

We will demonstrate how to use the display options Tables, BarChart, Histogram, Plots, and MovAverage. We will use the Statistical option Intervals and the Files option Append.

6.2.1 The Tables Option

The Tables option generates a table of the values in a specified data file. To use the Tables option, click on **Tables** under the display pop-up menu. This will open the Tables option dialog box shown in Figure 6.3. The fields are described as follows:

Field	Description	Default
Title	Title given to the table (20 characters) [Optional]	Blank
Independent Axis Label	Label given to the first column of values of the independent variable (16 characters) [Optional]	Time

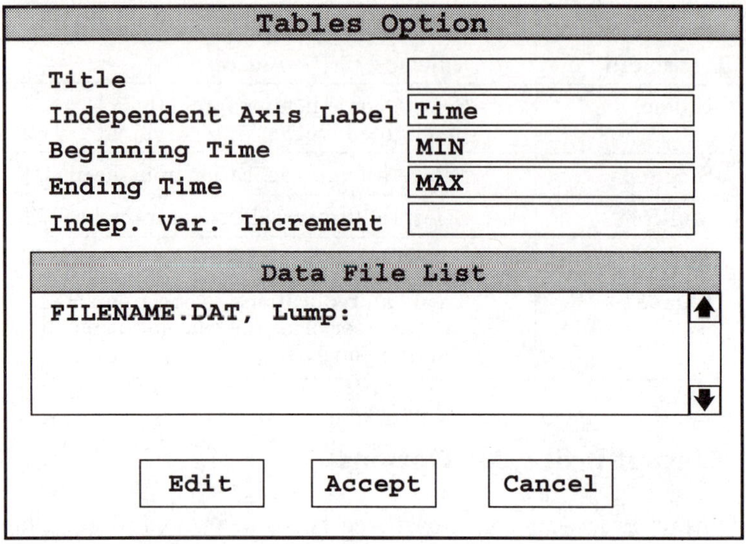

Tables Option

Title
Independent Axis Label: Time
Beginning Time: MIN
Ending Time: MAX
Indep. Var. Increment

Data File List

FILENAME.DAT, Lump:

Edit | Accept | Cancel

Figure 6.3 The Tables dialog box.

Field	Description	Default
Beginning Time	Beginning time or observation number [constant, MIN, or MIN*(Val)*, where *Val* is a constant]	MIN
Ending Time	Ending time or observation number [constant, MAX or MAX*(Val)*, where *Val* is a constant]	MAX
Independent Variable Increment	The increment to index the independent variable if all values are not desired [Optional]	—

The independent variable is either simulated time or the observation number. The *Data File List* is the list of files that will be used to generate the table. Multiple files can be displayed (maximum of three) if they share the same independent variable. The *Data File List* can be expanded or decreased by clicking on the **Edit** button.

6.2.2 The BarChart Option

The BarChart option generates a barchart of the variable specified in the data file. Click on **BarChart** under the display pop-up menu to use

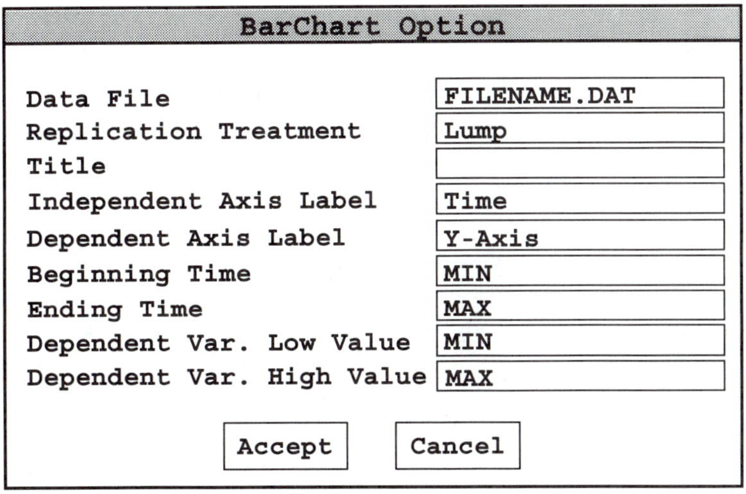

BarChart Option	
Data File	FILENAME.DAT
Replication Treatment	Lump
Title	
Independent Axis Label	Time
Dependent Axis Label	Y-Axis
Beginning Time	MIN
Ending Time	MAX
Dependent Var. Low Value	MIN
Dependent Var. High Value	MAX
Accept	Cancel

Figure 6.4 The BarChart Option window.

the BarChart option. This will open the BarChart option dialog box shown in Figure 6.4. The fields are described as follows:

Field	Description	Default
Data File	Input data file [FileID]	File highlighted in the New Data Group window
Replication Treatment	Replication treatment [integer, all, or lumped]	Selected in the New Data Group window
Title	The title of the barchart to be created (16 Characters) [Optional]	Blank
Independent Axis Label	The label given to the independent axis (X-axis) (16 Characters) [Optional]	Time
Dependent Axis Label	The label given to the dependent axis (Y-axis) (16 Characters) [Optional]	Y-axis
Beginning Time	Beginning time or observation number [constant, MIN, or MIN(*Val*), where *Val* is a constant]	MIN

Field	Description	Default
Ending Time	Ending time or observation number [constant, MAX, or MAX*(Val)*, where *Val* is a constant]	MAX
Dependent Variable Low	The lowest ordinate value of the dependent variable [constant, MIN, or MIN*(Val)*, where *Val* is a constant]	MIN
Dependent Variable High	The highest ordinate value of the independent variable [constant, MAX, or MAX*(Val)*, where *Val* is a constant]	MAX

6.2.3 The Histogram Option

The Histogram option generates a histogram for the variable in the specified data file. The Histogram option is executed by clicking on **Histogram** under the display pop-up menu. This opens the Histogram option dialog box shown in Figure 6.5. The fields are described as follows:

Field	Description	Default
Data File	Input data file [FileID]	File highlighted in the New Data Group window
Replication Treatment	Replication treatment [integer, all, or lumped]	Selected in the New Data Group window
Title	The title of the histogram to be created (16 Characters) [Optional]	Blank
Independent Axis Label	The label given to the independent axis (X-axis) (16 Characters) [Optional]	Blank
Number of Interior Cells	The number of cells used in the histogram [integer, selected automatically if left blank]	10–20

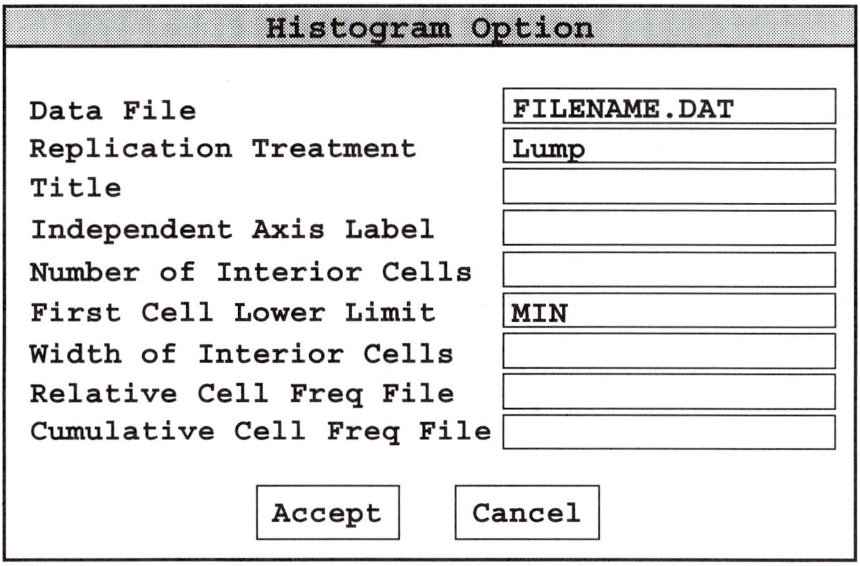

Histogram Option

Data File	FILENAME.DAT
Replication Treatment	Lump
Title	
Independent Axis Label	
Number of Interior Cells	
First Cell Lower Limit	MIN
Width of Interior Cells	
Relative Cell Freq File	
Cumulative Cell Freq File	

Accept   Cancel

Figure 6.5 Histogram Option window.

Field	Description	Default
First Cell Lower Limit	First interior cell lower limit [constant, MIN, or MIN(*Val*), where *Val* is a constant]	MIN
Width of Interior Cells	The width of the cells in the histogram [Constant]	(MAX-MIN)/ *Number of Interior Cells*
Relative Cell Frequency File	File to which the relative cell frequencies are written [FileID]	No save
Cumulative Cell Frequency File	File to which the cumulative cell frequencies are written [FileID]	No save

The fields *Number of Interior Cells*, *First Cell Lower Limit*, and *Width of Interior Cells* are usually defaulted. However, when these values are user specified, an open cell is added to each end of the histogram to tabulate the observations that do not fall into the interior cells.

6.2.4 The Plots Option

The Plots option is used to display the data in the form of a scatter plot. To use the Plots option, click on **Plots** under the display pop-up

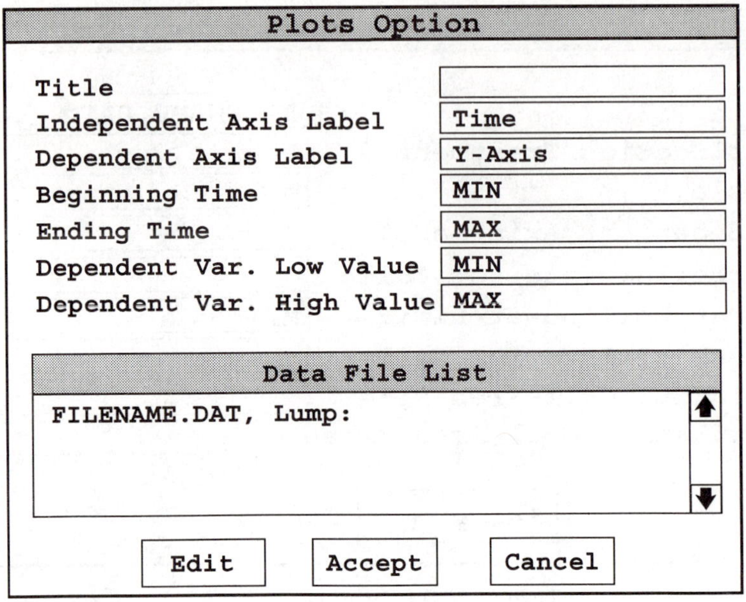

Plots Option	
Title	
Independent Axis Label	Time
Dependent Axis Label	Y-Axis
Beginning Time	MIN
Ending Time	MAX
Dependent Var. Low Value	MIN
Dependent Var. High Value	MAX

Data File List

FILENAME.DAT, Lump:

Edit Accept Cancel

Figure 6.6 The Plots Option dialog box.

menu. This will open the Plots option dialog box shown in Figure 6.6. The fields are described as follows:

Field	Description	Default
Title	The title of the plot to be created (16 Characters) [Optional]	Blank
Independent Axis Label	The label given to the independent axis (X-axis) (16 Characters) [Optional]	Time
Dependent Axis Label	The label given to the dependent axis (Y-axis) (16 Characters) [Optional]	Y-axis
Beginning Time	Beginning time or observation number [constant, MIN, or MIN*(Val)*, where *Val* is a constant]	MIN
Ending Time	Ending time or observation number [constant, MAX, or MAX*(Val)*, wher *Val* is a constant]	MAX

Field	Description	Default
Dependent Variable Low	The lowest ordinate value of the dependent variable [constant, MIN, or MIN*(Val)*, where *Val* is a constant]	MIN
Dependent Variable High	The highest ordinate value of the independent variable [constant, MAX, or MAX*(Val)*, where *Val* is a constant]	MAX

The independent variable is either simulated time or the observation number. The *Data File List* is the list of files that will be used to generate the table.

6.2.5 The MovAverage Option

The MovAverage option generates a plot of the moving average of the observations in the data file. This command smooths the data, making it easier to identify trends and underlying value changes. The MovAverage option is executed by clicking on **MovAverage** under the display pop-up menu. The MovAverage option dialog box is shown in Figure 6.7. The fields are described as follows:

Field	Description	Default
Data File	Input data file [FileID]	File highlighted in the New Data Group window
Replication Treatment	Replication treatment [integer, all, or lumped]	Selected in the New Data Group window
Title	The title of the moving average plot to be created (16 Characters) [Optional]	Blank
Type of Average	The type of average to be computed [MOV (moving), CUM (cumulative) or EXP (exponential smoothing)]	MOV
Type Value	MOV average: number of periods to include [integer]	10
	CUM average: minimum data points allowed [integer]	5
	EXP smooth: constant between 0 and 1 [constant]	0.1

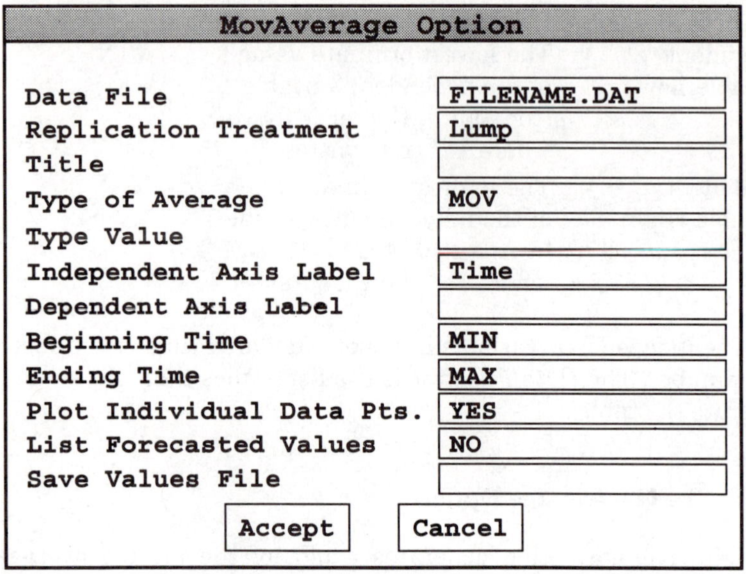

MovAverage Option

Data File	FILENAME.DAT
Replication Treatment	Lump
Title	
Type of Average	MOV
Type Value	
Independent Axis Label	Time
Dependent Axis Label	
Beginning Time	MIN
Ending Time	MAX
Plot Individual Data Pts.	YES
List Forecasted Values	NO
Save Values File	

Accept Cancel

Figure 6.7 The MovAverage Option dialog box.

Field	Description	Default
Independent Axis Label	The label given to the independent Axis (X-axis) (16 Characters) [Optional]	Time
Dependent Axis Label	The label given to the dependent Axis (Y-axis) (16 Characters) [Optional]	Element name or *Data File*
Beginning Time	Beginning time or observation number [constant, MIN, or MIN(*Val*), where *Val* is a constant	MIN
Ending Time	Ending time or observation number [constant, MAX, or MAX(*Val*), where *Val* is a constant	MAX
Plot Individual Data Points	Option to plot individual data points in addition to the moving average [YES/NO]	YES
List Forecasted Values	Option to list forecasted values before displaying moving average plot [YES/NO]	YES
Save Values	File to which forecasted values are to be written [FileID]	No save

The cumulative average (CUM) includes all data up to the current observation. The moving average (MOV) considers the k most recent observations, where k is specified by *Type of Values*. In addition, an exponentially weighted moving average (EXP) gives the specified weight to the current observation and the complement of that weight to prior observations.

6.2.6 The Intervals Option

The Intervals option generates a confidence interval for the data set stored in the specified output file. This option is accessed by clicking on **Intervals** under the Statistical pop-up menu. The Intervals Option dialog box is shown in Figure 6.8. The fields are described as follows:

Field	Description	Default
Title	The title given to the interval (20 characters) [Optional]	Blank
Scale Output	Automatic graphical scaling [YES/NO]	YES
Confidence Coefficient	The value of the confidence coefficient α [constant between 0.0 and 1.0]	0.05

The *Scale* field is used when generating multiple confidence intervals and will graphically display the multiple confidence intervals scaled with respect to the maximum and minimum values of all the data sets. As with the Tables option, multiple data sets can be analyzed simultaneously by adding files to the *Data File List*.

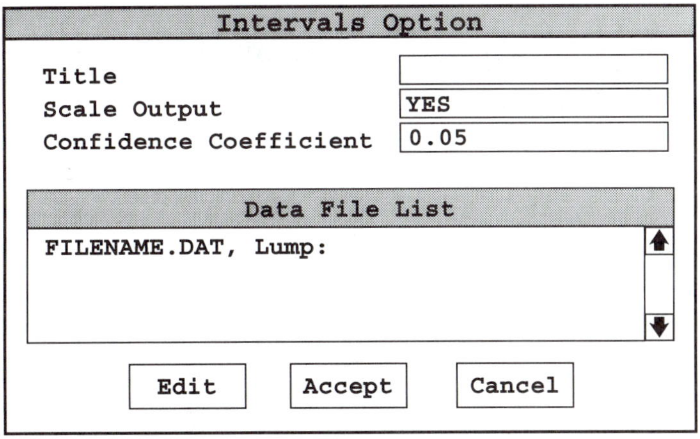

Figure 6.8 The Intervals Option dialog box.

The Intervals option assumes that the observations in the data file are independent, identically and normally distributed. The default value 0.05 for the confidence coefficient corresponds to a 0.95 or 95% confidence level.

6.2.7 The Append Option

The Append option is used to combine the data from two individual files into one file. The grouping of data allows output analysis of the data to continue with all the data in one data file. This is necessary when performing statistical analyses of the data. The Append option is accessed by selecting **Append** under the Files pop-up menu, with the corresponding Appends Source File dialog box shown in Figure 6.9 and the Append Option dialog box shown in Figure 6.10. The fields in the Appends Source File dialog box are described as follows:

Field	Description	Default
Directory	Current directory of files	C:\SIMAN
File Spec	File specification to be listed	*.DAT

The fields in the Append Option dialog box are described as follows:

Field	Description	Default
Source File	Filename of data files to be appended	File highlighted in New Data Group window
Replication Treatment	Replication treatment [integer, all, or lumped]	All
Destination File	Filename to receive appended files	File highlighted in New Data Group window

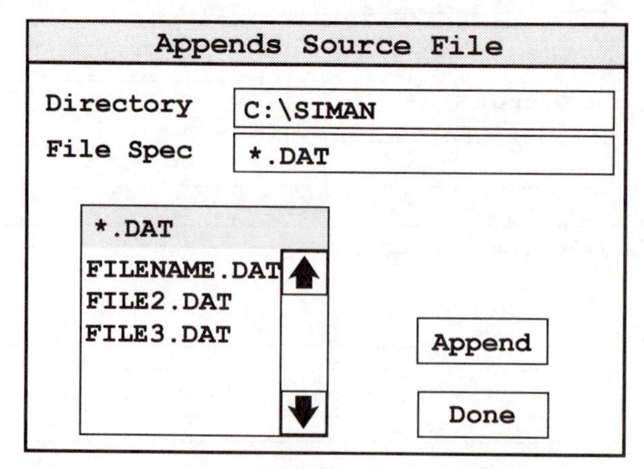

Figure 6.9 Append Source File Option dialog box.

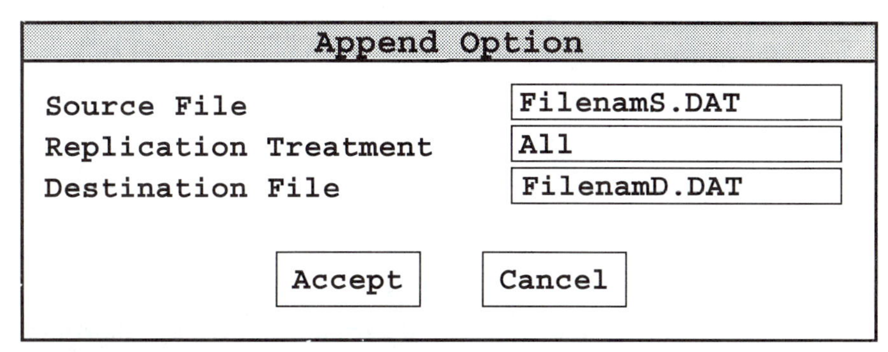

Figure 6.10 Append Option Dialog Box

The Append option adds data from the *Source File* to the *Destination File*. The *Destination File* is the file that is highlighted in the New Data Group window when the Append command is selected. To append another file to the *Destination File*, highlight the file to be appended on the file list in the Appends Source File dialog box and click on the **Append** button. This will open the Append Option dialog box. If the default values are acceptable, click on Accept. Otherwise, change the values as appropriate. The result will be one data file with the name specified by the *Destination File* that contains the data from both files. The various files shown in the data set list can be altered by changing the *Directory* and *File Spec* in the Append Source File dialog box.

Upon executing an Output Processor option, a window is created that contains the results from the option. These results are either textual and graphical or graphical only. Once a result window has been created, it can be printed as text by selecting **Print** under the File pop-up menu in a textual window, or graphically by selecting print that appears in the pop-up window upon clicking the slanted blue **A** in the upper left-hand corner of the window. Other options under the slanted blue A are close, shrink, or enlarge the window, push to back, and refresh.

EXAMPLE 6.1 Terminating System

Consider the three-machine system described in Example 3.1 with the following changes:

a. One hundred jobs are started and the system shuts down after they have been completed.

b. Processing times (in minutes):

Drill Time—Triangular (7.4, 8.6, 10.8)
Mill Time—Uniform (7, 9)
Grind Time—Discrete (0.25, 3, 0.65, 4, 1.0, 6.5)

c. Resource capacities:

Drill—2
Mill—2
Grinder—1

d. Queue capacities:

Drill Queue—5
Mill Queue—7
Grinder Queue—6

The objective of the simulation is to determine the average flowtime for parts through the system. Use random number stream 1 for all distributions. Perform the following:

a. Simulate the system for a sufficient number of days so that the average flowtime is within 5% of the sample mean with 95% confidence.
b. Determine the likelihood that the flowtime will be in the range of 5% less than the average to 10% above the average.

The system is terminating since the production area completes a fixed number of jobs each day, then shuts down and reopens the next day empty and idle.
Solution Strategy-Part (a)

1. Execute a pilot run of 10 one-day replications to generate a sample of average flowtimes that can be used to determine the final number of replications required. To accomplish this:
    a. Add the OUTPUTS element to the experiment frame to create an output file that contains the average flowtime for each replication.
    b. Use the Output Processor options Tables and Barchart to generate a table of the average flowtimes and the corresponding barchart using the data recorded in the output file. This step is for visual analysis of the data in order to gain a better understanding of the results.
    c. Use the Output Processor option Intervals to generate a confidence interval for the average flowtime.
2. If the confidence interval generated above is too large, determine the number of additional replications required to obtain an appropriate confidence interval using the following equation:

$$n^* = n\left(\frac{h}{h^*}\right)^2 \tag{6.1}$$

where:

n = number of replications in pilot run

n^* = total number of replications required

h = half-width of the confidence interval for the pilot run

h^* = half-width of the confidence interval for all replications (desired half-width)

3. If additional replications are required, the procedure is as follows:

a. Use the SEEDS element in the experiment frame to re-initialize the random number streams to create independent replications from those already performed.
b. Use a different output file to record the data from the additional replications.
c. Use the Output Processor option Appends to add the data of the additional replications to the data of the original replications.
d. Use the Output Processor to generate a table and the corresponding barchart for the average flowtime using the data from all replications (Tables and BarChart options). As above, this is for visual analysis of the data.
e. Use the Output Processor to generate a new confidence interval for the average flowtime using data from all replications (Intervals option).

4. Repeat Steps 2 and 3 if the confidence interval is still too large.

6.3 Experiment Frame Elements

In order to perform the above procedure, we will need to become familiar with two new elements, OUTPUTS and SEEDS, as described in Table 6.2.

Table 6.2 Elements Introduced in this Chapter

New Element	Purpose
OUTPUTS	Generates output data files for specific performance measures
SEEDS	Specifies initial seed for each random number stream

6.3.1 The OUTPUTS Element

The OUTPUTS element saves the value of a specified expression in a designated output file at the end of each replication. For example, if the expression of interest is the number of rejects for each production run, the OUTPUTS element will generate a file that lists the number of rejects for each replication performed. The format for the OUTPUTS element is as follows:

OUTPUTS: *Number, Expression, Output File, Report Label*: repeats;

The operands are described as follows:

Operand	Description	Default
Number	Output element number [Optional]	Consecutive integers beginning with 1
Expression	Expression to be displayed/recorded	—
Output File	Output file name	No saving
Report Label	Output label	Expression

The first operand, *Number*, is the output element number. This is usually defaulted. The second operand, *Expression*, is the value that will be written into the specified output file at the end of each replication. The expression is most often specified by the statistical summary variables shown in Table 6.3.

The third operand, *Output File*, is the name of the output file where the observations will be stored at the end of each replication.

Table 6.3 Statistical Summary Variables for OUTPUTS Element

Variable	Description
DAVG(DstatID)	Average Dstat value for DstatID
DMIN(DstatID)	Minimum Dstat value for DstatID
DMAX(DstatID)	Maximum Dstat value for DstatID
TAVG(TallyID)	Average Tally value for TallyID
TMIN(TallyID)	Minimum Tally value for TallyID
TMAX(TallyID)	Maximum Tally value for TallyID
NC(CountID)	Counter value for CountID

The *Output File* is a valid system filename, which is always enclosed in double quotes. Examples in DOS are: "Parts.dat", "A:\worker.dat", and "C:\results\queue1.dat". The observations will automatically be written into the data file indicated, and any existing data in the file will be erased. If the file does not exist at the beginning of the simulation, SIMAN V will automatically create the file. The last operand, *Report Label*, is the label used in the Output Processor and the summary report to identify the output observations.

An example of the OUTPUTS element is as follows:

```
OUTPUTS:   DAVG(InspUtil), "Inspect.dat", Inspector Utilization:
           TMIN(Flowtime)," Minflow.dat", Minimum Flowtimes:
           NC(Rejects), "Rejects.dat", Number of Rejects;
```

In this example, three different data files will be created. The first file, *"Inspect.dat"*, will store the average values for the *DstatID* named *InspUtil* at the end of each replication. This data will be labeled *Inspector Utilization*. The second file, *"Minflow.dat"*, will store the minimum values of the *TallyID* named *Flowtime* for each replication and label it *Minimum Flowtimes*. The last file, *"Rejects.dat"*, will store the values of the *CountID* named *Rejects* and label them *Number of Rejects*.

For Example 6.1, the value of interest is the average flowtime of the parts through the system. In order to store this value into an output data file, we need to include the following OUTPUTS element in the experiment frame:

```
OUTPUTS: TAVG(Flowtime), "FlwTime1.dat", Average Flowtime;
```

This will store the average value of the *TallyID* named *Flowtime* in the output data file *"FlwTime1.dat"*. This file will hold one observation for each replication performed in the simulation run. In our case, after the pilot simulation run, the file *"FlwTime1.dat"* will contain 10 observations. The data will be labeled *Average Flowtime* in the summary report and Output Processor.

6.3.2 The SEEDS Element

The SEEDS element is used to change the initial seed values of the random number streams used during the simulation. At the beginning of each simulation replication, a default number is used as the seed value of each random number stream. The SEEDS element is used to change that default seed value in order to generate independent replications for different simulation runs. Unless the SEEDS element specifies different seed values, the same random numbers will be used, thus creating identical output. To illustrate, if 10 replications are performed in one simulation run and then 25 additional replications are performed in a second simulation run without changing the initial random number seeds, then the first 10 replications of the 25 added will be identical to the initial 10 replications conducted. The format for the SEEDS element is as follows:

SEEDS: *Identifier, Seed Value, Initialize Option*: repeats;

Operand	Description	Default
Identifier	Number or name of stream to initialize [integer or symbol name]	—
Seed Value	Initial seed value [integer]	Machine dependent
Initialize	Re-initialize stream between replications	No
Option	[Yes, No, Common, Antithetic]	

The operand *Identifier* indicates the random number stream to be initialized. The *Seed Value* is the new seed value. The operand *Initialize Option* indicates how the random number streams are to be initialized between replications. If *Initialize Option* is Yes, then the random number stream will be re-initialized between each replication, thus each replication will use the same random number sequence; if No, then each replication will use different random number seeds. The values Common and Antithetic are used if variance reduction techniques are employed; however, they will not be discussed in this text.

An example of the SEEDS element is as follows:

```
SEEDS:  1, 79531:
        2, 52187, YES;
```

In this example, the seed value for random number stream 1 is initialized to 79531. The seed for stream 2 is 52187, and it will return to this value at the beginning of each replication.

At one point during the solution for Example 6.1, we will be using the following SEEDS element:

```
SEEDS:  1, 48311;
```

This will change the seed value of random number stream 1 to 48311.

EXAMPLE 6.1 Solution

Example 6.1 is based on the system modeled in Example 3.1. Modifying the model frame of Example 3.1 to include a TALLY for flowtime, limiting the number of parts created to 100, and altering the queue sizes, we have the model frame shown in Exhibit 6.1. Note: the changes are italicized.

The alterations to the experiment frame of Example 3.1 for Example 6.1 include the addition of the OUTPUTS and TALLIES elements, modification of the processing times and resource capacities, specification of the *ArrTime* attribute, and specification of 10 replications. The resulting experiment frame is shown in Exhibit 6.2. Note: the changes are italicized.

EXHIBIT 6.1 **Model frame for Example 6.1.**

```
BEGIN;
     CREATE:                 !Create jobs according to EXPRESSION
        ArrivalTime,100:     !that defines the arrival rate
        MARK(ArrTime);       Mark the arrival time
     QUEUE,
        DrillQ,              !Enter the drill queue and wait for a drill
        5,                   !Maximum capacity of the drill queue is 2
        Balk;                If queue is full, go to block labeled Balk
     SEIZE:
        Drill;               Capture an available drill
     DELAY:                  !Process according to the EXPRESSION
        DrillTime;           that defines the drilling process time
     RELEASE:
        Drill;               Free the drill when finished
     QUEUE,
        MillQ,               !Enter the mill queue and wait for a mill
        7,                   !Maximum capacity of the mill queue is 3
        Balk;                If queue is full, go to block labeled Balk
     SEIZE:
        Mill;                Capture an available mill
     DELAY:                  !Process according to the EXPRESSION
        MillTime;            that defines the milling process time
     RELEASE:
        Mill;                Free the mill when finished
     QUEUE,
        GrindQ,              !Enter grinder queue and wait for grinder
        6,                   !Maximum capacity of grinder queue is 2
        Balk;                If queue is full, go to block labeled Balk
```

EXHIBIT 6.1 *continued.*

```
    SEIZE:
          Grinder;              Capture an available grinder
    DELAY:                      !Process according to the EXPRESSION
          GrindTime;            that defines the grinding process time
    RELEASE:
          Grinder;              Free the grinder when finished
    TALLY:                      !Tally the job flowtime
          Flowtime, INT(ArrTime);
    COUNT:
          Jobs completed:       !Count completed jobs
          DISPOSE;              and destroy the entity
Balk COUNT:
          Number Balked         !Count jobs that balk
          DISPOSE;              and destroy the entity
```

EXHIBIT 6.2 Experiment frame for Example 6.1.

```
BEGIN;
PROJECT,          Three Machine, Team;
ATTRIBUTES:       ArrTime;
EXPRESSIONS:      ArrivalTime,   EXPO(5, 1)
                  DrillTime,     TRIA(7.4, 8.6, 10.8 1):
                  MillTime,      UNIF(7, 9, 1):
                  GrindTime,     DISC(0.25, 3, 0.65, 4, 1.0, 6.5, 1);
QUEUES:           1, DrillQ:
                  2, MillQ:
                  3, GrindQ:
RESOURCES:        1, Drill, 2:
                  2, Mill, 2:
                  3, Grinder, 1;
COUNTERS:         1, Jobs Completed:
                  2, Number Balked;
TALLIES:          Flowtime;
OUTPUTS:          TAVG(Flowtime), "FlwTime1.dat", Average Flowtime;
DSTATS:           1, NQ(DrillQ), Jobs in Drill Queue:
                  2, NQ(MillQ), Jobs in Mill Queue:
                  3, NQ(GrindQ), Jobs in Grinder Queue:
                  4, NR(Drill)/2, Drill Utilization:
                  5, NR(Mill)/2, Mill Utilization:
                  6, NR(Grinder)/2,Grinder Utilization;
REPLICATE, 10;
```

We now have all the necessary tools and components to begin our analysis for Example 6.1. The first step is to compile and run the simulation. This will generate two files: the normal results file and the specified OUTPUTS data file. Exhibit 6.3 displays a segment of the results file. Notice that there is an additional section in the results file called OUTPUTS. This section indicates the expressions and their corresponding values that were written to output data files. The OUTPUTS data file that was generated during the simulation run is "FlwTime1.dat", which was defined in the OUTPUTS element in the Experiment frame. This file contains the value of average flowtime for each of the 10 replications.

***EXHIBIT 6.3* Segment of the Results file for Example 6.1.**

Summary for Replication 1 of 10

Project: Three Machine   Run execution date : 8/24/1994
Analyst: Team   Model revision date: 8/24/1994
Relication ended at time: 559.002

TALLY VARIABLES

Identifier	Average	Variation	Minimum	Maximum	Observations
Flowtime	31.758	.29520	19.110	59.632	100

DISCRETE-CHANGE VARIABLES

Identifier	Average	Variation	Minimum	Maximum	Final Value
Jobs in Drill Queue	.82638	1.3502	.00000	4.0000	.00000
Jobs in Mill Queue	.00580	13.086	.00000	1.0000	.00000
Jobs in Grinder Queue	.89307	1.1008	.00000	5.0000	.00000
Drill Utilization	.78258	.42591	.00000	1.0000	.00000
Mill Utilization	.69415	.50588	.00000	1.0000	.00000
Grinder Utilization	.85564	.41075	.00000	1.0000	.00000

Counters

Identifier	Count	Limit
Jobs Completed	100	Infinite
Number Balked	0	Infinite

OUTPUTS

Identifier	Value
Average Flowtime	31.758

6.4 Output Analysis of a Terminating System

For Example 6.1, we will be using the following options:

Command	Function
Tables	Creates a listing of the values in a data file
BarChart	Generates a barchart for a specified variable
Intervals	Generates confidence intervals for a specified variable
Append	Joins two data files into one data file
Histogram	Generates a histogram for a specified variable

We now enter the Output Processor to begin our examination of the output data file by selecting **New** under the Output pop-up menu. ADD the file, "FLWTIME1.DAT," to the files under analysis and then select the Lumped data treatment by clicking on the **LumpRep** button on the screen. Note that even though capitals and lowercase letters were used in the experiment frame, all letters will appear as capitals in the Output Processor. The next step is to generate a table of the flowtime values in each replication using the Tables option under the display pop-up window with the following field values:

Title:	Average Flowtime
Independent Axis Label:	Replication #

All other fields are defaulted, which will result in Exhibit 6.4. Notice that there are 10 values, one for each replication, ranging from 27.9 to 38.9.

We now generate a barchart of the values of flowtime. Use the BarChart option with the following field values:

Title:	Average Flowtime
Independent Axis Label:	Replication #
Dependent Axis Label:	Flowtime

All other fields are defaulted. The Output Processor will generate the barchart shown in Exhibit 6.5.

The data shown in Exhibits 6.4 and 6.5 do not contain any outlying observations, that is, values that have a large difference in magnitude from the rest of the observations. With such a small sample size, one outlying point can skew the statistical results of the analysis.

EXHIBIT 6.4 Table of flowtimes for Example 6.1 (Tables option).

TABLE: AVERAGE FLOWTIME

REPLICATION #	AVERAGE FLOWTIME
1	31.8
2	35.5
3	36.6
4	31.5
5	36.9
6	27.9
7	28.9
8	38.9
9	29.5
10	33.5

EXHIBIT 6.5 Barchart of average flowtimes for Example 6.2 (BarChart option).

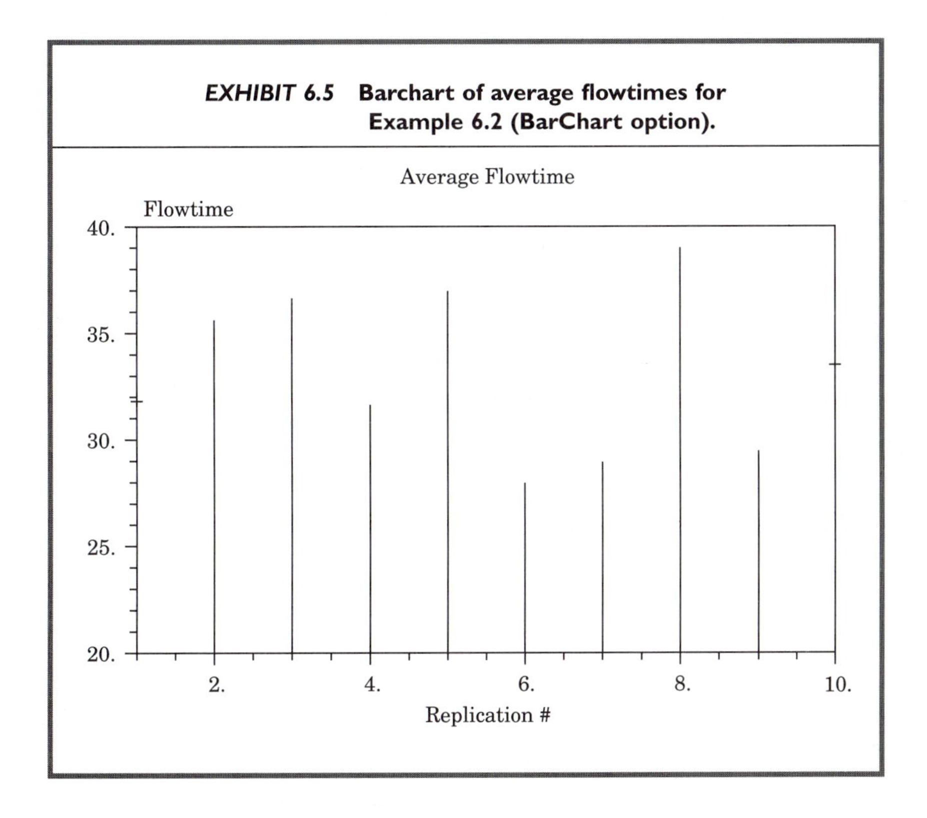

Next, we generate the confidence interval for the average flowtimes using the Intervals option with the following field value:

Title: Average Flowtime

With all other fields defaulted, two windows will be generated: a graphical representation of the confidence interval, shown in Exhibit 6.6; and a parameter report for the confidence interval, shown in Exhibit 6.7.

After generating the confidence interval, the next step in the procedure is to determine whether the confidence interval meets the requirements defined in part (a) of Example 6.1. The requirements are that the average flowtime is within 5% of the sample mean with 95% confidence. The half-width of the confidence interval must be less than or equal to 5% of the mean. Using the confidence interval generated, the sample mean is 33.1. Calculating 5% of this mean equals 1.66. The half-width of the initial confidence interval is 2.71. Since 2.71 is greater than 1.66, the initial confidence interval does not meet the necessary requirements, and since the confidence interval is too large, more replications are needed.

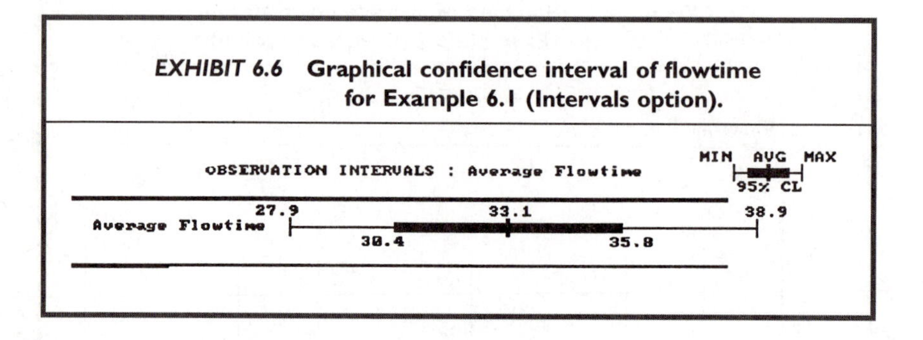

EXHIBIT 6.6 **Graphical confidence interval of flowtime for Example 6.1 (Intervals option).**

EXHIBIT 6.7 **Parameter report for confidence interval of flowtime for Example 6.1 (Intervals option).**

INTERVALS: AVERAGE FLOWTIME

IDENTIFIER	AVERAGE	STD DEV	.950 C.I. HALF-WIDTH	MIN VALUE	MAX VALUE	NUMBER OF OBS.
AVERAGE FLOWTIME	33.1	3.79	2.71	27.9	38.9	10

With $n = 10$, $h = 2.71$, and $h^* = 1.66$, using Equation 6.1, the total number of replications needed is calculated as follows:

$$n^* = 10(2.71/1.66)^2 = 26.7$$

Generally, it is wise to round the number up to the nearest five replications. Since n^* is only an estimate, rounding up provides a margin of safety since too many is better than too few. In this case, round up to 30 replications. Since we have already conducted 10 replications, it is necessary to conduct only 20 additional replications.

Now that we have determined the number of additional replications required, we can modify the experiment frame accordingly. First we change the REPLICATE element in the experiment file to produce 20 replications. Next, we add the SEEDS element to the experiment frame in order to obtain independent results between the two simulation runs. Finally, we specify a new OUTPUTS data file so that our original will not be overwritten. The revised experiment frame is shown in Exhibit 6.8,

EXHIBIT 6.8 Revised experiment frame for Example 6.1.

```
BEGIN;
PROJECT, Three Machine, Team;
ATTRIBUTES: ArrTime;
EXPRESSIONS:  ArrivalTime,  EXPO(5, 1)
              DrillTime,    TRIA(7.4, 8.6, 10.8, 1):
              MillTime,     UNIF(7, 9, 1):
              GrindTime,    DISC(0.25, 3, 0.65, 4, 1.0, 6.5, 1);
QUEUES:       1, DrillQ:
              2, MillQ:
              3, GrindQ;
RESOURCES:    1, Drill, 2:
              2, Mill, 2:
              3, Grinder, 1;
COUNTERS:     1, Jobs Completed:
              2, Number Balked;
TALLIES:      Flowtime;
OUTPUTS:      TAVG(Flowtime), "FlwTime2.dat", Average Flowtime;
SEEDS:        1, 48311;
DSTATS:       1, NQ(DrillQ),       Jobs in Drill Queue:
              2, NQ(MillQ),        Jobs in Mill Queue:
              3, NQ(GrindQ),       Jobs in Grinder Queue:
              4, NR(Drill)/2,      Drill Utilization:
              5, NR(Mill)/3,       Mill Utilization:
              6, NR(Grinder)/2,    Grinder Utilization;
REPLICATE, 20;
```

where the changes are shown in italics. The model frame has not been changed.

After making these changes, we can conduct the simulation run resulting in a total of 30 replications. This simulation run will create a second output file called "FlwTime2.dat", which contains the value of flowtime for the 20 additional replications.

After conducting the new replications, enter the Output Processor and add the original data file, "FLWTIME1.DAT". Next, click on the **Append** option under the Files pop-up window. With the file "FLWTIME2.DAT" highlighted, click the **Append** button on the screen. Then, click the **Accept** button in the Append Option dialog box. We now have one data file, "FLWTIME1.DAT", which contains all 30 replications.

Now, select the Lumped data treatment for "FLWTIME1.DAT" in the New Data Group window. As before, generate a table of the data, shown in Exhibit 6.9, and the corresponding barchart, shown in Exhibit 6.10. Visual analysis of the data indicates that we may proceed with our investigation.

EXHIBIT 6.9 **Table of flowtimes of all replications for Example 6.1—Revised (Tables option).**

TABLE: AVERAGE FLOWTIME

REPLICATION #	AVERAGE FLOWTIME
1	31.8
2	35.5
3	36.6
4	31.5
5	36.9
6	27.9
7	28.9
8	38.9
9	29.5
10	33.5
11	33.1
12	31.4
13	30.5
14	27.8
15	33.4
16	33.2
17	34.4
18	28.2

EXHIBIT 6.9 *continued.*

REPLICATION #	AVERAGE FLOWTIME
19	34.6
20	32.3
21	35.6
22	37.0
23	31.3
24	27.0
25	36.0
26	33.3
27	32.0
28	29.7
29	31.8
30	34.2

EXHIBIT 6.10 **Barchart of flowtimes of all replications for Example 6.2—Revised(BarChart option).**

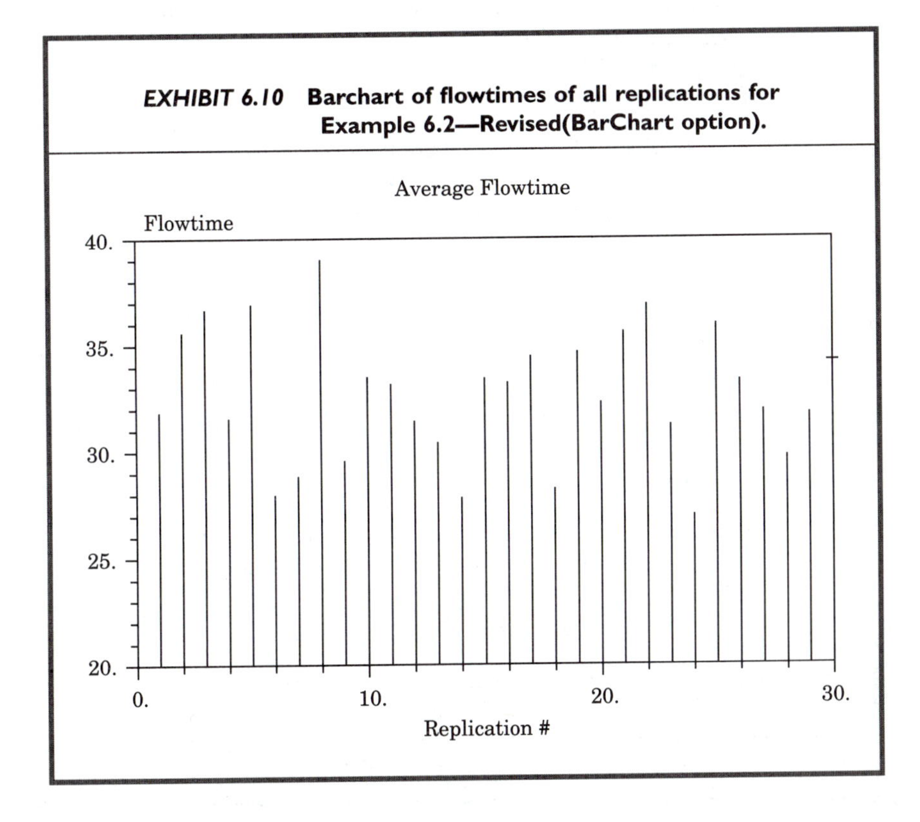

The final step is to generate a new confidence interval using data from all the replications. The resulting graphical confidence interval and parameter report are shown in Exhibits 6.11 and 6.12, respectively. Again, compare the half-width of the new confidence interval, computed as 1.15, to 5% of the mean (32.6), 1.63. This comparison shows that the confidence interval is acceptable since the half-width, 1.15, is less than 1.63. If this had not been the case, the number of replications required would have to be recalculated and the process would be repeated. Part (a) of Example 6.1 is now finished.

Example 6.1 part (b) asked for a calculation of the likelihood that the flowtime will be in the range from 5% less than the average to 10% above the average. The solution strategy to complete this task is as follows:

1. Use the data generated in part (a) to generate a histogram of the average flowtimes. (Histogram option)
2. Determine the expected value of flowtime.
3. Plot vertical lines at the desired likelihood values so that they meet the cumulative probability.
4. Plot horizontal lines back to the cumulative probability scale (Y-axis) from the point where the vertical lines meet the cumulative probability curve.

EXHIBIT 6.11 **Graphical confidence interval of flowtimes of all replications for Example 6.1—Revised (Intervals option).**

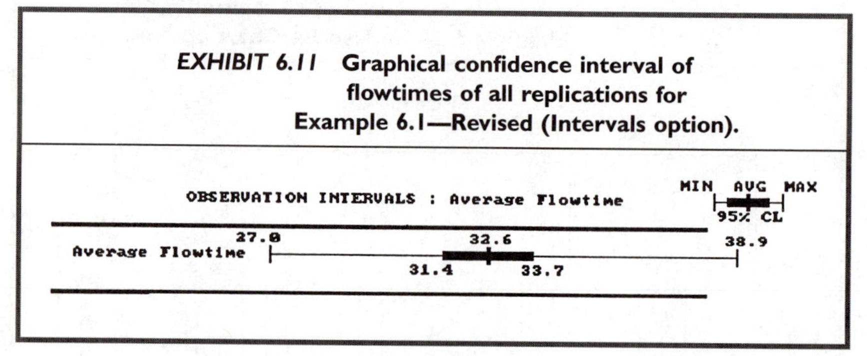

EXHIBIT 6.12 **Parameter report for confidence interval of flowtime for Example 6.1—Revised (Intervals option).**

INTERVALS: AVERAGE FLOWTIME

IDENTIFIER	AVERAGE	STD DEV	.950 C.I. HALF-WIDTH	MIN VALUE	MAX VALUE	NUMBER OF OBS.
AVERAGE FLOWTIME	32.6	3.08	1.15	27.0	38.9	30

To generate a histogram of the average flowtimes, use the Histogram option under the display menu. Using the file identified as "FLWTIME2.DAT", we generate a histogram of the values of average flowtime. Use the Histogram option with the following field values:

Title:	Average Flowtime
Independent Axis Label:	Flowtime

Default all other fields. The Output Processor will generate the graphical histogram, shown in Exhibit 6.13, and the corresponding numerical report, shown in Exhibit 6.14.

From the confidence interval that we found earlier, the average flowtime is 32.6 minutes. The values for −5% and +10% are 30.97 and 35.86, respectively. Now, we plot vertical lines at 30.97 and 35.86. Next, we plot horizontal lines from the intersections of the vertical lines and the cumulative probability curve. Mapping to the cumulative probability, we find that 30.97 and 35.86 have values of approximately 0.35 and 0.90, respectively. The probability of falling within 30.97

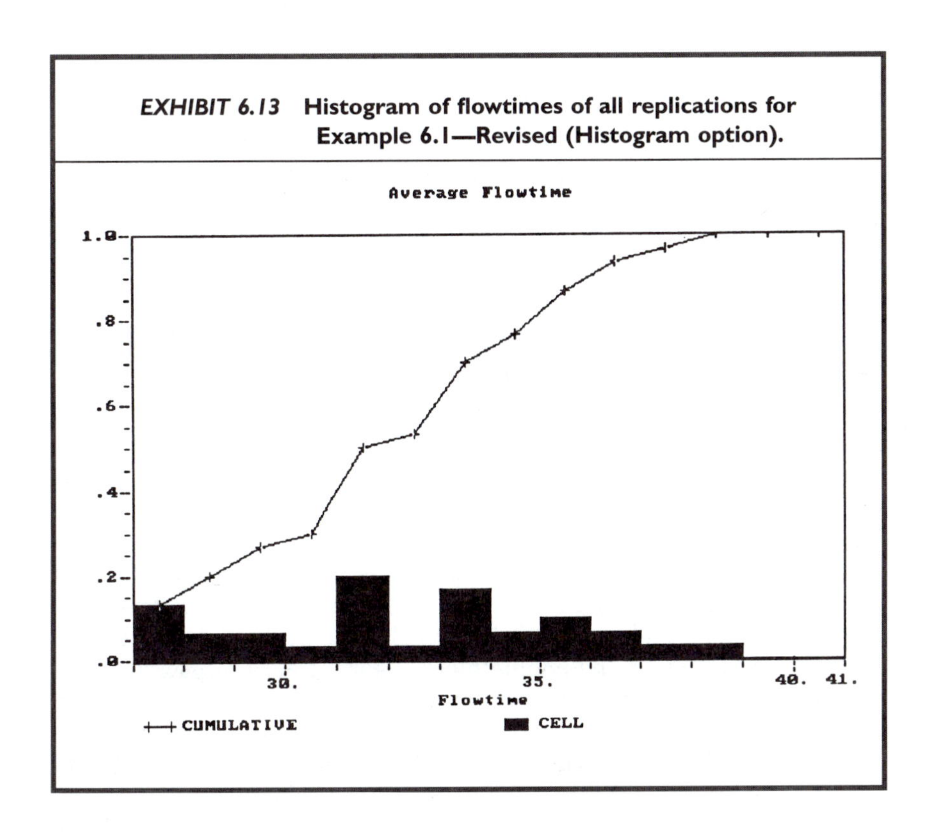

EXHIBIT 6.13 **Histogram of flowtimes of all replications for Example 6.1—Revised (Histogram option).**

EXHIBIT 6.14 **Numerical result for histogram of flowtimes for Example 6.1—Revised (Histogram option).**

	CELL LIMITS		ABSOLUTE FREQ.		RELATIVE FREQ.	
CELL NO.	FROM	TO	CELL	CUMUL.	CELL	CUMUL.
0	−INFINITY	28.000	4	4	.1333	.1333
1	28.000	29.000	2	6	.0667	.2000
2	29.000	30.000	2	8	.0667	.2667
3	30.000	31.000	1	9	.0333	.3000
4	31.000	32.000	6	15	.2000	.5000
5	32.000	33.000	1	16	.0333	.5333
6	33.000	34.000	5	21	.1667	.7000
7	34.000	35.000	2	23	.0667	.7667
8	35.000	36.000	3	26	.1000	.8667
9	36.000	37.000	2	28	.0667	.9333
10	37.000	38.000	1	29	.0333	.9667
11	38.000	39.000	1	30	.0333	1.000
12	39.000	40.000	0	30	.0000	1.000
13	40.000	+INFINITY	0	30	.0000	1.000

and 35.86 is computed by subtracting the cumulative probability of the lower limit from the cumulative probability of the upper limit. In this case, 0.90 − 0.35 = 0.55. This means that there is a 0.55 (55%) probability that flowtime will fall in the range given by 30.97 and 35.86.

EXAMPLE 6.2 Nonterminating System

Consider a distribution center where engine blocks are packed into transportation containers. There are five types of engines (1, 2, 3, 4, and 5) that flow into the center with an interarrival time of 1 minute. The distribution of engine types is as follows: 10% are engine Type 1, 10% are engine Type 2, 20% are engine Type 3, 30% are engine Type 4, and 30% are engine Type 5. Type 5 engine blocks must first go through a prepacking stage in which one worker must perform some special tasks on the engine block before it goes to packing. The prepacking process follows a uniform distribution between 3 and 6 minutes. Engine Types 1 to 4 flow directly to the packing process. Upon arrival at packing, one worker will be assigned to pack an engine block.

The times (in minutes) for one worker to pack one engine block are as follows:

Engine Type	Packing Times
1	Triangular(11,15,17)
2	Triangular(9,12,14)
3	Triangular(8,10,12)
4	Triangular(8,9,11)
5	Triangular(6,8,10)

After the worker finishes packing an engine block, another worker using a forklift places the engine block in a storage rack. The loading time is triangularly distributed (4.9, 5.95, 7.1) in minutes. There are two employees dedicated to prepacking, ten workers are in the packing area, and there are six forklifts with drivers. The distribution center runs continuously, 24 hours a day, 7 days a week. The objective is to simulate the system in steady state until the true value for the average time that an engine block spends in the system falls within 5% of the sample mean with 99% confidence. The solution strategy to accomplish this objective is as follows:

1. Execute three one-day pilot replications to determine the length of the transient (warm-up) phase. To accomplish this:
   - **a.** Modify the TALLIES element in the experiment frame to record time spent in the system by each engine block and store that value in an external data file.
   - **b.** Use the Plots option to generate a plot of the data from the pilot run.
   - **c.** Use the MovAverage option to generate a plot of the moving averages for time in the system.
   - **d.** Using the plot of the moving average, determine when the time in the system appears to stabilize. This point will represent the end of the transient (warm-up) phase.
2. Now that the transient phase has been determined, conduct a pilot run of 10 replications, excluding the statistics from the transient phase. The length of each replication should be equal to six times the length of the transient phase. To accomplish this:
   - **a.** Add the OUTPUTS element to the experiment frame to create an output file that contains the average for time in the system for each replication.
   - **b.** Use the Output Processor options Tables and BarChart to generate a table of the average time in the system and

the corresponding barchart using the data recorded in the output file. This step is for visual analysis of the data in order to gain a better understanding of the results.

c. Use the Output Processor option Intervals to generate a confidence interval on the average flowtime.

3. If the confidence interval generated above is too large, determine the number of additional replications required to obtain an appropriate confidence interval using Equation 6.1.
4. If additional replications are required, the procedure is as follows:
   a. Use the SEEDS element in the experiment frame to reinitialize the random number streams to create independent replications from those already performed.
   b. Use a different output file to record the data from the additional replications.
   c. Use the Output Processor option Append to add the data of the additional replications to the data of the original replications.
   d. Use the Output Processor to generate a table and the corresponding barchart for the average flowtime using the data from all replications (Tables and BarChart options). As already stated, this is for visual analysis of the data.
   e. Use the Output Processor to generate a new confidence interval for the average flowtime using data from all replications (Intervals option).
5. Repeat Steps 3 and 4 if the confidence interval is still too large.

6.5 The TALLIES Element Revisited

We need to understand how the system behaves over time to determine the length of the transient phase. Therefore, we need to capture data from a time-persistent statistic. Time-persistent statistics are gathered by the TALLIES and DSTATS elements. In this example, the performance measure is time in the system; therefore, we need to use the TALLIES element to capture the necessary data. If the performance measure was the number in a queue or resource utilization, the DSTATS element would be used. This section will discuss the use of the TALLIES element to collect time-persistent data. Refer to Appendix B for more details about the DSTATS element.

In Example 6.2, we will use the TALLIES element to capture and store data for the value of time in the system as a time-persistent statistic. The TALLIES element can be used to record the value for time in the system of each engine block. As described in Chapter 4, the

TALLIES element is used to define the TALLY variables. The format for the TALLIES element was indicated in Section 4.6, as follows:

TALLIES: *Number, Name, Output File*: repeats;

The Operands are described as follows:

Operand	Descriptions	Default
Number	Tally number [integer]	Sequential
Name	Name and identifier for labeling the statistics in the Summary Report	"Tally *Number*"
Output File	External file name of unit for storing observations	No saving

Previously, the tally values were not stored; however, for Example 6.2, these values need to be saved in an external data file. To create an external data file, we specify a filename enclosed in double quotes for the *Output File* operand. The file extension .dat is used to identify these data files. For example, the following TALLIES element would record each observation of *QueueTime* in the file *"QTime.dat"* and the values for *ExitTime* would be saved in *ExitTime.dat* on drive A.

```
TALLIES:    QueueTime, "QTime.dat":
            ExitTime, "A:\ExitTime.dat";
```

For Example 6.2, the TALLIES Element will be as follows:

```
TALLIES:    Time In System, "SysTime1.dat";
```

This will record each observation of *Time In System* in the data file *"SysTime1.dat"*.

6.6 The Model and Experiment Frames for Example 6.2

The model frame for Example 6.2 is shown in Exhibit 6.15. The experiment frame is shown in Exhibit 6.16.

Key areas to note in the experiment frame are the TALLIES and REPLICATE elements. The TALLIES element is used to write the data for the TALLY *Time In System* to the file *"SysTime1.dat"*. The

EXHIBIT 6.15 Model frame for Example 6.2.

```
BEGIN;
        CREATE:
             ArrivalTime:                  !Interarrival time
             MARK(ArrTime);                Mark arrival time attribute
        ASSIGN
             PartType=PartMix;             Assign part number 1 to 5
        BRANCH,1:
             IF, PartType==5, PrePack: !If part type is 5 then go to PrePack
             ELSE, Packing;                Else go to packing
PrePack QUEUE,
             PrePackQ;                     Wait for PrePack Worker
        SEIZE:
             PrePack Worker;               Capture PrePack Worker
        DELAY:
             PrePackTime;                  Delay for PrePackTime
        RELEASE:
             PrePackWorker;                Free PrePackWorker resource
Packing QUEUE,
             PackingQ;                     Wait for PackingWorker
        SEIZE:
             PackingWorker;                Capture PackingWorker
        DELAY:
             ED(PartType);                 Delay for PackTime
        RELEASE:
             PackingWorker;                Free PackingWorker
        QUEUE:
             ForkliftQ;                    Wait for Forklift
        SEIZE:
             Forklift;                     Capture Forklift
        DELAY:
             TravelTime;                   Delay for TravelTime
        RELEASE:
             Forklift;                     Free Forklift
        TALLY:                             !Tally the time in the system
             Time in System,INT(ArrTime);
        DISPOSE;                           Destroy entity
```

REPLICATE element indicates that the simulation is to run for three one-day replications of 1440 minutes each.

EXHIBIT 6.16 Experiment frame for Example 6.2.

```
BEGIN;
PROJECT,          Example 6.2, Team;
ATTRIBUTES:       ArrTime:
                  PartType;
QUEUES:           PrePackQ:
                  PackingQ,LVF(PartType):
                  ForkliftQ,LVF(PartType);
RESOURCES:        PrePackWorker, 2:
                  PackingWorker, 10:
                  Forklift, 6;
EXPRESSIONS:      1, PackTime1, TRIA(11, 15, 17, 1):
                  2, PackTime2, TRIA(9, 12, 14, 1):
                  3, PackTime3, TRIA(8, 10, 12, 1):
                  4, PackTime4, TRIA(8, 9, 11, 1):
                  5, PackTime5, TRIA(6, 8, 10, 1):
                     ArrivalTime, 1:
                     PartMix, DISC( 0.1,1, 0.2,2, 0.40,3, 0.70,4, 1.00,5, 1):
                     PrePackTime, UNIF(3, 6, 1):
                     TravelTime, TRIA(4.9, 5.95, 7.1, 1);
TALLIES:          Time In System, "SysTime1.dat";
DSTATS:           NR(PrePackWorker)/2, PrePackWorker Util:
                  NR(PackingWorker)/10, PackingWorker Util:
                  NR(ForkLift)/6, ForkLift Util;
REPLICATE, 3, 0, 1440;
```

EXAMPLE 6.2 Solution

A segment of the resulting output file is shown in Exhibit 6.17. The simulation run also created the data file *"SysTime1.dat,"* which stores the values of *Time In System* for each observation for the three replications.

For Example 6.2, we use the following Output Processor options:

Command	Function
Plots	Generates a scatter plot of the values in the data file
MovAverage	Generates a plot of the moving average of the data
Tables	Generates a table of the data in the data file

EXHIBIT 6.17 Segment of Output file for Example 6.2.

Summary for Replication 1 of 3

Project: Example 6.2 Run execution date: 8/24/1994
Analyst: Team Model revision date: 8/24/1994
Replication ended at time: 1440.0

TALLY VARIABLES

Identifier	Average	Variation	Minimum	Maximum	Observation
Time In System	21.678	.24388	13.900	43.412	1416

DISCRETE-CHANGE VARIABLES

Identifier	Average	Variation	Minimum	Maximum	Final Value
PrePackWorker Util	.67597	.54370	.00000	1.0000	.00000
PackingWorker Util	.97405	.07722	.00000	1.0000	1.0000
ForkLift Util	.98268	.10280	.00000	1.0000	1.0000

Command	**Function**
BarChart	Generates a barchart for a specified variable
Intervals	Generates confidence intervals for a specified variable
Append	Joins two data files into one file

The first step in the analysis will be to generate a scatter plot of the data in the file *"SysTime1.dat"*. After invoking the Output Processor, ADD the file *"SYSTIME1.DAT"* and select the SingRep data treatment and select replication 1. Next, use the Plots option with the following field values to generate a plot of the data:

Title:	Time In System
Independent Axis Label:	Simulation Time
Dependent Axis Label:	Time

Default the remaining fields. The resulting plot is shown in Exhibit 6.18. Examining the plot, we can see that the time in the system

EXHIBIT 6.18 Plot of pilot run data for Example 6.2 (Plots Option).

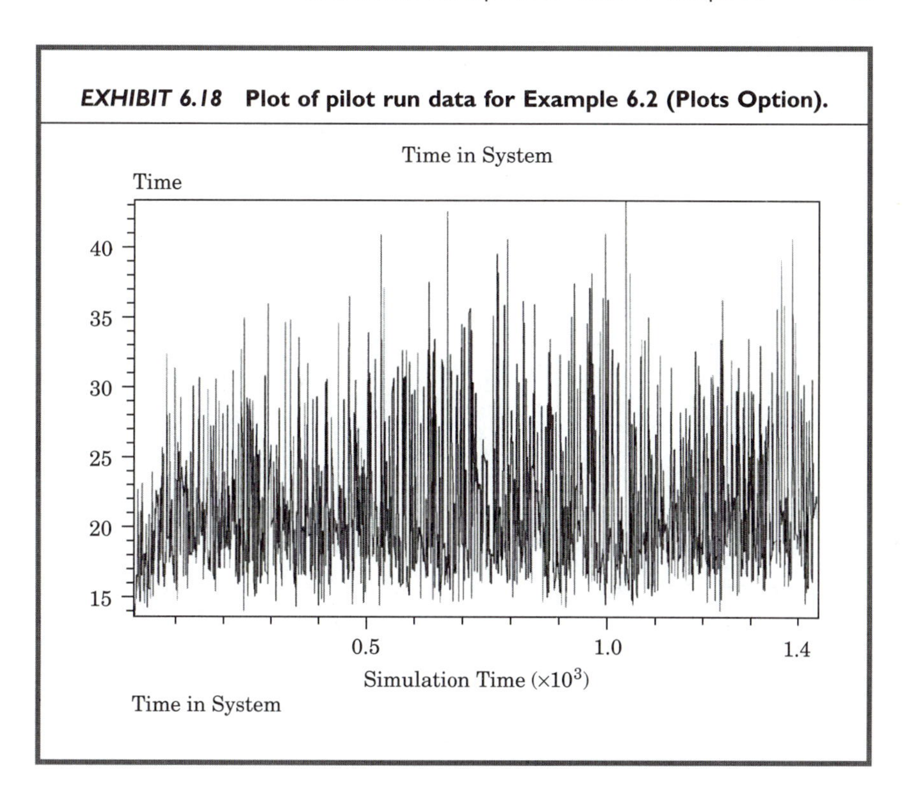

is quite variable, but it is not possible to determine any underlying patterns.

To gain a clearer understanding of the data, we will create a plot of the moving average of 25 observations. This is accomplished using the MovAverage command with the following field designations:

Title:	Time in System
Type of Average:	MOV
Type Value:	25
Independent Axis Label:	Simulation Time
Dependent Axis Label:	Time
Plot Individual Data Points:	No
List Forecasted Values:	No

Default the remaining fields. This will result in the plot shown in Exhibit 6.19.

Analyzing the plot of moving averages, we can see that the value reaches stability after approximately 300 minutes. However, one replication is not a good indicator. Therefore, we generate the plots of the

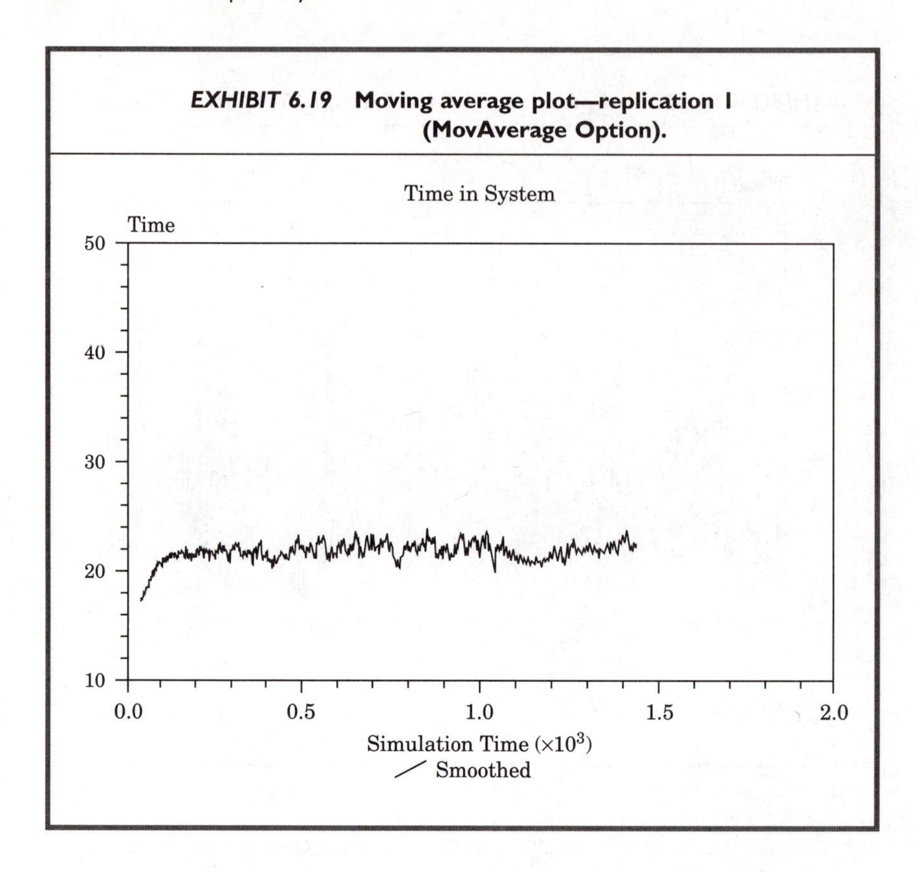

EXHIBIT 6.19 **Moving average plot—replication 1 (MovAverage Option).**

moving average for the other two replications by selecting replications 2 and 3, respectively, in the SingRep replication treatment. The resulting plots are shown in Exhibits 6.20 and 6.21. After analyzing all three plots, we will select 700 minutes as the length of the transient (warm-up) phase. Selection of 700 minutes is based on a conservative estimate of the transient phase length; it is better to estimate a longer transient phase than one that is too short.

Once the transient phase has been identified, the procedure is similar to that of a terminating system. The first step will be to conduct the pilot run of 10 replications, excluding the statistics from the first 700 minutes of each replication. A good rule of thumb is that the length of each replication should be six times the transient phase. Since we exclude the transient phase from the statistics, we are actually capturing data for a time period equal to five times the transient phase. For Example 6.2, the transient phase is 700 minutes; therefore, the length of the replication is 4200 minutes. A few modifications

EXHIBIT 6.20 Moving average plot—replication 2 (MovAverage Option).

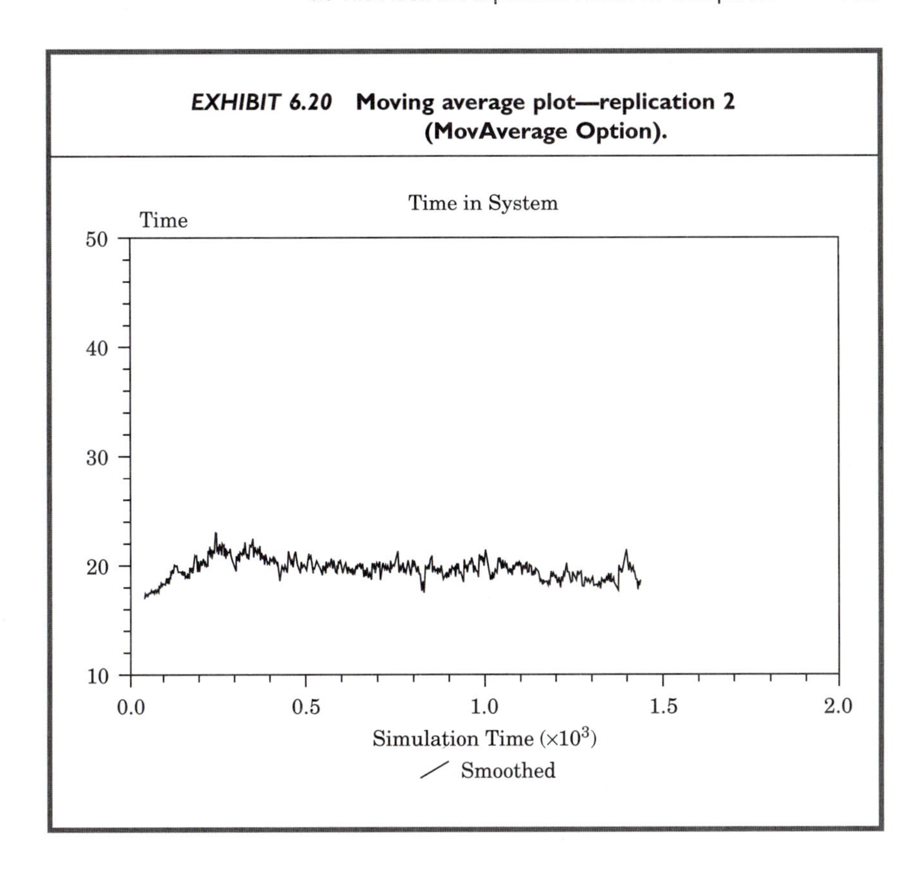

must be made to the experiment frame before conducting the pilot run. First, we must remove the data file operand from the TALLIES element. That was only required to determine the transient phase. Second, we need to add the OUTPUTS element so that we can capture the average time in the system for each replication. Finally, we have to change the REPLICATE element in order to conduct 10 replications with a length of 4200 minutes and a transient phase of 700 minutes. The modified experiment frame is shown in Exhibit 6.22. The changes are shown in italics.

We now execute the pilot run. In addition to the normal output file, we have also created a data file named "SysTime2.dat" that contains the value of the average time in the system for each of the 10 replications. Once again, we use the Output Processor to conduct the output analysis. First, Add the file "SYSTIME2.DAT" and select the Lumped replication treatment. Next, use the Tables option to generate

EXHIBIT 6.21 **Moving average plot—replication 3 (MovAverage Option).**

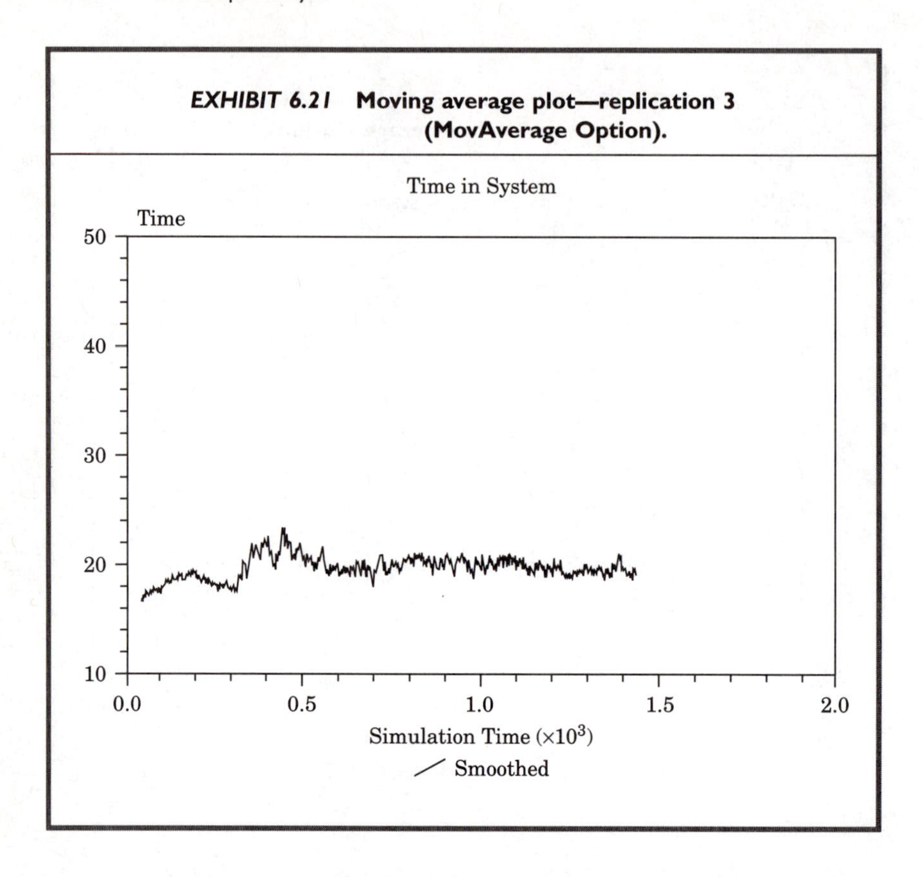

a list of the average time in the system for each of the replications. Use the following field values:

Title:	Time In System
Independent Axis Label:	Replication #

Default the remaining fields. The result is shown in Exhibit 6.23.

Next, generate a barchart of the data using the BarChart command. Use the following field values:

Title:	Time In System
Independent Axis Label:	Replication #
Dependent Axis Label:	Time

Default the remaining fields. The resulting barchart is shown in Exhibit 6.24. The table and barchart shown in Exhibits 6.23 and 6.24 are used to visually examine the data and identify any existing outliers. After examining the data, we find that everything is in order and we can proceed in creating the confidence interval.

EXHIBIT 6.22 Modified experiment frame for Example 6.2.

```
BEGIN;
PROJECT,         Example 6.2, Team;
ATTRIBUTES:      ArrTime:
                 PartType;
QUEUES:          PrePackQ:
                 PackingQ, LVF(PartType):
                 ForkliftQ, LVF(PartType);
RESOURCES:       PrePackWorker, 2:
                 PackingWorker, 10:
                 Forklift, 6;
EXPRESSIONS:     1, PackTime1, TRIA(11, 15, 17, 1):
                 2, PackTime2, TRIA(9, 12, 14, 1):
                 3, PackTime3, TRIA(8, 10, 12, 1):
                 4, PackTime4, TRIA(8, 9, 11, 1):
                 5, PackTime5, TRIA(6, 8, 10, 1):
                    ArrivalTime, 1:
                    PartMix, DISC(0.1,1, 0.2,2, 0.40,3, 0.70,4, 1.00,5, 1):
                    PrePackTime, UNIF(3, 6, 1):
                    TravelTime, TRIA(4.9, 5.95, 7.1, 1);
TALLIES:         Time In System;
OUTPUTS:         TAVG(Time In System),"SysTime2.dat",
                 Avg Time In System;
DSTATS:          NR(PrePackWorker)/2,PrePackWorker Util:
                 NR(PackingWorker)/10,PackingWorker Util:
                 NR(ForkLift)/6,ForkLift Util;
REPLICATE, 10, 0, 4200, Yes, Yes, 700;
```

EXHIBIT 6.23 Table of time in system for Example 6.2 (Tables option).

TABLE: TIME IN SYSTEM

REPLICATION #	AVG TIME IN SYST
1.00	22.6
2.00	22.9
3.00	25.3
4.00	23.7
5.00	25.3
6.00	24.2
7.00	24.1
8.00	25.4
9.00	24.7
10.0	23.8

EXHIBIT 6.24 Barchart of Time in System for Example 6.2 (BarChart option).

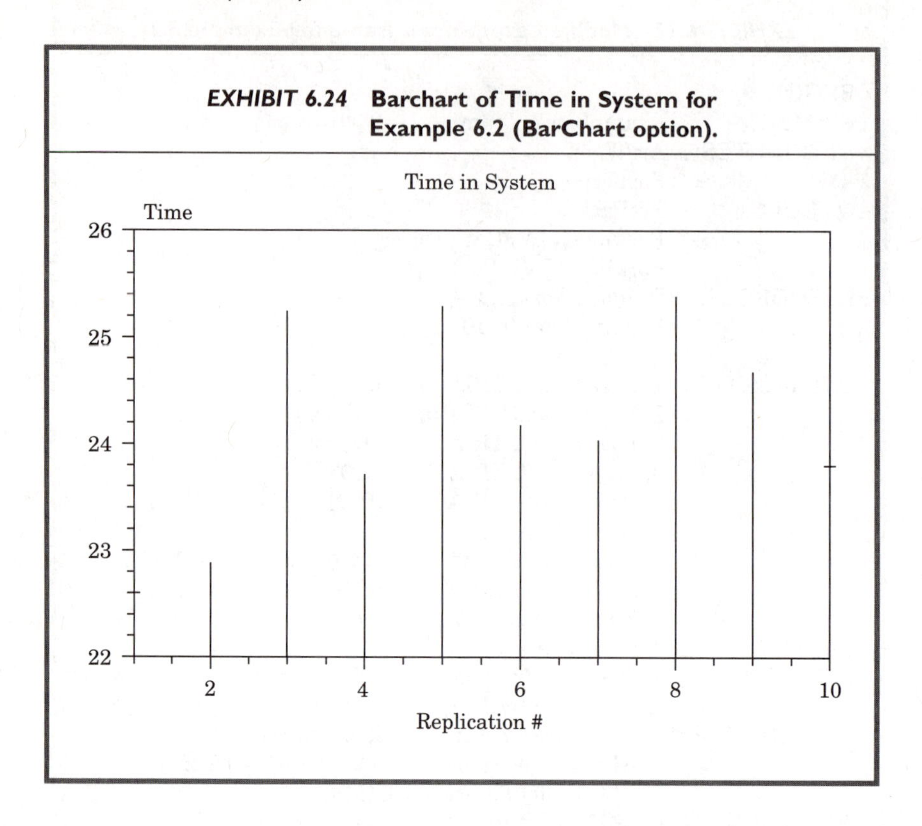

To generate the 99% confidence interval, we use the Intervals option with the following field values:

Title:	Time In System
Confidence Coefficient:	0.01

The resulting confidence interval is shown graphically in Exhibit 6.25 with the corresponding parameter report shown in Exhibit 6.26.

The results are as follows:

1. The average time in the system is 24.2 minutes.
2. With a 99% confidence, the half-width is 0.998 resulting in the confidence interval (23.2, 25.2).

Computing 5% of the average (24.2), we find that the desired half-width is 1.21. Since the actual half-width (0.998) is less than the desired half-width (1.21), we have arrived at a satisfactory answer. Thus, Example 6.2 is complete. If the confidence interval had been too large, we would have proceeded by estimating the number of

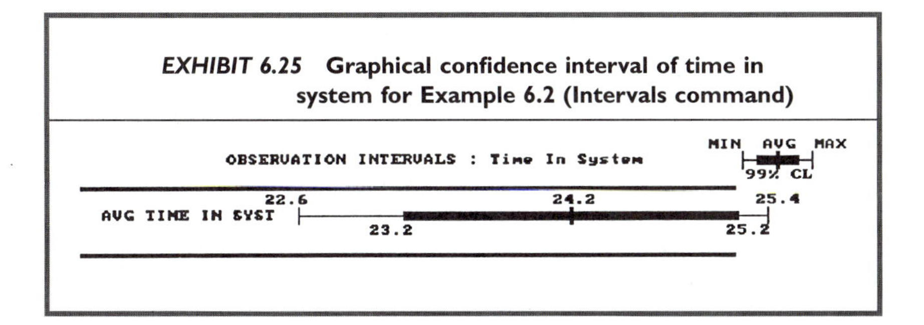

EXHIBIT 6.25 **Graphical confidence interval of time in system for Example 6.2 (Intervals command)**

EXHIBIT 6.26 **Parameter report for confidence interval of time in system for Example 6.2 (Intervals option).**

INTERVALS: JOB FLOWTIME

IDENTIFIER	AVERAGE	STD DEV	.990 C.I. HALF-WIDTH	MIN VALUE	MAX VALUE	NUMBER OF OBS.
JOB FLOWTIME	24.2	.971	.998	22.6	25.4	10

replications required using Equation 6.1. Next, we would perform the additional replications after adding the SEEDS element and changing the OUTPUTS data file. Finally, we would recompute the confidence interval and check to see if it fell within the desired limits.

6.7 SUMMARY

The real power of simulation is its ability to capture the variation of stochastic systems. However, due to the stochastic nature of the models, you have to be very careful when analyzing simulation results in order to properly determine the true nature of the system. Simulation runs must adequately capture the variation in order to accurately represent the values of the performance measures. The run length of the simulation defines the amount of variation captured. In a terminating system, the length of each replication is equal to system operation time. In a nonterminating system, the simulation runlength is equal to six times the transient phase, as a rule of thumb. In both cases, pilot runs are used to gain an understanding of system variation. The difference in the analysis procedures for terminating and nonterminating systems is that the transient phase on the nonterminating system must be identified and excluded from the statistics captured.

The Output Processor is used to perform the statistical analysis as well as to graphically represent the data gathered. Options such as Tables, BarChart, Plots, MovAverage, and Histogram are used to display the data gathered; whereas the Intervals option is used to analyze the statistical nature of the data. In this chapter, we have presented the methods for analyzing both terminating and nonterminating systems. When conducting any simulation project, these methods should be used to determine accurate interpretations of simulation results.

6.8 REFERENCE

Welch, P. D., "Statistical Analysis of Simulation Results," in *The Computer Performance Modeling Handbook*, S. S. Lavenberg, ed., pp. 268–328, Academic Press, New York (1983).

6.9 EXERCISES

E6.1 This assignment is a modification of the manufacturing system described in E3.1.

- **a.** Based on 10 replications of an 8-hour shift, generate a 95% confidence interval for the average utilization for the assembly process.
- **b.** Generate a 99% confidence interval for the average number of parts waiting for the inspector over an 8-hour shift. This interval should have a half-width less than 20% of the mean.

E6.2 Customers arrive at the Goldrush Bank following an exponential distribution with a mean of 0.9 minutes. These customers join a single waiting line that is served by four tellers. The service time for each teller is triangularly distributed (1, 4, 6) in minutes. The bank is open daily from 8:30 A.M. to 5:00 P.M. Determine a 95% confidence interval for the average number of customers waiting in line. This interval should have a half-width less than 7% of the mean.

E6.3 A manufacturing facility operates continuously, producing the same types of jobs as described in E4.1. Using 10 replications, determine a 95% confidence interval on the steady-state value of the time that a Type I job spends in the system.

E6.4 A soft drink manufacturer is designing a production line that molds and fills 64-ounce bottles. The bottles are produced at five independent molding machines. The completed bottles flow into a single buffer that feeds a filling machine with a capacity of 10. Bottle molding follows a normal distribution with a mean of 1.00 minute and a standard deviation of 0.02 minutes. Bottles are filled following a normal distribution with a mean of 2 minutes and a standard deviation of 0.04 minutes. The system will be operated on a continuous basis. Determine a 95% confidence interval for the number of bottles in the buffer. Base the results on 10 replications.

7

More Intermediate Resource Concepts

Chapters 3 and 4 discussed basic SIMAN blocks and elements. Whereas these blocks and elements may be found in most SIMAN models, they alone do not provide the flexibility needed to model systems of increasing complexity. In this chapter, we will examine new blocks and revisit some familiar ones in order to gain that required flexibility. This chapter covers methods for changing a resource's capacity during the simulation run, causing breakdowns of resources, creating decision branches and loops, grouping resources, and changing the flow of the model.

EXAMPLE 7.1 Problem Statement

Recall the model of Example 4.1. View this system running in 8-hour shifts, three shifts per day, 5 days each week. Consider the following new requirements and add them to the model. After the changes and additions, determine the average flowtime of each entity, and utilizations of each resource.

1. The inspector takes a 30-minute lunch break 4 hours (240 minutes) into each 8-hour shift.
2. During the first shift, the delay time of the grinders is 90% of the original delay time. During the third shift, the delay time is 110% of the original delay time. During the second shift, the delay time is 100% of the original delay time.
3. At the beginning of the second shift, one drill is taken off-line for 5 minutes for preventive maintenance. Increase the maximum queue before the drill to 5.
4. Model the system for 15 shifts of 8 hours each.

The new blocks and elements needed to address the examples in this chapter are given in Table 7.1.

Table 7.1 Blocks/Elements Introduced in this Chapter

New Block/Element	Purpose
SETS element	Groups related resources into a single class
STATESETS element	Allows defined states to be associated with a resource SET
FAILURES element	Used to define breakdowns or other forms of downtime for a resource SET
ALTER block	Increases or decreases the capacity of a particular resource
SCHEDULES element	Increases or decreases the capacity of a particular resource based on timed intervals
FREQUENCIES element	Used to obtain summary report information about resource states
PROCEED block	Used to define a gate or entity passage reference point
BLOCK block	Denies passage through a PROCEED block to an entity
UNBLOCK block	Reverses the effect of a BLOCK block
BLOCKAGES element	Defines blockage points and their initial conditions within a model
IF, ELSEIF, ELSE, ENDIF blocks	Used to create logical decision branches
WHILE, ENDWHILE blocks	Used to create a looping activity

7.1 Changing Resource Capacity

7.1.1 The SCHEDULES Element

The SCHEDULES element allows for modeling changes in resource availability (such as a lunch break or a shift change). This element may be used to fulfill some of the requirements of Example 7.1. It has the following format:

SCHEDULES: *Identifier, Resource Capacity*Capacity Duration,...* : repeats;

Operand	Description	Default
Indentifier	Schedule number or name [integer or symbol name]	—
Resource Capacity	Resource capacity [integer]	—
Capacity Duration	Capacity duration [expression]	Infinite

The SCHEDULES element creates a time-dependent schedule for capacity changes in resources. The *Resource Capacity* operand is the absolute capacity of a resource that is indicated in the RESOURCES element in the following manner:

RESOURCES: *ResourceName, SCHED(NumberName)*: repeats;

Consider the following example:

RESOURCES: Press, SCHED(1);
SCHEDULES: 1, 2*480, 3*480, 1*480;

The resource *Press* has a capacity of 2 for the first 480 time units, a capacity of 3 for the next 480 time units, and a capacity of 1 for another 480 time units. If the simulation time is longer than 1440 time units then the schedule will be recycled. If we simulate for only 1440 time units, then the SCHEDULES element may be restated in the following manner:

SCHEDULES: 1, 2*480, 3*480, 1;

Omitting the duration from the last specification indicates that only one press is available for the remainder of the simulation.

Below is part of the model frame for Example 4.1 with the changes required above. The changes are shown in italics.

```
RESOURCES:   Drill:
             Mill, 3:
             Plane, 3:
             Grind, 2:
             Inspect, SCHED(1):
             Shift, SCHED(2);
SCHEDULES:   1, 1*240, 0*30, 1*210:    !Lunch break
             2, 1*480, 2*480, 3*480;   Shift divider
```

The first requirement is to model a lunch break for the inspector. Think of this as a change in the capacity of the inspector from 1 to 0. Notice in the RESOURCES element that the resource *Inspect* has an additional operand *SCHED(1)*. This indicates that the resource *Inspect* should follow the instructions of SCHEDULE number 1. In the SCHEDULES element, the first schedule indicates that the capacity of *Inspect* should be 1 for a period of 240 minutes, 0 for a period of 30 minutes, and 1 for a length of 210 minutes. Then the schedule is recycled.

A new resource, *Shift*, has been added. This resource will never be seized by any entity, and its utilization is not important. This is a "dummy" resource that serves merely to help denote the beginning of each shift. The SCHEDULE attached to this resource indicates that its capacity is 1 for the first shift, 2 for the second, and 3 for the third. This property will be used to fulfill the second requirement regarding the grinder.

Below, in italics, are the model frame changes for the grinder.

```
Grinding QUEUE,                                  !Grinder queue, capacity
            GrinderQ, 2, EJCT;                    2, overflow to EJCT
         SEIZE:
            Grinder;                              Capture grinder
         BRANCH, 1:
            IF, MR(SHIFT)==1, Delay1: !First Shift
            IF, MR(SHIFT)==2, Delay2: !Second Shift
            ELSE, Delay3;
Delay1   DELAY:
            GrindTime * 0.9:                      !Delay for 90% GrindTime
            NEXT(Rel);
Delay2   DELAY:
            GrindTime:                            !Delay for GrindTime
            NEXT(Rel);
Delay3   DELAY:
            GrindTime * 1.1:                      !Delay for 110% GrindTime
            NEXT(Rel);
Rel      RELEASE:
            Grinder;                              Free the resource grinder
```

This particular section of SIMAN code contains a number of important changes. First is the use of a new SIMAN variable in the BRANCH block. This variable is MR. By using the name of a resource as its operand, MR returns the capacity of that resource. For example, MR(Inspect) will return the value "0" when the inspector is on lunch break, and "1" at all other times. MR is used here in the BRANCH block to indicate where the entity should be sent based on the current shift. If MR evaluates to “1”, then the entity is sent to *Delay1* where it is delayed by the grinder for 0.9 times *GrindTime*. After this delay, the entity is sent to *Rel* where the grinder is released. Similar logic follows for shifts 2 and 3. The shift resource is never truly utilized here, it merely indicates the shift.

The next requirement indicates a change in capacity for the drill at the beginning of the second shift. This could be accomplished using the SCHEDULES element. However, there is another block that could be used for this purpose—the ALTER block.

7.1.2 The ALTER Block

The ALTER block, when executed, changes the capacity of a resource by a positive or negative amount. The format is as follows:

ALTER: *Resource ID, Quantity to Change*: repeats;

Operand	Description	Default
Resource ID	Resource to be altered [expression truncated to an integer or symbol name]	—
Quantity to Change	Capacity change [expression truncated to an integer]	1

Resource ID is the number or symbol name of a resource. The operand *Quantity to Change* is the positive or negative change in the resource's capacity.

In Example 7.1, the capacity of the drill decreases by 1 at the beginning of the second shift. This may be accomplished with the block:

```
ALTER:
      Drill, -1;
```

One way to determine when this event should take place, without using a SCHEDULE, is by using a subprogram. Unlike a subroutine or function call in a conventional computer program, the SIMAN subprogram may be thought of as running simultaneously with the main program. In this subprogram, an entity will be created specifically to execute the ALTER block at the required moment. The subprogram is as follows:

```
          CREATE:
            480;
          BRANCH, 1:
            IF, MR(Shift) == 2, ChgDrill:
            ELSE, NORMAL;
ChgDrill  ALTER:
            Drill, -1;
          DELAY:
            5;
          ALTER:
            Drill, +1;
Normal    DISPOSE;
```

The CREATE block causes an entity to enter the system every 480 minutes (the beginning of each shift). If MR evaluates to "2", indicating that it is the second shift, then the entity executes a sequence of blocks that decreases the drill's capacity by 1, delays for 5 minutes, then increases the capacity by 1. The entity is then destroyed. If the entity is created during the first or third shift, it is immediately destroyed.

Because we are interested in the utilization of the drill, there is an important change that must be made to the DSTATS element in the experiment frame. In Example 4.2, the line

NR(1)/2, Drill Utilization

sufficed for calculating the utilization. However, the drill does not always remain at a capacity of 2. Once again, the SIMAN variable MR comes into play. By substituting the "2" with "MR(1)", it is assured that as the capacity of the drill changes at the beginning of the second shift, the correct utilization will be calculated.

The last two changes required by Example 7.1 are changing the size of *DrillQ* to 5 and changing the length of the simulation from 2400 to 7200 to represent 15 shifts of 8 hours each.

Exhibits 7.1, 7.2, and 7.3 show the model and experiment frames and the summary report of this simulation.

EXAMPLE 7.2 Problem Statement

Recall the manufacturing model of Example 4.1. We will build on this example by adding machine failures and changes in entity flow as described in the following paragraphs:

EXHIBIT 7.1 Model frame for Example 7.1.

```
!Model Frame for Example 7.1
!
BEGIN, NO;
           CREATE:                               !Create arriving jobs according to
             EX(1,1):                            !Exponential distribution and save
             MARK(ArrTime);                      Arrival times in attribute ArrTime
           ASSIGN:
             Jobtype = DP(2,1):                  !Type I or Type II job
             DrillTime = UN(3,1):                !Uniform drill time
             MillTime = TR(4,1):                 !Triangular mill time
             PlaneTime = TR(5,1):                !Triangular plane time
             GrindTime = DP(Jobtype+5, 1):       !Grind time is dependent on job type
             InspectTime = RN(8,1):              Normally distributed inspection time
Drilling   QUEUE,                                !Queue for drill with capacity 5
             DrillQ, 5, Eject;                   Send overflow to Eject
           SEIZE:
             Drill;                              Capture resource Drill
           DELAY:
             DrillTime;                          Process for DrillTime
           RELEASE:
             Drill;                              Free the resource Drill
           BRANCH,1:
             IF, JobType == 1, Milling:          !Send Type I to Milling
             ELSE, Planing;                      Send Type II to Planing
Milling    QUEUE,                                !Queue for mill with capacity 3
             MillQ, 3, Eject;                    Send overflow to Eject
           SEIZE:
             Mill;                               Capture resource Mill
           DELAY:
             MillTime;                           Process for MillTime
           RELEASE:
             Mill:                               Free the resource Mill
             NEXT(Grinding);
Planing    QUEUE,                                !Queue for plane with capacity 3
             PlaneQ, 3, Eject;                   Send overflow to Eject
           SEIZE:
             Plane;                              Capture resource Plane
           DELAY:
             PlaneTime;                          Process for PlaneTime
           RELEASE:
             Plane;                              Free the resource Plane
Grinding   QUEUE,                                !Queue for grinder with capacity 2
             GrinderQ, 2, Eject;                 Send overflow to Eject
           SEIZE:
             Grinder;                            Capture resource Grinder
           BRANCH,1:
             IF, MR(SHIFT) == 1, Delay1:
             IF, MR(SHIFT) == 2, Delay2:
             ELSE, Delay3;
Delay1     DELAY:
             GrindTime * 0.9:                    !Delay for .9 normal GrindTime
             NEXT(Rel);
```

EXHIBIT 7.1 *continued.*

```
Delay2    DELAY:
            GrindTime:                     !Delay for normal GrindTime
            NEXT(Rel);
Delay3    DELAY:
            GrindTime * 1.1:               !Delay for 1.1 normal GrindTime
            NEXT(Rel);
Rel       RELEASE:
            Grinder;                       Free the resource Grinder
Inspect   QUEUE,                           !Queue for inspector with capacity 4
            InspectorQ, 4, Eject;          Send the overflow to Eject
          SEIZE:
            Inspector;                     Capture resource Inspector
          DELAY:
            InspectTime;                   Process for InspectTime
          RELEASE:
            Inspector;                     Free the resource Inspector
          BRANCH, 1, 10:                   !Take only one branch
            WITH, 0.05, Rework:            !Rework 5% of the jobs
            WITH, 0.05, Scrap:             !Scrap 5% of the jobs
            ELSE, Done;                    Remaining 90% are complete
Done      TALLY:                           Tally the average flowtime
            Flowtime, INT(ArrTime);
          TALLY:
            Exit Period, BET;              !Tally average time between exits
          COUNT:
            JobType, 1:                    !Count number of jobs by type and
            DISPOSE;                       Destroy the entity
Rework    ASSIGN:                          !Reassign processing times
            DrillTime = UN(3,1):
            MillTime = TR(4,1):
            PlaneTime = TR(5,1):
            GrindTime = DP(Jobtype+5, 1):
            InspectTime = RN(8,1);
          COUNT:
            Reworks, 1:                    !Count number of reworked jobs
            NEXT(Drilling);                Send reworks back to Drill
Scrap     COUNT:
            Fails, 1:                      !Count number of failed jobs and
            DISPOSE;                       Destroy the entity
Eject     COUNT:
            Ejections:                     !Count number of jobs ejected
            DISPOSE;                       Destroy the entity
          CREATE:
            480;                           !Create an entity every 480 minutes
          BRANCH,1:
            IF, MR(SHIFT) == 2, ChgDrill:
            ELSE, Normal;
ChgDrill  ALTER:
            Drill, -1;                     Reduce Drill capacity by 1
          DELAY:
            5;                             Delay for 5 minutes
          ALTER:
            Drill, +1                      Increase Drill capacity by 1
Normal    DISPOSE;                         Destroy the entity
```

EXHIBIT 7.2 Experiment frame for Example 7.1.

```
!Experiment frame for Example 7.1 using SCHEDULES
!
BEGIN, NO;
PROJECT, Example 7.1, Team;
ATTRIBUTES:   ArrTime:
              JobTime:
              DrillTime:
              MillTime:
              PlaneTime:
              GrindTime:
              InspectTime;
RESOURCES:    Drill, 2:
              Mill, 3:
              Plane, 3:
              Grinder, 2:
              Inspector, SCHED(1):
              Shift, SCHED(2);            Shift resource
SCHEDULES:    1, 1*240, 0*30, 1*210:      !Inspector's lunch break
              2, 1*480, 2*480, 3*480;     Shift divider
QUEUES:       DrillQ:
              MillQ:
              PlaneQ:
              GrinderQ, LVF(JobType):
              InspectorQ;
COUNTERS:     Type I Job Count:
              Type II Job Count:
              Fails:
              Reworks:
              Ejections;
PARAMETERS:   1, 5:
              2, .7, 1, 1.0, 2:
              3, 6, 9:
              4, 10, 14, 18:
              5, 20, 26, 32:
              6, .1, 6, .75, 7, 1.0, 8:
              7, .1, 6, .35, 7, .65, 8, .9, 9, 1.0, 10:
              8, 3.6, .6;
TALLIES:      Flowtime:
              Exit Period;
DSTATS:       NQ(DrillQ), Drill Queue:
              NQ(MillQ), Mill Queue:
              NQ(PlaneQ), Plane Queue:
              NQ(GrinderQ), Grinder Queue:
              NR(1)/MR(1), Drill Utilization:
              NR(2)/3, Mill Utilization:
              NR(3)/3, Plane Utilization:
              NR(4)/2, Grinder Utilization:
              NR(5), Inspector Util.;
REPLICATE, 1, 0, 7200;
```

EXHIBIT 7.3 Summary report for Example 7.1.

Summary for Replication 1 of 1

Project: Example 7.1 Run execution date: 8/24/1994
Analyst: Team Model revision date: 8/24/1994
Replication ended at time: 7200.0

TALLY VARIABLES

Identifier	Average	Variation	Minimum	Maximum	Observations
Flowtime	48.210	.32503	26.995	164.68	1266
Exit Period	5.6679	.79281	1.8160	38.959	1265

DISCRETE-CHANGE VARIABLES

Identifier	Average	Variation	Minimum	Maximum	Final Value
Drill Queue	.96641	1.3917	.00000	5.0000	.00000
Mill Queue	.18499	2.6188	.00000	3.0000	.00000
Plane Queue	.10781	3.6029	.00000	3.0000	.00000
Grinder Queue	.31663	1.7926	.00000	2.0000	.00000
Inspect Queue	.49459	2.0072	.00000	4.0000	.00000
Drill Utilization	.76933	.46296	.00000	2.0000	.00000
Mill Utilization	.67172	.47667	.00000	1.0000	.66667
Plane Utilization	.53180	.64179	.00000	1.0000	.33333
Grinder Utilization	.73884	.47470	.00000	1.0000	1.0000
Inspector Util.	.69753	.65850	.00000	1.0000	.00000

COUNTERS

Identifier	Count	Limit
Type I Job Count	885	Infinite
Type II Job Count	381	Infinite
Fails	72	Infinite
Reworks	62	Infinite
Ejections	108	Infinite

Simulation run complete.

After a drill completes 25 consecutive jobs, it is taken off-line for 10 minutes to remove excess filings. Consequently, the drills have two states; drilling and cleaning. Determine the percentage of time that the drills spend in each state.

The operator of the drill will prefer to use drill 1 if possible because it is a quieter machine. The mill operator, in an effort to use all machines equally, tries to cycle the preference among the three machines.

In order to improve on quality, the manager has created more inspection points for each job, changing the distribution of inspection time to N(10, 1.6). The queue before the inspector will now hold 10 jobs.

If more than eight jobs are waiting in the inspection queue, then the operator of the third mill is sent to assist the inspector, and a block is placed before the grinder queue. When the number of jobs in the inspection queue falls below 5, the mill worker leaves and the blockage is removed.

The mills have four states: milling, broken, waiting, or borrowed. When a mill is working, it is in the milling state. Mills break down according to an exponential distribution with a mean of 80 minutes. The repair time is normally distributed with a mean of 10 minutes and a standard deviation of 1.2 minutes. The mills are in state waiting when they are idle. The third mill will enter the borrowed state when it is sent to assist the inspector. Determine the average time that each mill spends in each of its possible states.

The length of queues before each resource is as follows: 2 before the drills, 3 before the mills, 3 before the planes, 15 before the grinder, and 10 before the inspector.

7.2 The SETS Element

The SETS element is used to group resources. You may wish to establish such a group because the resources are all used during a particular shift, all used in a specific operation, or they all may be subject to a malfunction or blockage in the system. The SETS element has the following format:

SETS: *Number, Name, Member, Member, . . .* : repeats;

Operand	Description	Default
Number	Set number	Sequential
Name	Set name [symbolic]	—
Member	Name of resource to include in set [symbol name or ResIDLow..ResIDHigh]	—

Each resource of the set is called a member, and each member has a name of its own and a sequential index. When defining the members they may be specified as a range of resources from the RESOURCES element. For example,

```
RESOURCES:  Truck:
            Van:
            Flatbed:
            Car:
            Worker:
            Drill:
            Pickup;
SETS:       Motorpool, Truck..Car, Pickup;
```

The RESOURCES element defines four resources. In the SETS element, the range of resources, *Truck* through *Car* (this includes the *Van* and the *Flatbed*), is specified as part of the set *Motorpool*. The resource *Pickup* is included as part of *Motorpool*, but could not be included in the range of resources unless the *Worker* and *Drill* are also defined as part of Motorpool.

The following functions may be of use when grouping resources into sets:

NumMem*(Set ID)* Returns the number of members in the set *Set ID*.

For example, NumMem*(Motorpool)* would return a value of 5 because there are five members of the set *Motorpool: Truck, Van, Flatbed, Car,* and *Pickup*.

Member*(Set ID, Index)* Returns the resource number of the Index into the *Set ID*.

For example, Member *(Motorpool, Pickup)* would return a value of 7 because the resource *Pickup* is the seventh defined resource in the RESOURCES element.

MemIdx*(Set ID, Resource ID)* Returns the member index of resource *Resource ID*.

For example, MemIdx*(Motorpool, Pickup)* would return a value of 5 because it is the fifth member of the *Motorpool* set.

In subsequent sections, we will introduce additional operands of the RESOURCES element and the SEIZE and RELEASE blocks that apply to the use of the SETS element.

For Example 7.2 we define two SETS in the following manner:

```
SETS:  Drills, Drill1, Drill2:
       Mills, Mill1..Mill3;
```

Two SETS, *Drills* and *Mills,* are defined. The *Drills* set contains the resources *Drill1* and *Drill2*. The *Mills* set contains all resources between *Mill1* and *Mill3*.

7.3 The RESOURCES Element Revisited

In Section 3.3.5, we introduced the RESOURCES element that defines important information about each resource used in the model frame. This element has three additional operands: *StateSetID, Failure*

ID, and *Failure Entity Rule*. The expanded RESOURCES element includes the following:

RESOURCES: *Number, Name, Capacity Keyword, StateSetID, FAILURE(Failure ID, Failure Entity Rule)*: ... : repeats;

Operand	Description	Default
StateSetID	StateSet number or name which defines the states for this resource	No stateset
Failure ID	Failure identifier [integer or failure symbol name]	No failures
Failure Entity Rule	Rule identifier [IGNORE, WAIT, or PREEMPT]	Ignore

The *StateSetID* refers to the states defined for this resource in the STATESETS element. The *Failures* operand associates the resource to a type of failure described by the FAILURES element (discussed in detail in Section 7.6.2). The *Failure Entity Rule* may be specified with any of the following:

IGNORE At the scheduled time the resource's capacity is changed, but the resource becomes inoperative only after it has finished the entity it is processing. The schedule/failure begins immediately as defined, but the actual time that the resource is inoperative may be less than the specified downtime.

WAIT This rule dictates that a scheduled downtime begins only after the resource finishes processing the entity. The downtime will be of full duration, but may be postponed.

PREEMPT At the scheduled time the resource's capacity is changed and the resource becomes inoperative for the specified downtime, wresting control from the entity. If control cannot be taken from the entity, then the WAIT rule is used.

An example is as follows:

```
STATESETS:   WorkerStates, ...
RESOURCES:   Worker, 1, WorkerStates, FAILURE(Injured,
             PREEMPT):
             Forklift, 1, , FAILURE(Broken, PREEMPT);
```

In this example, two resources are defined: *Worker* and *Forklift*. The *Worker* has one failure state, *Injured,* and it occurs immediately. Additional states for the *Worker* are defined in the STATESETS element and are called *WorkerStates*. The *Forklift* has a failure state of *Broken*;

it has no additional states. Also, when a resource uses a SCHEDULE, any of the above three rules may be used to describe how the entity is affected by the SCHEDULE. An example is given by the following:

RESOURCES: Worker, SCHED(WorkSchedule, WAIT);

This code would cause the capacity of the *Worker* resource to change according to the SCHEDULE *WorkSchedule*. However, it would not change until the current entity finishes processing.

The following RESOURCES element is for Example 7.2. The differences between it and the corresponding RESOURCES element for Example 4.1 are italicized.

RESOURCES: *Drill1, 1, DrillStates, FAILURE(DrillClean, WAIT):*
Drill2, 1, DrillStates, FAILURE(DrillClean, WAIT):
Mill1, 1, MillStates, FAILURE(MillFail, PREEMPT):
Mill2, 1, MillStates, FAILURE(MillFail, PREEMPT):
Mill3, 1, MillStates, FAILURE(MillFail, PREEMPT):
Plane, 3:
Grinder, 2:
Inspector;

Here the resources for the SETS element have been defined. Notice that the drills and the mills have associated SETS *DrillStates* and *MillStates*, respectively, and associated FAILURES *DrillClean* and *MillFail*, respectively. The *DrillClean* failure waits until the current job completes processing, but the failure of the mill is immediate.

7.4 The STATESETS Element

Up to this point, we have limited our discussion of the *state* of a resource as either *busy* (if captured by an entity) or *idle* (if released and awaiting capture). Your own experience may suggest that the terms busy and idle, though accurate, are not adequate to describe the true state of a resource. For example, if the resource is a worker and he or she is not busy, then that worker could be on a break, moving between machinery stations, goofing off, ill, on personal time, or waiting for a repair to be done. The state "idle" simply does not cover every case. Therefore, SIMAN allows the user to define special-purpose states.

The STATESETS element has the following format:

STATESETS: *Number, Name, State Name(Type), . . .* : repeats;

Operand	Description	Default
Number	Stateset number [integer]	Sequential
Name	Stateset name [symbol name]	Blank
State Name	State name [symbol name]	Blank
Type	Autostate or FailurcID	—

Number is the numerical SIMAN identification. It does not need explicit definition. *Name* is a symbolic name for a particular set of states. The *State* operand may take one of three forms; user defined, autostate, or failure state.

A user-defined state requires only a symbol name and may be set using an ASSIGN block. An example follows:

```
ASSIGN:
    STATE(Drill) = Cleaning;
```

Observe the STATE keyword requires a resource symbol name as an argument.

An autostate uses one of four key words, IDLE, BUSY, INACTIVE, or FAILED, as an operand (e.g., Repair(INACTIVE)). When a resource becomes IDLE, BUSY, INACTIVE, or FAILED, then SIMAN sets the state of the resource to the user-defined state for that key word. Only one autostate key word may be referenced in a stateset (i.e., two different states may not both be defined as INACTIVE).

A failure state uses a *FailureID* defined in a RESOURCES element as an operand. When the resource experiences the failure *FailureID*, the resource state is set to the corresponding state name.

Note: The symbolic name for state may be used in multiple STATESETS, but it must be unique within a STATESET. For example, the *Worker* in the example in Section 7.3 may have the following defined states:

```
STATESETS:  WorkerStates,  Loading, Unloading,
                           Break(INACTIVE),
                           Loafing(IDLE);
```

This STATESETS element defines four states for the *Worker*. *Loading* and *Unloading* are user-defined states. These states must be set explicitly, for example,

```
ASSIGN:  STATE(Worker) = Loading
```

Break and *Loafing* are autostates. When resource *Worker* is INACTIVE or IDLE it is immediately set to state *Break* or *Loafing*, respectively.

Example 7.2 requires the following STATESETS:

STATESETS: DrillStates, Drilling(BUSY), Cleaning(DrillClean):
MillStates, Milling(BUSY), Broken(MillFail),
Waiting(IDLE), Borrowed(INACTIVE);

There are two DrillStates: *Drilling*, which corresponds to the autostate of BUSY, and *Cleaning*, which is dependent upon the *DrillClean* FAILURE.

The four MillStates are *Milling*, which corresponds to an autostate of BUSY, *Broken*, which corresponds to the *MillFail* FAILURE, *Waiting*, which serves as the IDLE state, and *Borrowed* which handles the INACTIVE autostate.

7.5 The SEIZE and RELEASE Blocks Revisited

SEIZE and RELEASE blocks, initially discussed in Chapter 3, are capable of using resources defined in sets. Recall the format of the SEIZE block:

SEIZE, *Priority: Resource ID, Number of Units*: repeats;

In earlier examples, *Resource ID* was either the number or name of a resource. The following forms for *Resource ID* can also be used:

SetName(Index)	*SetName* is the symbol name of the set and *Index* is the index into the set. The resource number will be returned.
Member*(Set Expression, Index)*	*Set Expression* evaluates to a set number from the SETS element and *Index* is the index into the set. This will return the resource number.
SELECT*(Set Expression [,rule][,Attribute ID])*	*Set Expression* evaluates to a number from the SETS element, *rule* is the optional Resource Selection Rule, shown in Table 7.2, and *Attribute ID* is the optional attribute used to save the selected set index.

Table 7.2 Resource Selection Rules

Resource Selection Rule (RSR)	**Description**
CYC	**Cyclic priority** Select the first available resource beginning with the successor of the last SEIZE block selected
LNB	**Largest number busy** Select the SEIZE block that has the largest number of busy resource units; break ties using the POR rule
LRC	**Largest remaining capacity** Select the SEIZE block that has the largest remaining resource capacity; break ties using the POR rule
POR	**Preferred order rule** Select the first SEIZE block for which the required resource units are available
RAN	**Random priority** Select randomly from the SEIZE blocks for which the required resource units are available
SNB	**Smallest number busy** Select the SEIZE block that has the smallest number of resource units busy; break ties using the POR rule
SRC	**Smallest remaining capacity** Select the SEIZE block that has the smallest remaining resource capacity; break ties using the POR rule

A similar addition is applicable to the RELEASE block. Recall the format of this block:

RELEASE: *ResourceID, Quantity to Release*: repeats;

The resource number or symbol name may be used in the *Resource ID* operand. The set functions *Resource ID(Index)* and *Member(Set Expression, Index)* may also be used. For the RELEASE block, the SELECT key word uses a slightly different form:

SELECT(SetName [,FIRST|LAST] [,Attribute ID])

In this form, the *SetName* is the symbol name of the set. The operational key words *FIRST* and *LAST* are used to RELEASE either the FIRST or LAST member of the set that was seized. The *Attribute ID* may be used to release a member of the set other than the FIRST or LAST.

As an example, consider the following elements:

```
RESOURCE:   Worker1:
            Worker2:
            Worker3:
            Worker4:
            Worker5;
SETS:       Loaders, Worker1..Worker5;
```

and the following blocks:

```
QUEUE,
  LoadQ;
SEIZE:
  SELECT(Loaders, POR, Index);
DELAY:
  LoadTime;
RELEASE:
  Loaders(Index);
```

In this example, five resources are defined, *Worker1* through *Worker5*, which all belong to the set called *Loaders*. In the model, when an entity executes the SEIZE block, it captures a resource from the *Loaders* set based on the Preferred Order Rule (POR). The POR rule dictates that the resources will be selected in order of their definition, if available. The index of the set member is stored in the attribute *Index* that is later used by the RELEASE block to free the specific resource.

In Example 7.2, the drilling and milling operations both SEIZE from the SETS of *Drills* and *Mills*, respectively. This is accomplished in the following way:

```
:
SEIZE:
  SELECT(1, POR, DrillNum);
DELAY:
  DrillTime;
RELEASE:
  Drills(DrillNum);
:
```

The SELECT causes the SEIZE block to choose from the first set *(Drills)* based on the Preferred Order Rule. The index of the member of the set seized is placed in the attribute *DrillNum*. In the RELEASE block, the attribute *DrillNum* is used to specify which member of *Drills* should be released.

Consider another example as follows:

```
:
SEIZE:
  SELECT(2, CYC, MillNum);
DELAY:
  MillTime;
RELEASE:
  Mills(MillNum):
NEXT(Grinding);
:
```

Here the SEIZE block chooses from the second set *(Mills)* in a cyclic order. The index of the member seized is placed in *MillNum*. *MillNum* is used by the RELEASE block to specify the member of *Mills* that should be released.

7.6 Downtimes

In Section 7.1, the SCHEDULES element and the ALTER block were introduced. They can be used to model shifts, schedule breaks, and downtime. An additional way that downtime can be simulated is described in the following sections.

7.6.1 The FAILURES Element

The FAILURES element defines characteristics of resource failures. It has the following format:

FAILURES: *Number, Name, Type(Between, Duration, State)*:
repeats;

Operand	Description	Default
Number	Failure number [integer]	Sequential
Name	Failure name [symbol name]	Blank
Type	COUNT or TIME	—
Between	Time or count between [expression truncated to an integer]	—
Duration	Time length of failure [expression]	—
State	State for which the Between time accumulates [time-based failures only]	All

Number is the sequential identifier. *Name* is a symbolic name for the failure. *Type* determines whether the failure is based on TIME or on a COUNT. The COUNT keyword causes a failure when the resource has been released the number of times specified in the operand *Between*. *Between* specifies the amount of time (time based) or the number of releases (count based) before the failure occurs. *Duration* is any valid SIMAN expression.

State is the state name for whatever STATESETS are associated with that particular resource. The *State* operand applies only to time-based failures and indicates the state that should be considered for the time between failures. If no *State* is specified, then SIMAN considers all states. In other words, the time between failures does not depend on any particular state but on the total simulation time.

Consider the following example:

```
FAILURES:   DrillCleaning, COUNT(25, 10):
            MillBroken, TIME(EXPO(80), EXPO(5), Milling);
RESOURCES:  Drill, 2, , DrillCleaning:
            Mill, 3, , MillBroken;
```

The first failure, *DrillCleaning*, is a count-based downtime that occurs every 25 releases of the *Drill*. The *Drill* is down for 10 minutes. The second failure, *MillBroken*, is associated with the *Mill*. Time between failures is based on an exponential distribution with mean of 80 minutes; its duration is exponential with mean 5 minutes, and the time between failures is based on how much time the *Mill* spends in state *Milling*.

7.6.2 The FREQUENCIES Element

The FREQUENCIES element collects time-persistent data on SIMAN variables or an expression. It uses the following format:

FREQUENCIES: *Number, Expression, Name, Output File*:
repeats;

Operand	Description	Default
Number	Frequency number [integer]	Sequential
Expression	SIMAN expression on which the FREQUENCIES statistics are to be recorded [expression or STATE(*ResourceID*) where *ResourceID* is a valid symbol name]	—

Operand	Description	Default
Name	Frequency name and identifier for labeling summary report [symbol name]	Expression
Output file	Output unit to which FREQUENCIES observations are written during the simulation run [unique integer number or system-specific filename enclosed in double quotes]	No save

Number is the sequential SIMAN identifier for each FREQUENCIES group. *Expression* is a valid SIMAN expression, such as NQ*(QueueID)*, or STATE*(ResourceID)*. The *Name* is an optional descriptor for the frequency statistic. If this is omitted, then the expression is used as the descriptor. The *Output File* is an optional operand that records each observation to a file or output device.

The FREQUENCIES element creates an additional section in the summary report. There are additional operands used to further define the format of the FREQUENCIES section in the summary report, but their functions are not relevant to this text.

In the experiment frame, the line:

```
FREQUENCIES:   STATE(Drill1)
```

produces the addition to the summary report shown in Exhibit 7.4. The *Identifier* is the same as the *Expression* in the FREQUENCIES element because no descriptor is specified. The report indicates that *Drill1* entered the Cleaning state 10 times and remained in that state for an average of 10 minutes. This translates to 3.85% of the total simulation time spent in the Cleaning state.

EXHIBIT 7.4 **FREQUENCIES section of the summary report.**

FREQUENCIES

Identifier	Category	Occurrences		Standard	Restricted
		Number	AvgTime	Percent	Percent
STATE(Drill1)	Drilling	126	15.869	76.91	76.91
	Cleaning	10	10.000	3.85	3.85
	Idle Autostate	118	4.241	19.25	19.25

7.7 Flow Control

In the previous section, ways to model equipment downtime were presented. During a period of downtime, the flow of entities through the model is interrupted or otherwise affected because the resources are restricted. In this section, new ways to affect the flow of entities due to conditions that can depend on environmental variables other than resources are explored.

This section discusses the BLOCKAGES element and the PROCEED, BLOCK, and UNBLOCK blocks. The QUEUE block is discussed in terms of flow control.

7.7.1 The BLOCKAGES Element

This element defines blockages used in the model, initializes the number of blockages, and sets the global priority for tie-breaking. The BLOCKAGES element has the following format:

BLOCKAGES: *Number, Name, Initial Blockages, GlobalPriority*: repeats;

Operand	**Description**	**Default**
Number	Blockage number [integer]	Sequential
Name	Blockage name [symbolic]	Blank
Initial Blockages	Initial number of blockages [expression truncated to an integer]	0
GlobalPriority	Global priority [QTIME, HVF(expression), LVF(expression)]	QTIME

There are two types of blockages: block-type and queue-type. The block-type blockages use the BLOCK, PROCEED, and UNBLOCK blocks to affect the flow of entities. Queue-type blockages use the QUEUE block's operands to affect entity flow. The queue-type blockages will be discussed in Section 7.7.5.

Specific blockages in the model are referenced by either their *Number* or *Name*. If the symbolic name is omitted from the BLOCKAGES element, then the trailing comma must be included. If the *Number* operand is omitted, then the compiler still assigns a sequential number for every named blockage. The *Name* of the blockage will be referred to as the *BlockageID* in the model frame.

Initial Blockages sets the number of block-type blockages for each specific *BlockageID*. The UNBLOCK block must be used to clear any of these blockages.

The operand *GlobalPriority* defines the tie-breaking rule for entities of the same priority waiting for the removal of a blockage. QTIME (length of time that the entity has been in the queue) is the default rule. The Highest Value First (HVF) and Lowest Value First (LVF) rules may also be used.

The number of blockages for any blockage *Number* or *Name* may be accessed using the SIMAN variable NB*(Number)* or NB*(Name)*. Examples of the BLOCKAGES element are as follows:

```
BLOCKAGES:  4;
```

This BLOCKAGES element defines four unnamed blockages, numbered 1 through 4, with no initial blockages, and a tie-breaking rule based on QTIME.

```
BLOCKAGES:  BlockDrill, 1:
            BlockPlane, , LVF(DueDate);
```

This BLOCKAGES element defines two blockages. *BlockDrill* has one initial blockage and uses QTIME to break any ties. *BlockPlane* begins with no blockages and breaks ties using a lowest value first rule based on the attribute *DueDate*.

7.7.2 The PROCEED Block

The PROCEED block provides a checkpoint for entities as they move through the model. When executed, the PROCEED block determines whether a blockage exists and holds the entity in an internal queue (if no QUEUE block precedes the PROCEED block) until the blockage is clear. If no blockage exists, then the entity executing the PROCEED block moves in an unhindered fashion.

The PROCEED block has the following format:

PROCEED, *Priority: BlockageID*: repeats;

Operand	Description	Default
Priority	Priority [expression truncated to an integer]	1
Blockage ID	Blockage identifier [expression truncated to an integer, blockage number or blockage symbol name]	—

In a case where the same *Blockage ID* is used at more than one PROCEED block, the *Priority* operand determines which entities may pass through these competing blocks. Entities at PROCEED blocks with lower *Priority* values have priority over those at PROCEED blocks with higher *Priority* values. In the event of a tie, the *GlobalPriority* in the BLOCKAGES element is used.

Examples of the use of the PROCEED block are as follows:

```
QUEUE,
  Bay1Q;
PROCEED:
  Bay1;
```

Entities wait in queue *Bay1Q* until all blockpoints have been cleared at blockage *Bay1*.

```
PROCEED, 1:
  Bay1;
QUEUE,
  WorkerQ;
SEIZE:
  Worker;
DELAY:
  LoadTime;
RELEASE:
  Worker;
QUEUE,
  DockQ;
PROCEED:
  Bay1;
QUEUE,
  InspectQ;
SEIZE:
  Inspect;
DELAY:
  InspectTime;
RELEASE:
  Inspect;
```

In this example, entities wait in an internal queue until all blockpoints have been cleared at *Bay1*. No entities may SEIZE the *Worker* until the blockages are cleared. Notice that, because of priorities, no entities may pass on to the *Worker* until the *DockQ* is empty and all blockage points have been cleared for *Bay1*.

7.7.3 The BLOCK Block

The BLOCK block has the following format:

BLOCK: *BlockageID, Number of Block Points*;

Operand	Description	Default
BlockageID	Blockage number or name [integer or symbol name]	—
Number of Block Points	Number of blockage points to add to the BlockageID [expression truncated to an integer]	1

Number of Block Points blockage points are added to *BlockageID*. If the total number of blockage points at a *BlockageID* is greater than zero, then no entities may pass through a PROCEED block using that same *BlockageID*. If, for some reason, the value *Number of Block Points* results in a negative number, SIMAN issues a warning, but it does not change the status of *BlockageID*.

An example of the BLOCK block is the following:

```
BLOCK:
  Bay1;
```

One blockage point is established at *Bay1*.

7.7.4 The UNBLOCK Block

The UNBLOCK block has the following format:

UNBLOCK: BlockageID, Number to Remove;

Operand	Description	Default
BlockageID	Blockage number or name [integer or symbol name]	—
Number to Remove	Number of blockage points to subtract from the BlockageID [expression truncated to an integer]	1

The UNBLOCK block is used to remove *Number to Remove* blockage points from a *Blockage ID*. If the total number of blockage points for *Blockage ID* is greater than zero, then no entities may pass through a PROCEED block using that *Blockage ID*. If *Number to Remove* is a negative number, SIMAN issues a warning but does not change the status of *BlockageID*.

When all blockage points are removed from a *Blockage ID* and multiple awaiting entities are cleared to pass through an associated PROCEED block, those entities are processed one at a time to account for the execution of other BLOCK blocks in the model.

An example of the UNBLOCK block is as follows:

```
        UNBLOCK:
          Bay1, 3;
```

Three blockage points are cleared from blockage *Bay1*.

```
        BLOCK:
          Bay1;
          :
        DELAY:
          WorkTime;
        BRANCH, 1:
          IF, NB(Bay1).GT.5, Clear:
          ELSE, Gate;
Clear   UNBLOCK:
          Bay1, NB(Bay1):
          NEXT(Gate);
Gate    PROCEED:
          Bay1;
        DELAY:
          ClearTime;
        :
```

In the above example, blockage points are accumulated for *Bay1*. If *Number of Blockage Points* at *Bay1* is greater than 5, then an entity is sent to execute an UNBLOCK block to remove all blockage points at *Bay1*; otherwise, entities wait in an internal queue at the PROCEED block. Once allowed to pass, there is a DELAY equal to *ClearTime*, and the entities continue their movement through the remainder of the model.

7.7.5 The QUEUE Block Revisited

The QUEUE block can also serve as a method to affect the flow of entities through the model. We will now introduce two additional operands to the QUEUE block's format:

QUEUE, *QueueID, Capacity, BalkLabel*: *Blockage Level, Blockage ID*: repeats;

The operands, *Blockage Level* and *Blockage ID* are used with the QUEUE block to control flow of entities.

Operand	Description	Default
Blockage Level	Triggers blockage point [increment or BLOCKWHEN(expression)]	—
Blockage ID	Blockage number or name [expression truncated to an integer or symbol name]	—

As indicated previously, the PROCEED block could be considered as a checkpoint. The *Blockage Level* operand of the QUEUE block can provide a similar type of checkpoint. If the number of entities waiting in the QUEUE block exceeds the *Blockage Level*, then the number of blockage points for *Blockage ID* is incremented by one, that is, NB*(Blockage ID)* is incremented by one. The *Blockage Level* can also be specified by BLOCKWHEN*(logical expression)*. If the expression evaluates to TRUE, then NB*(Blockage ID)* is incremented by one.

An example of the QUEUE block follows:

```
QUEUE,
  DockQ: 30, DockFull;
```

In this block, when there are 30 entities waiting in *DockQ*, the number of blockage points at *DockFull* is incremented by one.

Suppose that a shipping dock with three workers is being modeled. If two workers go on break simultaneously, further arrivals at the dock are to be blocked. This can be accomplished as follows:

```
QUEUE,
  DockQ: BLOCKWHEN(MR(Workers) < 2), DockFull;
PROCEED:
  DockFull;
QUEUE,
  WorkerQ;
SEIZE:
  Workers;
:
UNBLOCK:
  DockFull;
```

In this example, a blockage point is added to *DockFull* when the capacity of RESOURCE *Workers* dips below 2. Thus, until an UNBLOCK block is executed, entities cannot move through the PROCEED block and on through the model.

7.8 Logical Constructs

SIMAN V offers two types of logical constructs that are often found in conventional computer programming languages. These are the IF conditioning construct and the WHILE looping construct.

7.8.1 IF, ELSE, ELSEIF, and ENDIF Blocks

These four blocks work together to perform much the same function as a BRANCH block. The IF and ELSEIF blocks each have a single operand; ELSE and ENDIF have no operands. If the expression in the operand evaluates to a nonzero integer, then the following group of blocks is executed. The entities then jump to the ENDIF and continue. In Example 4.1, we used a BRANCH block to determine the job type. We could use the logical structures of this section to accomplish the same purpose, as follows:

```
IF: JobType = = 1;
        QUEUE,                  !Queue for Mill with capacity
           MillQ, 3, Eject;     3, send overflow to Eject
        SEIZE:
           Mill;                Capture Mill
        DELAY:
           MillTime;            Process for MillTime
        RELEASE:
           Mill;                Free the resource Mill
ELSE;
        QUEUE,                  !Queue for Plane with capacity
           PlaneQ, 3, Eject;    3, send overflow to Eject
        SEIZE:
           Plane;               Capture Plane
        DELAY:
           PlaneTime;           Process for PlaneTime
        RELEASE:
           Plane;               Free the resource Plane
ENDIF;
        QUEUE, 2                !Queue for Grinder with capacity
           GrindQ, 2, Eject;    2, send overflow to Eject
```

If the *JobType* is equal to 1, then the expression is a nonzero value and the group of blocks between the IF block and the ELSE block are executed. Then the entities jump to the ENDIF block and continue. If *JobType* does not equal 1, then the entity jumps to the following group of blocks at the next ELSEIF block where another expression is evaluated or to the ELSE block.

An example using the ELSEIF block follows:

```
IF:      Color == 1;
      QUEUE, Pack1;
      :
ELSEIF: Color == 2;
      QUEUE, Pack2;
      :
ELSEIF: Color == 3;
      QUEUE, Pack3;
      :
ELSE;
      DISPOSE;
ENDIF;
```

In this example, the logical construct is used to determine the color of the entity. Based on that information, a different set of operations is performed. If an entity is not of color 1, 2, or 3, then it is disposed. The ELSEIF block allows multiple decision branches within one IF...ENDIF grouping.

An ENDIF block must always be used in any IF/ELSEIF/ELSE combination. An ELSE block requires an IF block, but does not require an ELSEIF block.

7.8.2 WHILE and ENDWHILE Blocks

The WHILE and ENDWHILE constructs are used to create a looping activity within the simulation. WHILE has one operand, a logical or mathematical expression. ENDWHILE has no operand, but it must be used with the WHILE block.

When an entity enters the WHILE block, the expression is evaluated, and if it evaluates to a nonzero value (TRUE), then the group of blocks between the WHILE and ENDWHILE blocks are executed. If the expression evalutes to zero (FALSE), then the entities jump to the ENDWHILE block and do not execute any of the intervening blocks. Note: Be very careful not to create an infinite loop. This may happen if the expression never evaluates to FALSE.

In Example 4.1, after a part is inspected, it may pass inspection, fail inspection, or be sent for rework. We accomplish the rework loop with a BRANCH block. The same activity may be duplicated with the WHILE, ENDWHILE combination, shown as follows:

```
            ASSIGN:
            Done=0;
          WHILE: Done= =0;
            QUEUE,                      !Queue for Drill
              DrillQ,2,Eject;           Send overflow to Eject
            SEIZE:
              Drill;                    !Capture Drill
            DELAY:
              DrillTime;                !Process for DrillTime
            :
            RELEASE:
              Inspector;                Free resource Inspector
            BRANCH,1,10:                !Using seed 10
              WITH,.9,Pass:             !Pass 90% of jobs
              WITH,.05,Scrap:           !Scrap 5% of jobs
              ELSE,Rework;              Rework the rest (5%)
Pass        ASSIGN:
              Done=1:
              NEXT(Ew);
Scrap       ASSIGN:
              Done=2:
              NEXT(Ew);
Rework      ASSIGN:
              Done=0;
            COUNT:
              Reworks,1;                Count reworked jobs
Ew        ENDWHILE;
          IF:  Done= =1;
            TALLY:                      !Tally average flowtime of jobs
              Flowtime,INTERVAL(ArrTime);
            TALLY:                      !Tally average time between exits
              Exit Period,BETWEEN;
            COUNT:
              JobType,1:                !Count number of jobs by type
            DISPOSE;                    Destroy entity
          ELSE;
            COUNT:
              Fails,1:                  !Count number of failures
              DISPOSE;                  Destroy entity
          ENDIF;
Eject       COUNT:
              Ejections:                Count number of ejected jobs
              DISPOSE;                  Destroy entity
```

An attribute called *Done* has been added. At the start of the simulation, every entity is assigned a value of 0 to the *Done* attribute. Then the entities enter the WHILE/ENDWHILE loop. After inspection, if the entity passes inspection, its *Done* attribute is set to 1 and it exits the WHILE loop because the expression Done == 1 is no longer TRUE. A similar situation occurs with the failed jobs, except that the *Done* attribute is set to a value of 2. After the ENDWHILE block, the entities continue through the simulation.

EXAMPLE 7.2 Solution

The Model Frame

Below is the entire model frame for Example 7.2 with some explanation. Important additions to the code are italicized.

EXHIBIT 7.5 Model frame for Example 7.2.

```
!Model Frame for Example 7.2
!
BEGIN,NO;
          CREATE:                              !Create arriving jobs according to
            EX(1, 1):                          !Exponential distribution and save
            MARK(ArrTime);                     Arrival times in attribute ArrTime
          ASSIGN:                              !Store job type and delay times
            Jobtype = DP(2, 1):                !Type I or Type II job
            DrillTime = UN(3, 1):              !Uniform drill time
            MillTime = TR(4, 1):               !Triangular mill time
            PlaneTime = TR(5, 1):              !Triangular plane time
            GrindTime = DP(Jobtype+5, 1):      !Grind time is dependent on job type
            InspectTime = RN(8, 1);            Normally distributed inspection time
Drilling  QUEUE,                               !Queue for drill with capacity 2
            DrillQ, 25, Eject;                 Send overflow to Eject
          SEIZE:
            SELECT(1, POR, DrillNum);          !Capture resource drill from set
          DELAY:
            DrillTime;                         !Process for DrillTime
          RELEASE:
            Drills(DrillNum);                  !Free the drill by index
          IF: JobType == 1;
          QUEUE,                               !Queue for mill with capacity 3
            MillQ, 3, Eject;                   Send overflow to Eject
          SEIZE:
            SELECT(2, CYC, MillNum);           Capture resource mill from set
          DELAY:
            MillTime;                          Process for MillTime
```

EXHIBIT 7.5 *continued.*

```
RELEASE:
  Mills(MillNum);              Free the Mill by index
ELSE;
QUEUE,                         !Queue for Plane with capacity 3
  PlaneQ, 3, Eject;            Send overflow to Eject
SEIZE:
  Plane;                       Capture Plane
DELAY:
  PlaneTime;                   Process for PlaneTime
RELEASE:
  Plane;                       Free the resource Plane;
ENDIF;
QUEUE,
  BlockQ;
PROCEED:
  AtGrind;
QUEUE,                         !Queue for Grinder with capacity 15
  GrinderQ, 15, Eject;         Send overflow to Eject
SEIZE:
  Grinder;                     Capture grinder
DELAY:
  GrindTime;                   Process for GrindTime
RELEASE:
  Grinder;                     Free the resource grinder
IF: NQ(InspectorQ) > 8;
  IF : MR(Inspector) < 2;
  ALTER:
    Inspector, +1:
    Mill3, −1;
  ASSIGN:
    STATE(Mill3) = Borrowed;
  BLOCK:
    AtGrind;
  ENDIF;
ENDIF:
QUEUE,                         !Queue for Inspector with capacity 10
  InspectorQ, 10, Eject;       Send overflow to Eject
SEIZE:
  Inspector;                   Capture resource Inspector
DELAY:
  InspectTime;                 Process for InspectTime
RELEASE:
  Inspector;                   Free the resource Inspector
IF: NQ(InspectorQ) > 5:
  IF: MR(Inspector) > 1;
    IF: NR(Inspector) < 2;
    ALTER:
      Inspector, -1:
      Mill3, +1;
```

EXHIBIT 7.5 *continued.*

```
              UNBLOCK:
                AtGrind, NB(AtGrind);
              ENDIF;
            ENDIF;
            BRANCH, 1, 10:                   !Take only one branch using seed 10
              WITH, 0.05, Rework:            !Rework 5% of the jobs
              WITH, 0.05, Scrap:             !Scrap 5% of the jobs
            ELSE, Done;                      Remaining 90% are complete
Rework      ASSIGN:                          !Store job type and delay times
              JobType = DP(2, 1):            !Type I or Type II job
              DrillTime = UN(3, 1):          !Uniform drill time
              MillTime = TR(4, 1):           !Triangular mill time
              PlaneTime = TR(5, 1):          !Triangular plane time
              GrindTime = DP(JobType+5, 1):
                                             !Grind time is dependent on job type
              InspectTime = RN(8, 1);        Normally distributed Inspect time
            COUNT:
              Reworks, 1:                    !Count the number of jobs reworked
              NEXT(Drilling);                Send reworks back to Drill
Done        TALLY:
              Flowtime, INT(ArrTime);        Tally the average flowtime of jobs
            TALLY:
              Exit Period, BET;              Tally the average time between exits
            COUNT:
              JobType, 1:                    !Count number of jobs by type and
              DISPOSE;                       Destroy the entity
Scrap       COUNT:
              Fails, 1:                      !Count number of failed jobs and
              DISPOSE;                       Destroy the entity
Eject       COUNT:
              Ejections:                     !Count the number of jobs ejected
              DISPOSE;                       Destroy the entity
```

The first major difference between Examples 7.2 and 4.1 is in the SEIZE block. The entity executing the SEIZE block will choose from the first SETS definition, choose a member of the set using the Preferred Order Rule, and then the index of that set member is placed in the attribute *DrillNum*. The RELEASE block then uses *DrillNum* to free the correct resource.

The IF/ENDIF construct replaces the use of the BRANCH block in Example 4.1. The entity will then enter the *BlockQ* and the PROCEED block called *AtGrind*. This block is where the entity will wait when the *Inspector* queue becomes too large.

In the next section, the IF block is used to determine the length of *InspectorQ*. If it is greater than 8 and if the capacity of the *Inspector* is less than 2 (meaning that the *Inspector* has not already called for help), then the capacities of the *Inspector* and *Mill3* are increased and decreased by 1, respectively. A block is then placed at *AtGrind*.

The section that follows determines whether the size of the *InspectorQ* has decreased below 5. If it has, then the capacities of the *Inspector* and of *Mill3* are returned to their initial status and the block is removed from *AtGrind*. The entity then completes the remaining blocks in the model.

The Experiment Frame

Exhibit 7.6 contains the entire experiment frame for Example 7.2. Important additions are italicized.

EXHIBIT 7.6 Experiment frame for Example 7.2.

```
!Experiment Frame for Example 7.2
!
BEGIN,NO;
PROJECT, Example 7.2, Team;
ATTRIBUTES:  ArrTime:
             JobType:
             DrillTime:
             MillTime:
             PlaneTime:
             GrindTime:
             InspectTime:
             Drillnum:
             Millnum:
STATESETS:   DrillStates, Drilling(Busy), Cleaning(DrillClean):
             MillStates, Milling(Busy), Broken(MillFail),
             Waiting(Idle), Borrowed(INACTIVE):
FAILURES:    DrillClean, COUNT(25, 10):
             MillFail, TIME(EXPO(80), NORM(10, 1.2), Milling);
RESOURCES:   Drill1, 1, DrillStates, FAILURE(DrillClean, WAIT):
             Drill2, 1, DrillStates, FAILURE(DrillClean, WAIT):
             Mill1, 1, MillStates, FAILURE(MillFail, PREEMPT):
             Mill2, 1, MillStates, FAILURE(MillFail, PREEMPT):
             Mill3, 1, MillStates, FAILURE(MillFail, PREEMPT):
             Plane, 3:
             Grinder, 2:
             Inspector;
SETS:        Drills, Drill1, Drill2:
             Mills, Mill1..Mill3;
```

EXHIBIT 7.6 *continued.*

```
FREQUENCIES:      STATE(Drill1):
                  STATE(Drill2):
                  STATE(Mill1):
                  STATE(Mill2):
                  STATE(Mill3);
BLOCKAGES:        AtGrind;
QUEUES:           BlockQ:
                  DrillQ:
                  MillQ:
                  PlaneQ:
                  GrinderQ, LVF(JobType):
                  InspectorQ;
COUNTERS:         Type I Job Count:
                  Type II Job Count:
                  Fails:
                  Reworks:
                  Ejections;
PARAMETERS:       1, 5:
                  2, 0.7, 1, 1.0, 2:
                  3, 6, 9:
                  4, 10, 14, 18:
                  5, 20, 26, 32:
                  6, 0.1, 6, 0.75, 7, 1.0, 8:
                  7, 0.1, 6, 0.35, 7, 0.65, 8, 0.9, 9, 1.0, 10:
                  8, 10, 1.6;
TALLIES:          Flowtime:
                  Exit Period;
DSTATS:           NQ(DrillQ), Drill Queue:
                  NQ(MillQ), Mill Queue:
                  NQ(PlaneQ), Plane Queue:
                  NQ(GrinderQ), Grinder Queue:
                  NQ(InspectorQ), Inspect Queue:
                  NR(1), Drill1 Utilization:
                  NR(2), Drill2 Utilization:
                  NR(3), Mill1 Utilization:
                  NR(4), Mill2 Utilization:
                  NR(5), Mill3 Utilization:
                  NR(6)/MR(6), Plane Utilization:
                  NR(7)/MR(7), Grinder Utilization:
                  NR(8)/MR(8), Inspector Util.;
REPLICATE, 1, 0, 2400;
```

Discussion of the Summary Report

The summary report in Exhibit 7.7 yields the answers to the questions asked in the problem statement of Example 7.2.

EXHIBIT 7.7 Summary report for Example 7.2.

Summary for Replication 1 of 1

Project: Example 7.2 Run execution date: 8/24/1994
Analyst: Team Model revision date: 8/24/1994
Replication ended at time: 2400.0

TALLY VARIABLES

Identifier	Average	Variation	Minimum	Maximum	Observations
Flowtime	138.64	.34143	34.300	405.48	332
Exit Period	7.1117	.59540	.09652	30.895	331

DISCRETE-CHANGE VARIABLES

Identifier	Average	Variation	Minimum	Maximum	Final Value
Drill Queue	.46478	1.4925	.00000	2.0000	2.0000
Mill Queue	.66092	1.4444	.00000	3.0000	1.0000
Plane Queue	.12648	3.0891	.00000	3.0000	.00000
Grinder Queue	4.7830	1.0672	.00000	15.000	11.000
Inspect Queue	5.5195	.60986	.00000	10.000	6.0000
Drill1 Utilization	.75844	.56436	.00000	1.0000	1.0000
Drill2 Utilization	.70582	.64559	.00000	1.0000	.00000
Mill1 Utilization	.71156	.63668	.00000	1.0000	1.0000
Mill2 Utilization	.66734	.70604	.00000	1.0000	.00000
Mill3 Utilization	.29639	1.5407	.00000	1.0000	1.0000
Plane Utilization	.58850	.57195	.00000	1.0000	.33333
Grinder Utilization	.63414	.73944	.00000	1.0000	1.0000
Inspector Util.	.97400	.16337	.00000	1.0000	1.0000

COUNTERS

Identifier	Count	Limit
Type I Job Count	224	Infinite
Type II Job Count	108	Infinite
Fails	24	Infinite
Reworks	20	Infinite
Ejections	127	Infinite

EXHIBIT 7.7 *continued.*

FREQUENCIES

Identifier	Occurrences Category	Number	AvgTime	Standard Percent	Restricted Percent
STATE(Drill1)	Drilling	118	15.425	75.84	75.84
	Cleaning	9	10.000	3.75	3.75
	IDLE	110	4.4522	20.41	20.41
STATE(Drill2)	Drilling	91	18.615	70.58	70.58
	Cleaning	9	9.0391	3.39	3.39
	IDLE	84	7.4365	26.03	26.03
STATE(Mill1)	Milling	64	26.683	71.16	71.16
	Broken	18	10.079	7.56	7.56
	Waiting	46	11.104	21.28	21.28
STATE(Mill2)	Milling	73	21.939	66.73	66.73
	Broken	26	10.166	11.01	11.01
	Waiting	48	11.126	22.25	22.25
STATE(Mill3)	Milling	43	14.916	26.73	26.73
	Broken	6	8.6726	2.17	2.17
	Waiting	26	12.238	13.26	13.26
	Borrowed	16	86.771	57.85	57.85

Simulation run complete.

1. The *drills* spend the following percentages in their respective states:

Resource	Drilling	Cleaning	Idle
Drill1	75.84%	3.75%	20.41%
Drill2	70.58%	3.39%	26.03%

2. The *mills* spend the following average times in each of their states (minutes):

Resource	Milling	Broken	Waiting	Borrowed
Mill1	71.16%	7.56%	21.28%	0.00%
Mill2	66.73%	11.01%	22.25%	0.00%
Mill3	26.73%	2.17%	13.26%	57.85%

7.9 SUMMARY

SETS and STATESETS elements were introduced to amplify use of the RESOURCES element and the SEIZE and RELEASE blocks. SETS provides a means of grouping resources when they perform related functions or are associated with each other in some way. By using STATESETS, a means for studying the use of resources is accomplished. The ALTER block and the SCHEDULES and FAILURES elements are used to change the availability or the state of resources. FAILURES used with STATESETS offers an explicit way of defining downtime for sets of resources. The ALTER block may be effectively used to change the capacity of single resources. SCHEDULES offers a means of affecting resources based on a regular routine.

In order to model changes in the flow of entities through the model, the BLOCK and UNBLOCK blocks and the PROCEED block were introduced. The PROCEED acts as a checkpoint or gate through which entities must pass. The BLOCKAGES element defines changes in flow while the BLOCK and UNBLOCK blocks create or remove them from the system.

For those familiar with logical constructs in conventional programming languages such as BASIC, FORTRAN, Pascal, or C, SIMAN offers a parallel. The IF, ELSEIF, ELSE, and ENDIF blocks provide a means for logical comparisons and decision branching. These constructs are a valuable augmentation to the BRANCH block. Also, the WHILE and ENDWHILE constructs may be used to create conditional loops within the model.

7.10 EXERCISES

E7.1 Every 60 seconds, 50 parts arrive to be sorted by color. The parts are colored red, blue, yellow, and green with each color equally likely. Simulate the sortation of 1000 parts and show the resulting distribution.

- **a.** Using a BRANCH block
- **b.** Using IF-THEN-ELSE logic

E7.2 A part arrives every second. Every 100th part is sent to inspection. Simulate the arrival of 10,000 parts and show the number that went to inspection and the number that did not.

- **a.** Using a BRANCH block
- **b.** Using IF-THEN-ELSE logic

E7.3 Consider the production process described in E4.1. However, in this assignment, we assume that operators will be needed. There are 4 mill operators, 2 plane operators, 3 drill operators, and 1 inspection machine operator. In actual operation, a job SEIZEs an operator and then SEIZEs a machine. When a part is ready for drilling, the drilling operator will use the machine with the largest remaining capacity. Additionally, the drilling machines suffer from breakdowns. Assume that a breakdown occurs only when the machine is idle or when it finishes a job. The time between breakdowns is uniformly distributed between 6 and 10 hours. A repair takes between 5 minutes and 30 minutes, uniformly distributed. A drill operator will perform the repair with priority over other activity. The operators have the following queue capacities before them:

Mill	8
Plane	8
Drill	6
Inspector	3

In this assignment, the capacity of the queue before the machines is 1. The system runs for three 8-hour shifts per day, 5 days per week. Simulate 1 week of production after a 1-day warm-up of the system. Use random number stream 10 for any BRANCH blocks and random number stream 1 for all other purposes. Answer the following questions:

- **a.** How many good jobs are made of each type?
- **b.** How many reworks occur?
- **c.** How many jobs are scrapped?
- **d.** What are the average system times for both types of job?
- **e.** What is the utilization of each class of machine and operator?
- **f.** How many jobs are ejected from the system in front of each class of machine?
- **g.** How much time do the drills spend in each of their states?

E7.4 Castings arrive at a machining area according to a normal distribution whose mean is 90 seconds and standard deviation is 5 seconds. An inspector examines the castings for cracks. Approximately 10% of the castings are rejected and scrapped at this point. There are four automatic millers that are used for the remaining castings. Two loaders load the mills in a time that is uniformly distributed from 65 to 105 seconds. The loaders will always try to load a casting into an empty mill before waiting for a mill to become available. The milling operation requires 180 seconds. A worker removes the milled casting in a time that is uniformly distributed from 30 to 50 seconds. After a mill processes 25 castings, it is shut down for a uniformly distributed time between 10 and 20 minutes for cleaning. An operator will complete processing of a casting before shutting down the mill for cleaning. Emergency orders occur according to a uniform distribution whose minimum is zero and whose maximum is 2 hours apart. An emergency order occupies the worker and mill for the same time as a regular order. There is a buffer in front of the operators who load the mill. Four hours into the simulation, the worker takes a 30-minute lunch break. During this time, the worker is replaced by an alternate worker who takes twice as long to perform the removal operation. Simulate the operation of the system until 500 castings are machined.

a. What is the average number in the buffer?

b. What is the average time in the system (excluding the inspection)?

c. What time did the work day finish if it started at 8:00 A.M.?

d. How many emergencies occurred?

e. How many castings were rejected by the inspector?

f. What is the utilization of the loaders, the worker, and the alternate worker?

g. What percentage of time does each mill spend in cleaning operations?

E7.5 Mechanics arrive at a tool crib according to an exponential distribution with a mean time between arrivals of 3 minutes. There are three tool crib attendants. The time to serve a mechanic is normally distributed with a mean of 7.5 minutes and a standard deviation of 1.5 minutes. Compare the following servicing methods. Be as efficient as possible with respect to the number of blocks used for modeling.

a. Mechanics form a single queue, choosing the next available tool crib attendant.

b. Mechanics enter the shortest queue (three queues from which to choose).

c. Mechanics choose one of three queues at random.

Collect data over a 26-hour period, clearing the system after the first 2 hours.

E7.6 Consider the intersection of a north–south road and an east–west road. Automobiles going east–west arrive at the intersection according to an exponential distribution with mean 10 seconds, whereas the north–south traffic arrives according to an exponential distribution with mean 28 seconds (this is a secondary road). In the intersection, there is a traffic light that controls the flow of traffic. The light switches from red to green and from green to red at a constant rate. Because of a local traffic ordinance, only two cars are allowed in the intersection at the same time—one in each lane. All cars take 4 seconds to cross the intersection. Determine the best length of time between switches so that no more than six cars must wait in the east–west direction.

E7.7 A machine manufactures parts in exactly 1.5 minutes. The machine requires a tool change according to an exponential distribution with a mean of 12 minutes between occurrences. The tool change time is also exponentially distributed with a mean of 3 minutes. Parts are started every 2 minutes. Determine the average time in the queue before the machine for three cases as follows:

- **a.** Taking the randomness into account as stated above
- **b.** Ignoring the randomness (treating the breakdown data as deterministic)
- **c.** Assuming that the machine has no breakdowns, but that the process time is 1.875 minutes

Simulate the system for 8 hours.

E7.8 Reconfigure Exercise E7.7 so that parts arrive in batches of 30 at the beginning of each hour. Explain the counterintuitive results.

8

Station Submodels

You may have noticed that certain sequences of code seem to appear frequently in a model. For example, the combination QUEUE-SEIZE-DELAY-RELEASE appears repeatedly in models beginning with Example 3.1. If you thought, Is there a way to be more efficient?, your time has arrived. Models can usually be decomposed into separate *submodels*, reducing the overall model size and decreasing the complexity of the models at the same time, a win–win situation. A submodel is represented by a station in SIMAN. A *station* is a collection of blocks describing an area where entities are processed. For identification purposes, a station name or station number is used. The special-purpose attribute M is reserved for this identification. Individual stations having the same logic can then be expressed in terms of a single generic station submodel. Examples will best describe the station concept.

EXAMPLE 8.1 Problem Statement

Example 4.1 had five QUEUE-SEIZE- DELAY-RELEASE combinations requiring 20 blocks. Other blocks were required to complete the model. Example 4.1 is an excellent candidate for the station concept, in that each of the five similar logical flows will be represented with a single-station submodel. In addition, the special-purpose attribute, M, will be used to specify individual workstations for each entity. New blocks and elements introduced in applying the station concept are shown in Table 8.1.

8.1 The STATION and ROUTE Blocks

The STATION block represents a point in the model to which entities are transferred. The format for the block is as follows:

STATION, *Beginning Station ID-Ending Station ID;*

Table 8.1 Blocks/Elements Introduced in this Example

New Block/Element	Purpose
STATION block	Used to represent an entry point into the station submodel
ROUTE block	Used to transfer jobs from one station to another
ASSIGN block	Used to store the job types and delay times
QUEUES element	Used to specify queue names
STATIONS element	Used to specify station names
RESOURCES element	Used to specify the resource capacities
SETS element	Used to index the resources

The operands are described as follows:

Operand	Description	Default
Beginning Station ID	Lower limit of station range [integer station number or station symbol name]	—
Ending Station ID	Upper limit of station range [integer station number or station symbol name]	No range

For Example 8.1, the STATION block is as follows:

STATION, SDrill-SInspect;

This block indicates that the station submodel begins with a station called *SDrill* and ends with a station called *SInspect*, corresponding to drilling and inspection. Between these two stations, we will also have stations *SMill, SPlane,* and *SGrind*, corresponding to milling, planing, and grinding. These stations will all be introduced in the STATIONS element.

Stations do not have to be entered in order in the model. For example, we may have

STATION, 5;
:
STATION, 1-4;

In this example, Station 5 is introduced, and later in the model, Stations 1 through 4 are introduced. These station numbers may also be represented by symbol names in the STATIONS element. Either the integer station number or symbol name may be used.

The ROUTE block transfers the entity to the indicated destination station. The format for the block is as follows:

ROUTE: *Duration, Destination;*

The operands are described as follows:

Operand	Description	Default
Duration	Routing time delay [expression]	0.0
Destination	Destination station [expression truncated to an integer station number, station symbol name, or SEQ]	SEQ

When an entity enters a ROUTE block, SIMAN evaluates the *Destination* operand and sets the entity's station attribute M to its value. After setting M, the entity is sent to the STATION specified by *Destination* for the delay specified by *Duration*. The next block entered is the STATION block. Any blocks between the ROUTE and STATION are not executed by this entity. The only modifier permitted is MARK. The use of SEQ is discussed in Section 8.5. Examples of the ROUTE block include the following:

ROUTE: 10, 4;

The entity is assigned an M value of 4, that is, the block routes the entity to Station 4 and is scheduled to arrive at that station in 10 time units.

ROUTE: MoveTime, SWeld;

This block routes the entity to the station named *SWeld* in *MoveTime* time units. *MoveTime* is an attribute or global variable.

For Example 8.1, one of the ROUTE blocks is:

ROUTE: 0, SDril;

This block indicates that the entity is going from its current location to the Station *SDril*, and that the time to travel is zero.

At this point in the presentation of SIMAN, travel times can be thought of as "in the air," that is, no resource is required to move entities from one location to another. In situations that have ample resources for material handling, or where material handling is not being simulated, such treatment of travel time is acceptable.

For success in simulation, models must be grown. They should start simply and be enhanced only after the logic is acceptable. For example, we could start with only the processing of entities, then we could add equipment failures, then add shift schedules, next add the transporters, then add conveyors, and finally include advanced logic. In essence, this text follows that same outline, and using ROUTE blocks prior to inserting transporters and conveyors is a step along that progression.

8.2 The Station Attribute M

M is a special-purpose attribute that stores the current station number for each entity. Initially, M = 0 for each new entity. If you followed the IRC examples quite closely, you may recall seeing M = 0 on the command "VIEW ENTITY." The meaning was not described at that time since the subject was beyond the intermediate modeling concepts in Chapter 4. As mentioned above, the value of M is automatically updated to the number of the destination station when an entity leaves the ROUTE block. (The station attribute M is also used with transporters and conveyors, as indicated in the next three chapters.)

8.3 Indexing of Queues and Resources

The key to using the station concept and radically reducing the size and complexity of models that have repeated logic is through indexing queues and resources. The queues can have varying limits, and the resources can have varying capacities. The station attribute M identifies the specific queue and resource required when moving through the blocks in a station submodel. Expressions involving M can also be used to identify specific queues and resources. The last two sentences are used in the following sequence of blocks that appear in the model frame for Example 8.1:

```
QUEUE,    M, V(M), Eject;
SEIZE:    Machine(M);
DELAY:    A(M+2);
RELEASE:  Machine(M);
```

The appropriate queue is given by the station attribute M, and the variable V(M) defines the upper limit of the number that can be in the queue. If entities arrive after the limit is reached, they are sent to the block labeled *Eject*.

8.4 The STATIONS Element

The STATIONS element identifies the stations that will be used in the model. The format for the element is:

STATIONS: *Number, Name*: repeats;

The operands are described as follows:

Operand	Description	Default
Number	Station number [integer]	Sequential
Name	Station name [symbol name]	Blank

An example is:

```
STATIONS:  10;
```

This element defines 10 stations, none of which are named.

In Example 8.1, the STATIONS element is given as follows:

```
STATIONS:  SDrill:
           SMill:
           SPlane:
           SGrinder:
           SInspect;
```

This element names Stations 1 through 5 as *SDrill, SMill, SPlane, SGrinder,* and *SInspect,* respectively.

The *Ending Station ID* must follow the *Beginning Station ID*. An alternate version of the STATIONS element is:

```
STATIONS:  1, SDrill:
           2, SMill:
           3, SPlane:
           4, SGrinder:
           5, SInspect;
```

The RESOURCES and SETS elements were described in Chapters 3 and 7. In Example 8.1, the RESOURCES and SETS elements are given as follows:

```
RESOURCES:   Drill, 2:
             Mill, 3:
             Plane, 3:
             Grinder, 2:
             Inspector;
SETS:        Machine, Drill..Inspector;
```

There are five different resources that are identified, with capacities as indicated. The capacity of *Inspect* is defaulted to 1. These resources will be identified in the set *Machine* so that *Machine*(1) is *Drill* with capacity 2 and so on.

EXAMPLE 8.1 Solution Using Logic

The model is broken down into several components including the following:

1. Creating arriving jobs
2. Workstation submodel
3. Routing to the next station
4. Collecting statistics

These will be presented and described in the following paragraphs:

1. Creating arriving jobs
   The statements to create the arriving jobs, assign the appropriate values to attributes, and route the jobs to the first station are:

```
BEGIN;
  CREATE:                           !Create arriving jobs according to
  EX(1,1):                          !Exponential distribution and save
  MARK(ArrTime);                    Arrival times in attribute ArrTime
ASSIGN:                             !Store the job type and process times
  JobType = DP(2,1):                !Type I or Type II job
  DrillTime = UN(3,1):              !Uniform Drill time
  MillTime = TR(4,1):               !Triangular Mill time
  PlaneTime = TR(5,1):              !Triangular Plane time
  GrindTime = DP(JobType+5,1):      !Grind time dependent on job type
  InspectTime = RN(8,1);            Normally distributed Inspection time
ROUTE:
  0, SDrill;                        Route jobs to SDrill station
```

2. Workstation submodel

```
STATION, SDrill-SInspect;  Stations are from SDrill to SInspect
QUEUE,                     !Queue number is given by M
    M, V(M), Eject;        Queue capacity is given by V(M)
SEIZE:
    Machine(M);            Capture indexed resource M
DELAY:                     !Process times by Attributes
    A(M+2);                (Attributes 1 and 2 are reserved)
RELEASE:
    Machine(M);            Free indexed resource M
```

3. Routing to the next station

```
          BRANCH, 1, 10:          !Take one branch only
            IF, M.EQ.SMill.OR.M.EQ.SPlane, GoGrind:
                                  !If at Mill or Plane, go to Grinder
            IF, M.EQ.SGrinder, GoInspect:
                                  !If at Grinder, go to Inspect
            IF, M.EQ.SDrill.AND.JobType.EQ.1, GoMill:
                                  !JobType I goes from Drill to Mill
            IF, M.EQ.SDrill.AND.JobType.EQ.2, GoPlane:
                                  !JobType II goes from Drill to Plane
            WITH, 0.05, Rework:   !5% are reworked
            WITH, 0.05, Scrap:    !5% are scrapped
            ELSE, Done;           The remaining jobs are complete
GoMill    ROUTE:
            0, SMill;             Destination is Mill
GoPlane   ROUTE:
            0, SPlane;            Destination is Plane
GoGrind   ROUTE:
            0, SGrinder;          Destination is Grinder
GoInspect ROUTE:
            0, SInspect;          Destination is Inspect
```

4. Collecting statistics

```
Rework  COUNT:
          Reworks;                          Count number of reworks
        ASSIGN:                             !New process times
          DrillTime = UN(3, 1):             !Uniform Drill time
          MillTime = TR(4, 1):              !Triangular Mill time
          PlaneTime = TR(5, 1):             !Triangular Plane time
          GrindTime = DP(JobType+5, 1):     !Time depends on job type
          InspectTime = RN(8,1);            Normally distributed Inspect
        ROUTE:
          0, SDrill;                        Destination is Drill
```

```
Done   TALLY:
          Flowtime, INT(ArrTime);   Tabulate Flowtime
       TALLY:
          Exit Period, BET;         Tally time between exits
       COUNT:
          JobType, 1:               !Count completed jobs
          DISPOSE;                  Destroy entity
Scrap  COUNT:
          Fails:                    !Count jobs scrapped
          DISPOSE;                  Destroy entity
Eject  COUNT:
          Ejections:                !Count of jobs ejected
          DISPOSE;                  Destroy entity
```

The entire experiment frame for this problem is given in Exhibit 8.1.

EXHIBIT 8.1 Experiment frame for Example 8.1

```
!Experiment Frame for Example 8.1
!
BEGIN;
PROJECT, First Station Example, Team;
ATTRIBUTES:   ArrTime:                     !Att 1 marked at arrival
              JobType:                     !Att 2 contains Job Type
              DrillTime:                   !Att 3 assigned Drill time
              MillTime:                    !Att 4 assigned Mill time
              PlaneTime:                   !Att 5 assigned Mill time
              GrindTime:                   !Att 6 assigned Grind time
              InspectTime;                 !Att 7 assigned Inspect time
VARIABLES:    1, MDrill, 2:                !Max queue at Drill is 2
              2, MMill, 3:                 !Max queue at Mill is 3
              3, MPlane, 3:                !Max queue at Plane is 3
              4, MGrinder, 2:              !Max queue at Grinder is 2
              5, MInspect, 4;              !Max queue at Inspect is 4
STATIONS:     SDrill:                      !Drilling station
              SMill:                       !Milling station
              SPlane:                      !Planing station
              SGrinder:                    !Grinding station
              SInspect;                    !Inspection station
RESOURCES:    Drill, 2:                    !2 Drills
              Mill, 3:                     !3 Mills
              Plane, 3:                    !3 Planes
              Grinder, 2:                  !2 Grinders
              Inspector;                   1 Inspect
SETS:         Machine,Drill..Inspector;    Machine(1) thru Machine(5)
```

EXHIBIT 8.1 *continued.*

```
QUEUES:       DrillQ:                              !Buffer before Drill
              MillQ:                               !Buffer before Mill
              PlaneQ:                              !Buffer before Plane
              GrinderQ:                            !Buffer before Grinder
              LVF(JobType):                        !Priority to JobType 1
              InspectorQ;                          Buffer before Inspect
COUNTERS:     Type I Job Count:
              Type II Job Count:
              Reworks:
              Fails:
              Ejections;
PARAMETERS:   1, 5:                                !Interarrival time
              2, .7, 1, 1.0, 2:                    !Job Type
              3, 6, 9:                             !Drilling time
              4, 12, 16, 20:                       !Milling time
              5, 20, 26, 32:                       !Planing time
              6, .25, 6, .75, 7, 1.0, 8:           !Grinding time Type I jobs
              7, .1, 6, .35, 7, .65, 8, .9, 9, 1.0, !Grinding time Type II jobs
              10:
              8, 3.6, .6;                          Inspect time
TALLIES:      Flowtime:
              Exit Period;
DSTATS:       NQ(DrillQ), Drill Queue:
              NQ(MillQ), Mill Queue:
              NQ(PlaneQ), Plane Queue:
              NQ(GrinderQ), Grinder Queue:
              NQ(InspectorQ), Inspect
              Queue:
              NR(1)/2, Drill Utilization:
              NR(2)/3, Mill Utilization:
              NR(3)/3, Plane Utilization:
              NR(4)/2, Grinder Utilization:
              NR(5), Inspect Utilization;
REPLICATE,    1, 0, 2400;
```

The output summary for Example 8.1 is shown in Exhibit 8.2. The reader can compare Exhibit 8.2 to the solution for Example 4.1 shown in Section 4.3. The differences are very minor, and are caused by the difference in the way some of the random numbers are used. The big difference in the two solutions is in the model frames. The model frame in Example 8.1 contains only 19 blocks. Compare this to the model frame for Example 4.1. The experiment frame for Example 8.1 is slightly larger than that of Example 4.1 by two elements to define the stations and the set. However, the overall savings in model size and complexity are enormous using stations. Furthermore, material

EXHIBIT 8.2 **Summary report for Example 8.1**

Summary for Replication 1 of 1

Project: First Station Example   Run execution date: 8/24/1994
Analyst: Team   Model revision date: 8/24/1994

Replication ended at time: 2400.0

TALLY VARIABLES

Identifier	Average	Variation	Minimum	Maximum	Observations
Flowtime	45.173	.25948	30.553	111.39	416
Exit Period	5.7047	.61742	2.2045	35.461	415

DISCRETE-CHANGE VARIABLES

Identifier	Average	Variation	Minimum	Maximum	Final Value
Drill Queue	.37844	1.6719	.00000	2.0000	1.0000
Mill Queue	.31966	2.0590	.00000	3.0000	2.0000
Plane Queue	.11064	3.5539	.00000	3.0000	.00000
Grinder Queue	.19932	2.2550	.00000	2.0000	.00000
Inspect Queue	.16792	2.2612	.00000	2.0000	.00000
Drill Utilization	.74430	.48388	.00000	1.0000	1.0000
Mill Utilization	.72298	.41624	.00000	1.0000	1.0000
Plane Utilization	.52615	.63507	.00000	1.0000	.33333
Grinder Utilization	.71106	.48503	.00000	1.0000	.50000
Inspect Utilization	.69810	.65761	.00000	1.0000	1.0000

COUNTERS

Identifier	Count	Limit
Type I Job Count	288	Infinite
Type II Job Count	128	Infinite
Reworks	19	Infinite
Fails	29	Infinite
Ejections	40	Infinite

Simulation run complete.

handling (using transporters, conveyors, and guided transporters in the SIMAN context) is only possible using the station concept.

8.5 Station Visitation Sequences

In Example 8.1, there are two types of entities visiting five stations. However, all the entities do not visit the stations in the same order.

Type I jobs go in the order Drill to Mill to Grinder to Inspect with a possible return to the Drill. Type II jobs go in the order Drill to Plane to Grinder to Inspect with a possible return to the Drill. In the model frame for Example 8.1, we were able to use logic to determine the next station for the jobs.

There is another very convenient method to accomplish the same purpose. SIMAN allows a separate station visitation sequence for each entity type. The sequence set number (NS) specifies a particular sequence for each entity type. For each NS, there is an index into the sequence set, IS. Thus, NS = 2 and IS = 3 indicates the third station visited in the second station visitation sequence. Initially, NS = 0 and IS = 0 for each new entity. The value of IS is automatically updated when an entity is transferred according to its sequence.

EXAMPLE 8.2 Using Station Visitation Sequences

Reconsider Example 8.1. In this example, we will use the concept of Station Visitation Sequences to solve the problem. As in Example 8.1, each of the five similar logical flows will be represented by a single-station submodel and the special-purpose attribute, M, will be used to assign individual workstations for each entity. Another special-purpose attribute, NS, will be used to specify a particular visitation sequence for each job. Another use for a previous block and one new element required in using station visitation sequences are shown in Table 8.2.

Table 8.2 Blocks/Elements Introduced in this Example

New Block/Element	Purpose
ROUTE block	Used to transfer entities from one station to another within each visitation sequence
SEQUENCES element	Used to define each visitation sequence, and to assign values to variables

8.6. The ROUTE Block and SEQUENCES Element

As discussed previously in Section 8.1, the ROUTE block sends the entity from one station to another station. The format for the block is as follows:

ROUTE: *Duration, Destination;*

The operands are described as follows:

Operand	Description	Default
Duration	Routing time delay [expression]	0
Destination	Destination station [expression truncated to an integer station number, station symbol name, or SEQ]	SEQ

When a station visitation sequence is used, the destination station in the ROUTE block must be stated as SEQ (or defaulted). The sequence set number, NS, determines which sequence is followed. NS must be assigned before the entity enters the ROUTE block. Finally, when the entity leaves the ROUTE block, the sequence set index, IS, will automatically be incremented by 1. Also, the station number, M, will automatically be updated to the next station in the sequence. In Example 8.2, the ROUTE block will be

ROUTE: 0, SEQ;

The format for the SEQUENCES element is given by the following:

SEQUENCES: *Number, Name, Station ID, Variable=Value &,*
...: repeats;

The operands are described as follows:

Operand	Description	Default
Number	Sequence set Number [integer]	Sequential
Name	Sequence set Name [symbol name]	Blank
Station ID	Next station in visitation sequence [integer station number or station symbol name]	—
Variable	Attributes or variables to be assigned *Value* upon entering station	—
Value	Values assigned to *Variables* upon entering station	—

For Example 8.2, the SEQUENCES element is as follows:

```
SEQUENCES:  1,TypeIPart,SDrill,ProcessTime=ED(3),&
            SMill,ProcessTime=ED(4),&
            SGrinder,ProcessTime=ED(6),&
            SInspect,ProcessTime=ED(8),PRedo=.05,PFail=.05&
            SDone:
            2,TypeIIPart,SDrill,ProcessTime=ED(3),&
            SPlane,ProcessTime=ED(5),&
            SGrinder,ProcessTime=ED(7),&
            SInspect,ProcessTime=ED(8),PRedo=.05,PFail=.05&
            SDone;
```

All parts entering the system will begin at station *SArrive*. After that, separate sequences are identified for Part Type I and Part Type II. The first station for Part Type I is *SDrill* at which the attribute *ProcessTime* is set to a value from a distribution given in expression set 3 and using random number stream 1. Part Type I continues through *SMill, SGrinder*, and *SInspect* in a similar fashion. The only difference is that at *SInspect* the values for attributes *PRedo* and *PFail* are each assigned the value 0.05. Note: more than one assignment can be made. The entity then goes to the exit station *SDone*. The sequence for Part Type II is explained in a similar fashion.

The model frame using the SEQUENCES element is shown in Exhibit 8.3. The portions that are different than in Example 8.1

EXHIBIT 8.3 Model frame for Example 8.2

```
!Model Frame for Example 8.2
!
BEGIN;
          CREATE:                        !Create arriving jobs according to
            ED(1):                       !distribution ED(1) and save their
            MARK(ArrTime);               Arrival times in attribute ArrTime
          ASSIGN:
            JobType = ED(2):             !Define job as Type I or Type II
            NS = JobType:                !Define SEQUENCE by JobType
            M = SArrive;                 Define current station as SArrive
          ROUTE:
            0, SEQ;                      Route jobs according to SEQUENCE
Process   STATION, SDrill-SInspect;      Stations are from SDrill to SInspect
          QUEUE,                         !Queue number is given by M
            M, V(M), Eject;              Queue capacity is given by V(M)
          SEIZE:
            Machine(M);                  Capture indexed resource M
          DELAY:
            ProcessTime;                 Process job - in SEQUENCE element
```

EXHIBIT 8.3 *continued.*

```
          RELEASE:
            Machine(M);               Free indexed resource M
          BRANCH, 1, 10:              !Take one branch only
          IF, M.LE.4, Move_On:        !If at first 4 stations continue SEQ
                                      !else, when at Inspection station,
            WITH, PRedo, Rework:      !Job is reworked with probability PRedo
            WITH, PFail, Scrap:       !Job is scrapped with probability PFail
            ELSE, Done;               !Job is complete - continue SEQUENCE
Move_On   ROUTE:
            0, SEQ;                   Route jobs according to SEQUENCE
Done      STATION; SDone;             SDone station for completed jobs
          TALLY:
            Flowtime,INT(ArrTime);    Tabulate Flowtime
          TALLY:
            Exit Period,BET;          Tabulate time between exits
          COUNT:
            Job Type,1:               !Count jobs by type and
          DISPOSE;                    destroy entity
Rework    COUNT:
            Reworks;                  Count number of jobs reworked
          ASSIGN:
            IS = 0;                   Restart SEQUENCE
          ROUTE:
            0, SEQ;                   Route jobs according to SEQUENCE
Scrap     COUNT:
            Fails:                    !Count number of jobs scrapped and
          DISPOSE;                    destroy entity
Eject     COUNT:
            Ejections:                !Count number of jobs ejected and
          DISPOSE;                    destroy entity
```

are shown by italics. These differences are explained as follows: The ASSIGN block results in the *JobType* being placed in the NS attribute. The ROUTE block makes use of the SEQUENCES element in the experiment frame. The STATION block is given a label as a destination for the reworked jobs. The DELAY block receives its action times from the SEQUENCES element.

The BRANCH block is rewritten to use the SEQUENCES element. Thus, if M is 1, 2, 3, or 4, then the entity proceeds to the ROUTE block labeled *Move_On*. If M is 5, then 5% are reworked, 5% are scrapped, and those remaining are good. Those that are to be reworked go to the block with label *Rework*.

At the following block, the position of the sequence is reset (IS = 0) in order for the job to be routed back to the beginning of the processing sequence, *SDrill*. Those that are scrapped go to the block with the label *Scrap*. Those that are good go to the block with the label *Done*.

The experiment frame using the SEQUENCES element is shown in Exhibit 8.4. The portions that are different than in Example 8.1 are shown in italics. These differences are explained as follows: The ATTRIBUTES element is changed as the first three values are given by the SEQUENCES element. The STATIONS element is the same except that Stations *SArrive* and *SDone* are added. The SEQUENCES element is being used.

EXHIBIT 8.4 Experiment frame for Example 8.2

```
!Experiment Frame for Example 8.2
!
BEGIN;
PROJECT, Second Station Example, Team;
ATTRIBUTES:   ProcessTime:              !First attribute in SEQUENCES
              PRedo:                    !Second attribute in SEQUENCES
              PFail:                    !Third attribute in SEQUENCES
              ArrTime:                  !Att 4 marked with arrival time
              JobType;                  Att 5 marked with JobType
VARIABLES:    1, MDrill, 2:             !Max queue at behind Drill is 2
              2, MMill, 3:              !Max queue at Mill is 3
              3, MPlane, 3:             !Max queue at Plane is 3
              4, MGrinder, 2:           !Max queue at Grinder is 2
              5, MInspect, 4;           !Max queue at Inspect is 4
STATIONS:     SDrill:                   !Drilling station
              SMill:                    !Milling station
              SPlane:                   !Planing station
              SGrinder:                 !Grinding station
              SInspect:                 !Inspection station
              SArrive:                  !Arrival station
              SDone;                    Exit station
RESOURCES:    Drill, 2:                 !2 Drills
              Mill, 3:                  !3 Mills
              Plane, 3:                 !3 Planes
              Grinder, 2:               !2 Grinders
              Inspector;                1 Inspector
SETS:         Machine,Drill..Inspector; Machine(1) thru Machine(5)
QUEUES:       DrillQ:                   !Buffer before Drill
              MillQ:                    !Buffer before Mill
              PlaneQ:                   !Buffer before Plane
              GrinderQ:                 !Buffer before Grinder
              LVF(JobType):             !Priority to JobType 1
              InspectorQ;               Buffer before Inspect
```

EXHIBIT 8.4 *continued.*

```
COUNTERS:     Type I Job Count:                 !Count Type I jobs completed
              Type II Job Count:                !Count Type II jobs completed
              Reworks:                          !Count number of jobs reworked
              Fails:                            Count number of jobs scrapped
              Ejections;                        Count number of jobs ejected
SEQUENCES:    1, TypeIPart,
              SDrill, ProcessTime = ED(3), &
              SMill, ProcessTime = ED(4), &
              SGrinder, ProcessTime = ED(6), &
              SInspect, ProcessTime = ED(8), PRedo = .05, PFail = .05 &
              SDone:
              2, TypeIIPart,
              SDrill, ProcessTime = ED(3), &
              SPlane, ProcessTime = ED(5), &
              SGrinder, ProcessTime = ED(7), &
              SInspect, ProcessTime = ED(8), PRedo = .05, PFail &
              SDone;
EXPRESSIONS:1, Dist1, EXPO(5, 1):               !Interarrival time
              2, Dist2, DISC(0.7, 1, 1.0, 2, 1): !Job Type distribution
              3, Dist3, UNIF(6, 9, 1):          !Drilling time
              4, Dist4, TRIA(12, 16, 20, 1):    !Milling time
              5, Dist5, TRIA(20, 26, 32, 1):    !Planing time
              6, Dist6, DISC(0.25, 6, 0.75, 7, 1.0, 8, 1):
                                                !Grinding time for Type I jobs
              7, Dist7, DISC(0.1, 6, 0.35, 7, 0.65, 8, 0.9, 9, 1.0, 10, 1):
                                                !Grinding time for Type II jobs
              8, Dist8, NORM(3.6, 0.6, 1);      Inspect time
TALLIES:      Flowtime:
              Exit Period;
DSTATS:       NQ(DrillQ), Drill Queue:
              NQ(MillQ), Mill Queue:
              NQ(PlaneQ), Plane Queue:
              NQ(GrinderQ), Grinder Queue:
              NQ(InspectorQ), Inspect Queue:
              NR(1)/2, Drill Utilization:
              NR(2)/3, Mill Utilization:
              NR(3)/3, Plane Utilization:
              NR(4)/2, Grinder Utilization:
              NR(5), Inspect Utilization;
REPLICATE,    1, 0, 2400;
```

The output for Example 8.2 is given in Exhibit 8.5. On comparing Exhibit 8.5 to Exhibit 8.2, differences will be observed. These differences occur even though the models are supposedly the same. The differences occur because the order in which the random numbers are used is not the same. Thus, Exhibits 8.2 and 8.5 will not have the same results.

EXHIBIT 8.5 Summary report for Example 8.2

Summary for Replication 1 of 1

Project: Second Station Example Run execution date: 8/24/1994
Analyst: Team Model revision date: 8/24/1994

Replication ended at time: 2400.0

TALLY VARIABLES

Identifier	Average	Variation	Minimum	Maximum	Observations
Flowtime	45.578	.25764	29.295	120.24	416
Exit Period	5.7026	.74327	1.9640	36.447	415

DISCRETE-CHANGE VARIABLES

Identifier	Average	Variation	Minimum	Maximum	Final Value
Drill Queue	.49308	1.4380	.00000	2.0000	2.0000
Mill Queue	.30267	1.9654	.00000	3.0000	1.0000
Plane Queue	.05178	4.7995	.00000	2.0000	.00000
Grinder Queue	.26655	1.9921	.00000	2.0000	1.0000
Inspect Queue	.19826	2.0456	.00000	2.0000	.00000
Drill Utilization	.74904	.48082	.00000	1.0000	1.0000
Mill Utilization	.72418	.44283	.00000	1.0000	1.0000
Plane Utilization	.51500	.66567	.00000	1.0000	.00000
Grinder Utilization	.70726	.51358	.00000	1.0000	1.0000
Inspect Utilization	.69637	.66032	.00000	1.0000	1.0000

COUNTERS

Identifier	Count	Limit
Type I Job Count	291	Infinite
Type II Job Count	125	Infinite
Reworks	19	Infinite
Fails	29	Infinite
Ejections	65	Infinite

Simulation run complete.

8.7 SUMMARY

This chapter describes the extremely important SIMAN concept of a station. Stations are necessary for all models involving material handling concepts such as transporters, conveyors, and guided transporters. New constructs introduced include the STATION block to represent an entry point into the station submodel, and the ROUTE block to transfer entities from one station to another. The special-purpose attribute M was introduced to provide for very efficient modeling. The STATIONS element specifies the station names. The same example was solved in two ways: The first solution method used logic and the second solution method used the concept of station visitation sequences. To use the latter method, the SEQUENCES element was introduced. Understanding of the two examples in this chapter and working exercises from the set below using both the logical method and the station visitation sequence method are necessary checkpoints to proceed with Chapters 9, 10, and 11.

8.8 EXERCISES

E8.1 Parts enter the system under study at a staging area for transfer to the milling station. After milling by an operator, the parts are transferred to the cleaning station where a second operator cleans the part. After cleaning, the parts are transferred to the painting station. After painting by a third operator, the parts are sent to the packaging station where they are packaged by a fourth operator. After packaging, the parts are sent to the shipping area where they exit the system.

The time between arrival of parts to be processed is exponentially distributed with a mean of 30 minutes. The time for milling is uniformly distributed between 12 and 40 minutes. Cleaning time is triangularly distributed between 10 and 50 minutes, with a most likely value of 27 minutes. Painting time is uniformly distributed between 10 and 42 minutes. Lastly, packaging time is normally distributed with a mean of 26 minutes and a standard deviation of 4 minutes. Transfer times between stations are exponentially distributed with a mean of 3 minutes. This time occurs from staging to each of the operations and also from the last operation to the shipping area. Make one replication of this system for the completion of 200 parts and determine the following:

a. The utilization at each workstation

b. Statistics about the queue before each workstation

c. Average part flowtime

Model this system, using logic in an efficient manner.

E8.2 Solve E8.1, using the SEQUENCES element in an efficient manner.

E8.3 A production operation has six workstations. Stations 1, 2, and 3 are in parallel, and stations 4, 5, and 6 are in series. A queue may form behind each station. Jobs arrive uniformly between 0 and 20 minutes apart to be processed. From the receiving station, it takes 2 minutes to travel to one of the parallel stations. A job selects one of Stations 1, 2, or 3 at random. The processing time at the parallel stations is exponential with a mean of 27 minutes. It takes 1 minute to transfer from the parallel to series stations and 1 minute to transfer between the series stations and to the exit station. The processing times for Stations 4, 5, and 6 are given by normal distributions with means of 7, 8, and 9 minutes, respectively, and standard deviations of 1 minute in all cases. Simulate the operation of the system for 40 hours after initializing the system for 8 hours. Be efficient in programming using logic.

a. Provide queueing and resource statistics for every station

b. Tally the time in the system for all jobs

E8.4 A production operation has six workstations. Stations 1, 2, and 3 are in parallel, and stations 4, 5, and 6 are in series. A queue may form behind each station. Jobs arrive uniformly between 0 and 20 minutes apart to be processed. Some 60% of the jobs are gadgets, and the remaining are widgets. At each station, widgets have priority over gadgets. From the receiving station it takes 2 minutes to travel to one of the parallel stations. Jobs are sent to stations at random according to the following probability distribution:

Station	Probability
1	.40
2	.35
3	.25

The processing times at the parallel stations are exponentially distributed with the following means:

Station	Gadgets	Widgets
1	21	23
2	25	22
3	35	36

It takes 1 minute to transfer from the parallel to series stations and 1 minute to transfer between the series stations and the exit station. The processing times at the series stations are normally distributed with the following means and standard deviations:

Station	Gadgets	Widgets
4	(7, 1)	(9, 1.5)
5	(8, 1)	(7, 1.1)
6	(9, 1)	(8, 1.2)

Simulate the operation of the system for 40 hours after initializing the system for 8 hours. Be efficient in programming using logic then sequences.

a. Provide queuing and resource statistics for every station

b. Tally the time in the system for gadgets

c. Tally the time in the system for widgets

d. Tally the time in the system for all jobs

E8.5 Consider the production process described in E4.1. Assume that movement between stations requires times that follow a triangular distribution with parameters (0.5, 1.0, 1.5) (in minutes). Movement includes the following:

Arrival to Milling	or	Arrival to Planing
Milling to Drilling	or	Milling to Eject
Planing to Drilling	or	Planing to Eject
Drilling to Inspection	or	Drilling to Eject
Inspection to Exit	or	Inspection to Drilling or Inspection to Eject

The data for arrivals and processing is as follows:

Arrivals	Exponential	Mean = 2 minutes
Milling	Triangular	(8, 12, 14)
Planing	Uniform	(6, 10)
Drilling	Normal	Mean = 5, Standard deviation = 1
Inspection	Uniform	(1, 2)
Breakdown	Uniform	(5, 30)
Repair time	Uniform	(360, 600)
Route time	Triangular	(0.5, 1.0, 1.5)

Sixty percent of the jobs are type I, the others are type II. Type I jobs are served at milling, then move to drilling, and finally to inspection. Type II jobs go to planing, then drilling, and then inspection. At drilling, Type I jobs have higher priority than Type II jobs. There is no priority system at inspection. Jobs fail inspection with probability 0.10. Fifty percent of the failed jobs are returned to drilling, and the remainder is scrapped.

Answer the following questions:

- **a.** How many good jobs are made of each type?
- **b.** How many reworks occur?
- **c.** How many jobs are ejected?
- **d.** How many jobs are scrapped?
- **e.** What is the average system times for both job types?
- **f.** What is the utilization of each class of machine and operator?
- **g.** How many jobs are ejected from the system in front of each class of machine?

Simulate 1 week of production after a 1-day warm-up of the system using logic or sequences.

9

Transporters

Initially, the examples presented in this text make the assumption that after an entity finishes its processing at one resource it is instantaneously transferred to the next queue. With the introduction of the STATIONS concept and the ROUTE block, the examples could have accounted for the time associated with physically moving an entity from one location to another. The ROUTE block, however, is very general because it assumes that the transfer device is immediately available at all times. It cannot account for delays or obstructions in the path. This chapter introduces several new blocks that improve the realism of transporting an entity between locations.

9.1 SIMAN's View of the Transporter

A transporter is analogous to a mobile resource. It must be requested, used, and then freed. It may have multiple units that correspond to the number of transfer devices present in the system. In SIMAN, a transporter is simply one or more identical devices that may be allocated to an entity in order to move entities between stations. In the situation where there is more than one transporter unit, SIMAN monitors each unit as though it were a separate device and monitors variables that define its current or destination station, travel velocity, operational status (active or inactive), and its allocation station (idle or busy).

In SIMAN, there are two types of transporter systems: the *free-path* transporter and the *guided* transporter. The guided transporter is constrained to a network of paths and may be delayed by obstructions or congestion, for example, Automated Guided Vehicles (AGVs). This type of system will be discussed in Chapter 11. Free-path transporters are not constrained to a network of paths and may choose an alternate path or move around an obstruction. The modeling of free-path transporters is the focus of this chapter.

EXAMPLE 9.1 Problem Statement

Let us now modify Example 8.2 to include transportation of jobs. Assume that a job traveling between any two stations must be carried by a transporter at a rate of 80 feet/minute, loaded or unloaded. The closest transporter will respond for service. Below are the distances between stations:

	Mill	Plane	Grinder	Inspect	Exit	Arrive
Drill	60’	60’	120’	140’	200’	60’
Mill		120’	60’	80’	140’	120’
Plane			60’	80’	140’	120’
Grinder				20’	80’	180’
Inspect					60’	200’
Exit						260’

Distances along the diagonal are 0 feet, and the distances between two stations are the same in either direction. For example, the distance from the mill to the plane is the same as from the plane to the mill. Figure 9.1 provides a schematic view of the system. All transporters begin at the arrival station and are active. In this example, we are interested in determining the number of transporters needed in this system so that their utilization is within the 0.50–0.60 range. We will replicate the simulation five times for each change in the number of transporters and use the average transporter utilization to make the decision.

Table 9.1 lists the new blocks and elements used in modeling Example 9.1.

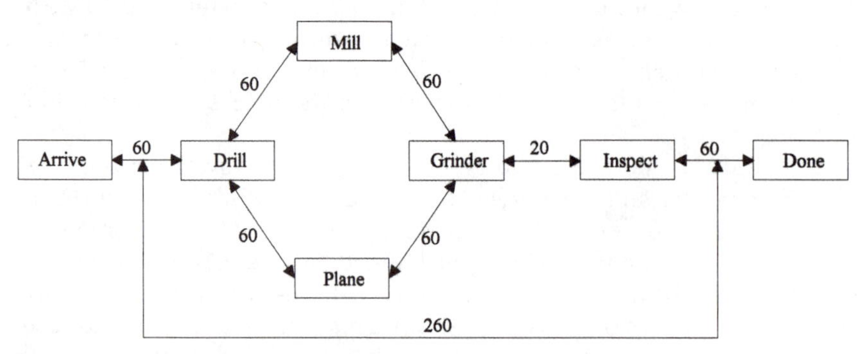

Figure 9.1 Schematic of Example 9.1.

Table 9.1 Blocks/Elements Introduced in this Chapter

New Block/Element	Purpose
REQUEST block	Used to call a transporter according to a selection rule or by a specific index
TRANSPORT block	Used to "move" an entity from one station to another
FREE block	Used to relieve the transporter from carrying an entity
TRANSPORTERS element	Used in the experimental frame to define default parameters for transporters
DISTANCES element	Used to define the distances between stations

9.2 Model Frame Transport Blocks

9.2.1 The REQUEST Block

The REQUEST block holds an arriving entity in the preceding QUEUE block until a transporter is available. The entity may REQUEST a specific transporter unit or any unit of a particular class of transporters. If the transporter is available (idle and active) its status is altered to busy and active, and it is sent to pick up the entity that waits in the REQUEST block. If, for some reason, the requested transporter is not available (busy or inactive), the REQUEST block holds the entity in the queue until the transporter is available. The format of the REQUEST block is

REQUEST, *Priority, Storage ID*: *Transporter Name(Unit), Velocity, EntityLocation;*

Operand	Description	Default
Priority	REQUEST block priority [expression truncated to an integer]	1
StorageID	Storage associated with the REQUEST block [expression truncated to an integer storage number, or storage symbol name]	No storage

Operand	Description	Default
Transporter Name	Symbol name for the transporter	—
Unit	Transporter index	1
Velocity	Velocity for REQUEST	Current velocity
Entity Location	Location of the entity (*StationID*)	Entity M value

The *TransporterName*(*Unit*) operand defines the class of transporter unit called. If the unit distinction is omitted, SIMAN defaults the unit to 1. Alternatively, a selection rule may be defined rather than a specific unit. The rule sets up a criterion by which a transporter is chosen to answer the request. A very common rule is to choose the transporter that is closest to the station (called SDS). Another possibility is to request the transporter that is the most distant from the station (called LDS). Now, consider a system that uses three forklifts, Fork(1), Fork(2), and Fork(3). You may wish to request these forklifts using a special priority. By using a cyclic priority, each forklift is given the highest priority in order. When Fork(2) has highest priority, the REQUEST block will attempt to allocate it to an entity before considering Fork(1) or Fork(3). In another instance, you may wish to assign priority to each unit randomly rather than in order. Finally, you may assign a preferred order of selection such that the REQUEST block can attempt allocation of Fork(1) first, Fork(2) second, and Fork(3) last. Table 9.2 shows the SIMAN codes for each selection rule.

Velocity is the speed used to calculate the delay in moving the empty transporter to the awaiting entity. A default value for this operand can be defined in the experiment frame. Although the units for velocity (feet/second, miles/hour) are not explicitly defined, they must be consistent with the velocity and distance definitions in the experiment frame.

For Example 9.1, the REQUEST block is used as follows:

REQUEST: Forklift(SDS);

Table 9.2 Transporter Selection Rules

CYC	Cyclic priority
LDS	Largest distance to station
POR	Preferred order rule
RAN	Random priority
SDS	Smallest distance to station

Thus, the closest *Forklift* to the requesting entity is called. If the *Forklift* is active and idle it will be allocated to the entity and sent to pick up the entity. Other examples follow:

REQUEST: Cart;

This example requests the first unit of the transporters in class *Cart*.

REQUEST: Cart(1);

This second example makes the same request, but here the value specified for the unit is the same as the default.

REQUEST: Cart(3);

The next example requests the third unit of the transporters in class *Cart*.

REQUEST, Type: Truck, 10;

Here, the first unit of class *Truck* is called, and it will move with a velocity of 10 feet per second. Entities requesting this transporter will be prioritized according to the attribute *Type*. Thus an entity of *Type* = 1 will have priority over entities of *Type* = 2.

REQUEST: Truck(RAN, TruckID), 10;

Here a *Truck* is selected randomly by the requesting entity. The *Unit* of the *Truck* is stored in the attribute *TruckID*, and the velocity is set to 10 feet per second.

REQUEST, Type: Cart(SDS);

In this final example, entities prioritized by *Type* request a cart to be selected by the Smallest Distance to Station rule.

9.2.2 The TRANSPORT Block

Now that a transporter has been successfully allocated to an entity, the model must account for the delay in moving the entity to the new location. That activity is handled by the TRANSPORT block. Its format is as follows:

TRANSPORT: *TransporterName, EntityDestination, Velocity*;

Operand	Description	Default
Transporter Name	Name of the transporter	Unit the entity was allocated
Entity Destination	Destination station for the entity [station number, station symbol name, or SEQ]	
Velocity	Velocity for the transporter [expression]	Current velocity

Transporter Name is the specific transporter unit allocated to the entity. An entity executing a TRANSPORT block must be allocated to a transporter unit using the REQUEST block. SIMAN V will allow explicit referencing of this operand. However, models built under SIMAN V do not require explicit referencing of the transporter unit, and this value may be defaulted. This feature enables models built under earlier versions of SIMAN to still be run using SIMAN V.

In Example 9.1, the TRANSPORT block is used as follows:

```
TRANSPORT: Forklift, SEQ;
```

Here the travel velocity is defaulted to 80 feet per minute (defined in the experiment frame), and the entity destination is determined by the SEQUENCES element.

Following are examples of the TRANSPORT block:

```
TRANSPORT;
```

In this first example, all operands are defaulted. The TRANSPORT block uses the transporter allocated to the entity in the REQUEST block, determines the destination from the SEQUENCES element in the experiment frame, and uses the default velocity also defined in the experiment frame.

```
TRANSPORT: Cart;
```

In this example, the *Unit* is defaulted;

```
TRANSPORT: Cart(3);
```

The third example explicitly references the third unit of *Cart*. If *Cart(3)* is allocated to the entity in the REQUEST block, then all three examples will produce identical results.

```
TRANSPORT: Truck(TruckID), Shipping, 25;
```

In this example, all operands have been specified. A previously selected transporter of class *Truck* with its *Unit* stored in *TruckID* will transport an entity to the *Shipping* station at a velocity of 25.

9.2.3 The FREE Block

Once the transporter has reached its destination with the entity, it must be freed for further use much in the same way that a resource is released. This is accomplished by using the FREE block that has the following format:

FREE: *TransporterName(Unit)*;

Operand	Description	Default
Transporter Name	Transporter unit to free	Transporter that the entity was allocated
Unit	Transporter index	1

The FREE block's one operand is the name of the transporter unit. Like the *TransporterName* operand of the TRANSPORT block, it may be explicitly referenced or defaulted to the transporter allocated to the entity.

In Example 9.1, the FREE block is used as follows:

```
FREE:
  Forklift;
```

9.3 Experiment Frame Transport Elements

9.3.1 The TRANSPORTERS Element

This element defines characteristics of SIMAN free-path and guided transporters. The format for the free-path TRANSPORTERS element is as follows:

TRANSPORTERS: *Number, Name, Number of Units, System Map, Velocity, Position - Status, repeats*: repeats;

Operand	Description	Default
Number	Transporter number [integer]	Sequential
Name	Transporter name [symbol name]	—
Number of Units	Number of available units [integer]	1
System Map	System map [integer distance set for DISTANCES element]	1
Velocity	Velocity	1.0
Position	Initial Position [*StationID* or STATION(*StationID*)]	See following examples
Status	Initial status [active or inactive]	Active

Consider the following examples:

```
TRANSPORTERS:     1, Cart, 3, 1, 10, 1 - A, 2 - A, 3 - A:
                  2, Truck, 2, 2, 30, 1 - I:
                  3, Fork, 3, 1, 10, 1 - A, 2 - I;
```

Transporter 1, the *Cart*, has three units. The *System Map* is the first set in the DISTANCES element; the velocity is 10 feet per second; the first unit begins the simulation run at Station 1 with an active status; the second unit begins at Station 2 with an active status; the third at Station 3, also in active status (active and inactive may be abbreviated using A or I, respectively). The two transporters in class *Truck* reference the second DISTANCES set and move at 30 feet per second. The first unit is inactive at Station 1 when the simulation begins, and the second unit also begins at Station 1, but in an active status. The third transporter, the *Fork*, has three units that follow the same DISTANCES set as the *Carts*. They move at 10 feet per second and one is found active at Station 1 when the simulation begins; the second begins at Station 2 in an inactive status; and the third begins at Station 1 in active status. Using the default for the element results in all transporters set to an active status at the first station listed in the DISTANCES element.

9.3.2 The DISTANCES Element

In the TRANSPORTERS element, the number in the *System Map* operand references a set of instructions found here in the DISTANCES element. The instructions provide all of the travel distances that the transporter will need during the simulation. The format for the DISTANCES element is shown as follows:

DISTANCES: *Identifier, Starting Station ID - Ending Station ID-Distance repeats...*: repeats;

The *Identifier* operand, the number referenced by the TRANSPORTERS element, and the range of stations that the transporter will visit, *Starting Station ID - Ending Station ID*, are required. The *Distance* is assigned explicity for each from–to pair, and only the travel paths used by the transporter can be entered. An example is given as follows:

```
DISTANCES:     1, Arriv-Drill-60, Drill-Press-60,
               Press-Oven-20, Oven-Dryer-80,
               Dryer-Oven-120, Dryer-Exit-150,
               Exit-Arriv-210;
```

Note: we are able to define two different distances for the same pair, *Oven - Dryer* and *Dryer - Oven*. Unless explicitly defined, as in the previous example, SIMAN will assume the same distance for the reverse pair of stations. For example, 60 feet is assumed for both *Arriv - Drill* and *Drill - Arriv* even though only one pair is defined.

EXAMPLE 9.1 Solution

When dealing with a stationary resource, such as the drill in Example 3.1, an entity must pass through the following sequence of blocks: QUEUE, SEIZE, DELAY, and RELEASE. An entity to be transported between stations must go through a similar sequence. This sequence is QUEUE, REQUEST, TRANSPORT, and FREE. The QUEUE block holds the entities awaiting transport, REQUEST determines the availability of transporters and assigns one to the awaiting entity, TRANSPORT accounts for the delay while moving the entity, and the FREE block deallocates the transporter and resets the status.

In the remainder of this subsection, the solution for Example 8.1 is used as the base. Any new or altered blocks or elements are in italics.

The Model Frame

Exhibit 9.1 shows the entire model frame for Example 9.1. You will notice that in comparison to Example 8.2, the only significant addition is a section of code that begins with the label, *Move_Job*. This small section acts as a subroutine that handles all transporters much in the same way that the STATION section handles the processing sequence for each resource.

EXHIBIT 9.1 Model frame for Example 9.1.

```
!Model Frame for Example 9.1
!
BEGIN;
          CREATE:                          !Create arriving jobs according to
            ED(1):                         !distribution ED(1) and save their
            MARK(ArrTime);                 Arrival times in attribute ArrTime
          ASSIGN:
            JobType = ED(2):               !Define job as Type I or Type II
            NS = JobType:                  !Define SEQUENCES by JobType
            M = SArrive;                   Define current station as SArrive
Move_Job  QUEUE:
            ForkLiftQ;                     Wait for available ForkLift
          REQUEST:
            ForkLift(SDS);                 Request ForkLift (Shortest Dist Rule)
          TRANSPORT:
            ForkLift, SEQ;                 Transport according to SEQUENCES
Process   STATION,SDrill-SInspect;         Stations are from SDrill to SInspect
          FREE:
            ForkLift;                      Free ForkLift transporter
          QUEUE,                           !Queue number is given by M
            M,V(M), Eject;                 Queue capacity is given by V(M)
          SEIZE:
            Machine (M);                   Capture indexed resource M
          DELAY:
            ProcessTime;                   Process job - defined in SEQUENCES
          RELEASE:
            Machine(M);                    Free indexed resource M
          BRANCH,1,10:                     !Take one branch only
            IF, M.LE.4, Move_Job:          !If at first 4 stations continue SEQ
                                           !else, when at Inspect station,
            WITH, PRedo, Rework:           !Job is reworked with probability PRedo
            WITH, PFail, Scrap:            !Job is scrapped with probability PFail
            ELSE, Move_Job;                !Job is complete - continue SEQ
Done      STATION, SDone;                  SDone station for completed jobs
          FREE:
            ForkLift;                      Free ForkLift transporter
          TALLY:
            Flowtime,INT(ArrTime);         Tabulate Flowtime
          TALLY:
            Exit Period,BET;               Tabulate time between exits
          COUNT:
            JobType, 1:                    !Count jobs by type and
            DISPOSE;                       destroy entity
Rework    COUNT:
            Reworks;                       Count number of jobs reworked
          ASSIGN:
            IS=0:                          !Restart SEQUENCES
            NEXT(Move_Job);                Move job to Drilling Station
```

EXHIBIT 9.1 *continued.*

```
Scrap    COUNT:
           Fails:                 !Count number of jobs scrapped and
           DISPOSE;               destroy entity
Eject    COUNT:
           Ejections:             !Count number of jobs ejected and
           DISPOSE;               destroy entity
```

Experiment Frame

The experiment frame contains special definitions that allow for the computation of transporter travel times and for the default values for velocity. Pay particular attention to the use of the DISTANCES and TRANSPORTERS elements. Also, note how the SIMAN variable NT is used to calculate the utilization of the transporters.

EXHIBIT 9.2 **Experiment frame for Example 9.1.**

```
!Experiment Frame for Example 9.1
!
BEGIN;
PROJECT, Example 9.1, Team;
ATTRIBUTES:        ProcessTime:        !First attribute in SEQUENCES
                   PRedo:              !Second attribute in SEQUENCES
                   PFail:              !Third attribute in SEQUENCES
                   ArrTime:
                   JobType:
VARIABLES:         1, MDrill, 2:
                   2, MMill, 3:
                   3, MPlane, 3:
                   4, MGrinder, 2:
                   5, MInspect, 4;
STATIONS:          SDrill:
                   Smill:
                   SPlane:
                   SGrinder:
                   SInspect:
                   SArrive:
                   SDone;
```

EXHIBIT 9.2 *continued.*

```
RESOURCES:      Drill,2:
                Mill, 3:
                Plane, 3:
                Grinder, 2:
                Inspector;
SETS:           Machine, Drill..Inspector;
QUEUES:         DrillQ:
                MillQ:
                PlaneQ:
                GrinderQ, LVF(JobType):
                InspectorQ:
                ForkLiftQ;          Buffer for ForkLift requests
DISTANCES:      1, SArrive-Sdrill-60, SArrive-SMill-120,
                SArrive-SPlane-120, SArrive-SGrinder-180,
                SArrive-SInspect-200, SArrive-SDone-260,
                SDrill-SMill-60, SDrill-SPlane-60,
                SDrill-SGrinder-120, SDrill-SInspect-140,
                SDrill-SDone-200, SMill-SPlane-120,
                SMill-SGrinder-60, SMill-SInspect-80,
                SMill-SDone-140, SPlane-SGrinder-60,
                SPlane-SInspect-80, SPlane-SDone-140,
                SGrinder-SInspect-20, SGrinder-SDone-80,
                SInspect-SDone-60;
TRANSPORTERS:   ForkLift, 2, 1, 120, SArrive - ACTIVE;
COUNTERS:       Type I Job Count:
                Type II Job Count:
                Reworks:
                Fails:
                Ejections;
SEQUENCES:      1, TypeIPart, SDrill, ProcessTime = ED(3), &
                SMill, ProcessTime = ED(4), &
                SGrinder, ProcessTime = ED(7), &
                SInspect, ProcessTime = ED(8), PRedo = .05, PFail = .05 &
                SDone;
                2, TypeIIPart, SDrill, ProcessTime = ED(3), &
                SPlane, ProcessTime = ED(5), &
                SGrinder, ProcessTime = ED(6), &
                SInspect, ProcessTime = ED(8), PRedo = .05, PFail = .05 &
                SDone;
EXPRESSIONS:    1, , EXPO(5, 1):              !Interarrival time
                2, , DISC(0.7, 1, 1.0, 2, 1): !Job Type distribution
                3, , UNIF(6, 9, 1):           !Drilling Time
                4, , TRIA(12, 16, 20, 1):     !Milling time
                5, , TRIA(20, 26, 32, 1):     !Planing time
                6, , DISC(0.25, 6, 0.75, 7, 1.0, 8, 1):
                                              !Grinding time for Type I jobs
                7, , DISC(0.1, 6, 0.35, 7, 0.65, 8, 0.9, 9, 1.0, 10, 1):
                                              !Grinding time for Type II jobs
                8, , NORM(3.6, 0.6, 1);       Inspection time
```

EXHIBIT 9.2 *continued.*

```
TALLIES:        Flowtime:
                Exit Period;
DSTATS:         NQ(DrillQ), Drill Queue:
                NQ(MillQ), Mill Queue:
                NQ(PlaneQ), Plane Queue:
                NQ(GrinderQ), Grinder Queue:
                NQ(InspectorQ), Inspect Queue:
                NR(1)/2, Drill Utilization:
                NR(2)/3, Mill Utilization:
                NR(3)/3, Plane Utilization:
                NR(4)/2, Grinder Utilization:
                NR(5), Inspect Utilization:
                NT(1)/2, ForkLift Utilization;
REPLICATE, 5, 0, 2400;
```

The first and last summary reports only are shown in Exhibit 9.3. The objective was to determine the utilization of the transporters over the five runs. These are provided at each replication with the following results:

0.56427
0.53486
0.52259
0.50476
0.58647

EXHIBIT 9.3 First and last summary reports for Example 9.1.

Summary for Replication 1 of 5

Project: Example 9.1 Run execution data: 8/24/1994
Analyst: Team Model revision data: 8/24/1994
Replication ended at time: 2400.0

TALLY VARIABLES

Identifier	Average	Variation	Minimum	Maximum	Observations
Flowtime	51.904	.22398	34.920	125.85	413
Exit Period	5.7397	.60772	.34708	27.845	412

EXHIBIT 9.3 *continued.*

DISCRETE-CHANGE VARIABLES

Identifier	Average	Variation	Minimum	Maximum	Final Value
Drill Queue	.45356	1.5364	.00000	2.0000	.00000
Mill Queue	.28814	2.0686	.00000	3.0000	.00000
Plane Queue	.08224	3.7409	.00000	3.0000	3.0000
Grinder Queue	.21568	2.2243	.00000	2.0000	.00000
Inspect Queue	.18338	2.1588	.00000	2.0000	.00000
Drill Utilization	.74142	.49122	.00000	1.0000	1.0000
Mill Utilization	.72034	.44602	.00000	1.0000	.00000
Plane Utilization	.50857	.67696	.00000	1.0000	1.0000
Grinder Utilization	.70429	.50682	.00000	1.0000	.50000
Inspect Utilization	.68668	.67549	.00000	1.0000	1.0000
ForkLift Utilization	.56427	.71179	.00000	1.0000	.00000

COUNTERS

Identifier	Count	Limit
Type I Job Count	290	Infinite
Type II Job Count	123	Infinite
Reworks	19	Infinite
Fails	29	Infinite
Ejections	60	Infinite

Summary for Replication 5 of 5

Project: Example 9.1 Run execution date: 8/24/1994
Analyst: Team Model revision date: 8/24/1994
Replication ended at time: 2400.0

TALLY VARIABLES

Identifier	Average	Variation	Minimum	Maximum	Observations
Flowtime	52.813	.26529	34.102	151.73	426
Exit Period	5.5651	.57183	.15582	23.098	425

DISCRETE-CHANGE VARIABLES

Identifier	Average	Variation	Minimum	Maximum	Final Value
Drill Queue	.41362	1.6086	.00000	2.0000	.00000
Mill Queue	.32251	1.8215	.00000	3.0000	.00000
Plane Queue	.06194	5.1543	.00000	3.0000	.00000
Grinder Queue	.24962	2.0522	.00000	2.0000	.00000
Inspect Queue	.20677	2.0240	.00000	2.0000	.00000
Drill Utilization	.75854	.44968	.00000	1.0000	1.0000
Mill Utilization	.76837	.37215	.00000	1.0000	.66667
Plane Utilization	.49676	.67684	.00000	1.0000	.33333
Grinder Utilization	.72297	.48021	.00000	1.0000	.50000
Inspect Utilization	.70776	.64258	.00000	1.0000	.00000
ForkLift Utilization	.58647	.68470	.00000	1.0000	.50000

EXHIBIT 9.3 *continued.*

COUNTERS

Identifier	Count	Limit
Type I Job Count	303	Infinite
Type II Job Count	123	Infinite
Reworks	24	Infinite
Fails	25	Infinite
Ejections	61	Infinite

Simulation run complete.

The average of these five values is 0.54259, which lies between 0.50 and 0.60. Therefore, only two transporters are needed to meet the specifications.

9.4 Other Transporter Blocks

The basic blocks required to use a transporter are REQUEST, TRANSPORT, and FREE. As models become more advanced, greater control over the transporters may be needed.

9.4.1 The HALT Block

The HALT block has the following format:

HALT: *Transporter Name(Unit)*;

Operand	Description	Default
Transporter Name	Transporter name [symbol name]	—
Unit	Transporter index [expression or constant]	1

When an entity moves through a HALT block, the transporter unit *Transporter Name(Unit)* is set to inactive and is unavailable for allocation. If the *Transporter Name(Unit)* is idle, the status is simply set to inactive. Should the transporter be busy when halted, it is both busy and inactive. It may not be allocated to any new entities and may only be controlled by the entity in possession of the transporter. If this entity executes a FREE block, then the status of *Transporter Name(Unit)* is changed to inactive and idle and must await the execution of an ACTIVATE block before it may be allocated to another entity. An example of the HALT block is given by

```
HALT:
  ForkLift(3);
```

The operational status of *ForkLift(3)* is changed from active to inactive.

9.4.2 The ACTIVATE Block

The ACTIVATE block has the following format:

ACTIVATE: *Transporter Name (Unit)*;

Operand	Description	Default
Transporter Name	Transporter name [symbol name]	—
Unit	Transporter index [expression or constant]	1

Recall that a transporter's operational status may be active or inactive. Inactivity may be the result of a HALT block or because the transporter was defined as inactive in the TRANSPORTERS element. The ACTIVATE block sets the operational status of the transporter *Transporter Name(Unit)* from inactive to active. The entity moving through the block does not take control of the transporter, however. The block makes that transporter available to entities awaiting it at a REQUEST block.

The SIMAN variable IT(*Transporter Name, Unit*) stores the current status of a transporter unit. *Transporter Name* is the transporter name, and *Unit* is the index. A 0 value indicates that the transporter unit is idle, 1 indicates that the unit is busy, and a 2 value shows that the transporter unit is inactive. The variable MT(*Transporter Name*) monitors the number of active units of *Transporter Name*. Thus, when an entity arrives at an ACTIVATE block, MT(*Transporter Name*) is incremented by 1. The value of IT(*Transporter Name, Unit*) is altered from 2 to 0, if no entities await a transporter, or to 1, if requesting entities exist.

Consider the following example:

```
ACTIVATE:
    ForkLift(3);
```

This block changes the status of *ForkLift(3)* from inactive to active. Another example is given by

```
HALT:
     Truck;
DELAY:
     LoadTime;
ACTIVATE:
     Truck;
```

In this example, we see a basic use of the HALT and ACTIVATE blocks. The *Truck* is made inactive during loading and is reactivated after loading. Statistics on individual transporter units may be gathered by using the FREQUENCIES element on the SIMAN variable IT(*Transporter Name*, Unit). This element was described in Section 7.6.2.

9.4.3 The ALLOCATE Block

The ALLOCATE block assigns a transporter to an entity without moving the transporter from its current location. It has the following format:

ALLOCATE, *Priority, Alternate Path: Transporter Name(Unit),*
Entity Location;

Operand	Description	Default
Priority	Allocate block priority	1
Alternate Path	Alternate path	Follow system map
Transporter Name	Transporter symbol name	—
Unit	Transporter index [expression truncated to an integer]	1
Entity Location	Entity Location [*StationID*, STATION(*StationID*)]	Entity M value

The ALLOCATE block, unlike the REQUEST block, does not require a preceding QUEUE block. Entities that are unable to allocate a transporter are held in an internal QUEUE until a transporter becomes available. At that time, the status of transporter *Transporter Name(Unit)* is set to busy, and the entity proceeds to the next block.

As an example, consider the following:

```
ALLOCATE:
     Forklift;
```

In this example, entities awaiting allocation of the *Forklift* transporter are held in an internal queue. Consider another example given by

```
QUEUE,
     M;
ALLOCATE:
     Cart(M+2);
```

Entities are held in queue M while they await allocation of a *Cart* with index M+2.

9.4.4 The MOVE Block

The MOVE block has the following format:

MOVE, *Storage ID, Alternate Path: Transporter Name(Unit),*
Destination, Velocity;

Operand	Description	Default
Storage ID	Storage associated with the MOVE block (expression or constant or storage symbol name)	No storage
Alternate Path	Alternate path	Not used
Transporter Name	Transporter unit to move	Unit the entity was allocated
Unit	Transporter index	1
Destination	Destination of transporter	Entity M value
Velocity	Velocity for MOVE [expression]	Current velocity

To use the MOVE block, an entity must first have been allocated *Transporter Name(Unit)* by an ALLOCATE block. Since SIMAN monitors such assignments, this operand may be defaulted. The time needed to MOVE the entity to the *Destination* is determined by the *Velocity* operand and the distance between *Destination* and the entity's location at the time the MOVE block was executed. For guided transporters, there are a number of different ways to specify the *Destination* and

Alternate Path; for our free-path transporters, *StationID* will suffice for *Destination*, and *Alternate Path* should be defaulted.

Storage ID is used for gathering statistics on the number of entities in the MOVE block. It is also used in a Cinema animation to display the entity's symbol during the time delay. Consider the following example:

```
MOVE: ForkLift(3), Shipping, 20;
```

In this example, *ForkLift(3)* is moved to station *Shipping* at a velocity of 20. If the default velocity is used, we may restate this block in the following way:

```
MOVE: , Shipping;
```

Since *ForkLift(3)* must already have been allocated to the entity executing the MOVE block, *Transporter Name(Unit)* is defaulted.

Observe how the ALLOCATE and MOVE blocks are used in the following set of blocks:

```
QUEUE,
  Platform;
ALLOCATE:
  ForkLift(SDS);
QUEUE,
  Dispatch;
SEIZE:
  Operator;
MOVE:
  ForkLift(A(1)), Loading;
```

Entities await allocation of a *ForkLift* at the *Platform* queue. Only when an operator is available does the *ForkLift* move to station *Loading*. Here we have modeled the delay caused by waiting for a *ForkLift* operator to become available. The REQUEST block alone would not allow for this activity.

9.5 SUMMARY

The concepts presented in this chapter are particularly important to modeling of material handling systems. Here we have discussed generic transporters used to move entities between stations. Transporters are called by using the REQUEST block. The TRANSPORT block describes the movement between stations, and the FREE block releases the transporter from the entity. The TRANSPORTERS element is used to describe each transporter explicitly, whereas the DISTANCES element is used to provide location information. Like the resources of Chapter 7, the HALT, ACTIVATE, ALLOCATE, and MOVE blocks may be used to simulate breakdowns, obstructions, or other unexpected abnormalities.

9.6 EXERCISES

E9.1 Using Example 9.1, alter the number of transporters in the TRANSPORTERS element. At what point are additional transporters of no appreciable help to the system? Hint: Have the summary report list the utilization of the transporters.

E9.2 Modify Exercise E8.2 using carts as transporters. Use the matrix below to determine the distances between each station. The numbers represent distances in feet.

	Milling	**Cleaning**	**Painting**	**Packaging**	**Shipping**
Staging	10	40	70	110	135
Milling		30	60	100	125
Cleaning			30	70	95
Painting				40	65
Packaging					25

Simulate this system until 200 completed jobs have been shipped. The carts begin the simulation in an active state. The carts move at 10 feet per minute.

a. How many carts are needed for an average utilization of the carts between 0.60 and 0.65? (Base your decision on five replications.)

b. How many times are the carts called during the simulation?

c. How much did the flowtime increase when using transporters compared to using the ROUTE block?

E9.3 The system of E9.1 has two carts that break down at an exponential rate with a mean of 120 minutes. The repair time is triangular with a minimum of 5, a median of 17, and a maximum of 20 minutes. The carts begin the simulation in an active state, and they move at 10 feet per minute.

a. How many breakdowns occur for the two carts?

b. How is the flowtime affected by the breakdowns compared with the output of E9.1 using two carts?

c. How is the utilization of the carts affected by the breakdowns compared with the output of E9.1 using two carts?

E9.4 Add forklift transporters to the system of Exercise E8.2. The forklifts begin the simulation in an active state at the exit station. Use the following distances in feet:

	P1	**P2**	**P3**	**S1**	**S2**	**S3**	**SExit**
SReceiving	40	50	70	70	97	110	130
P1		30	80	30	70	70	90
P2			50	47	47	87	87
P3				90	50	109	90
S1					40	40	59
S2						40	40
S3							40

Model the system under the following options:

two forklifts that move at 15 feet per minute
five forklifts that move at 15 feet per minute
three forklifts that move at 25 feet per minute

Compare the flowtime, the utilization of the forklifts, and the number of completed jobs for each of the three situations.

E9.5 Jobs arrive at an input station according to an exponential distribution with a mean of 10 minutes. The jobs are transported by a worker to the mill, a distance of 100 feet. After milling, the jobs are transported by a worker to a lathe, a distance of 40 feet. From the lathe, a worker takes the job to the exit station, a distance of 60 feet. There are two workers. One of the workers is required to operate each machine. Mill and lathe times are distributed normally with a mean of 16 minutes and a standard deviation of 2 minutes. Workers move at the rate of 100 feet per minute. There is an unlimited queue at the input station, but queues before the mill and lathe are limited to two jobs in each instance. Simulate the operation of this system for 100 hours. Develop measures that indicate the utilization of the two workers according to what task they were performing. Also, indicate the time in the system for jobs and the average time spent waiting for transportation.

E9.6 Trucks arrive at a loading dock according to an exponential distribution with a mean of 30 minutes. Each truck carries five pallets of stock. There are three dock workers. When a truck arrives, a forklift is called, and the workers unload the truck. Once completely unloaded, the workers then load the stock onto the forklift. Unloading a truck takes a uniformly distributed amount of time between 6 and 14 minutes. Loading a forklift takes a uniformly distributed amount of time between 2 and 5 minutes. The forklift carries its load into the warehouse and deposits it in a staging area for later placement. The distance from the dock to the staging area is 240 feet. The forklift moves at a rate of 70 feet per minute. The dock can only handle 20 pallets at a time. There are currently two forklifts available. Determine if the dock manager should order a new forklift to help keep the dock clear.

10

Conveyors

In the previous chapter, the mobile transporter was used for moving parts between locations. Another common method of transport is the conveyor, a fixed-path device for moving objects from point to point. In this chapter, we will discuss the different types of conveyors that can be modeled and the required corresponding SIMAN V blocks and elements.

10.1 The Conveyor Concept

A conveyor is a type of material handling device in which entities are moved from one location to another, following a fixed path with specific load and unload points. A conveyor can be thought of as a continuous flow of cells that are used to transport parts, packages, or other discrete entities. Examples of some common categories of conveyors are belt, roller, bucket, chain, and power-and-free. A conveyor system can utilize many individual conveyors to perform the material handling function.

In SIMAN V, a conveyor is defined in terms of its velocity, load-size, and load/unload points. Turns, inclines, or declines do not affect the modeling strategy. The conveyor specifications and path specifications are provided in the experiment frame. In the model frame, the movement of the entities follows this pattern: The entities are loaded onto a conveyor, conveyed to their destination, and unloaded from the conveyor. In addition, the operational status (active or inactive) of a conveyor can be controlled in the model frame. In the experiment frame, conveyor specifications are made and the path is defined. The path consists of segments between load and unload points. Each segment is defined as a continuous section between each transfer point.

In SIMAN V, there are two types of conveyors that can be modeled: accumulating and nonaccumulating. Both are unidirectional, that

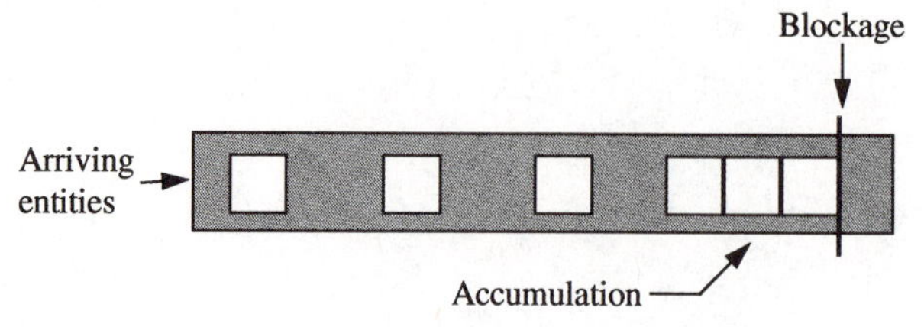

Figure 10.1 Entities on an accumulating conveyor.

is, they can only move in one direction and cannot be reversed. An accumulating conveyor runs continuously. If the forward progress of an entity is halted while on an accumulating conveyor, slippage occurs allowing the entity to remain stationary. As depicted in Figure 10.1, when a part is stopped on the conveyor, it creates a local blockage that prevents all other parts from being loaded at, or passing through, that section of the conveyor. Therefore, parts accumulate behind that point. When parts accumulate, any space between adjacent parts behind the blockage will disappear.

Once the blockage has been removed, the accumulated parts will begin moving again. For example, consider a belt conveyor that continuously moves parts from production into a warehouse. The parts on the conveyor can back up at the end of the conveyor until they are unloaded.

On nonaccumulating conveyors, the spacing between adjacent parts remains fixed. When a part is loaded onto this type of conveyor, the entire conveyor is stopped (disengaged) until the command to convey that part is given. Similarly, when a part arrives at its final destination, the entire conveyor is stopped again until that part is unloaded from the conveyor. An example is a bucket conveyor that moves a group of parts simultaneously. When a bucket reaches its destination, the entire conveyor stops, the desired part(s) are unloaded, and the conveyor resumes operation. To model a nonstop moving conveyor that does not allow for accumulation, model the conveyor as a nonaccumulating conveyor with a zero time delay for loading and unloading.

EXAMPLE 10.1 Problem Statement

Example 8.2 is now modified to include conveyors that will transport the jobs between the stations. The conveyor layout for this example is shown in Figure 10.2. The ArrConv and ExitConv conveyors are point-

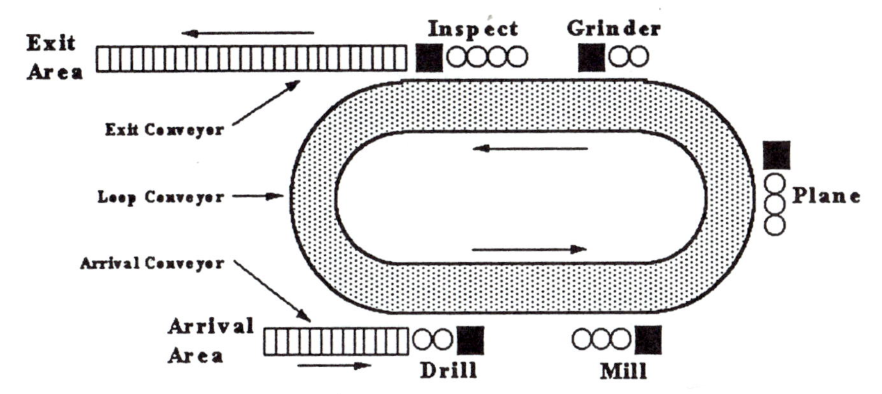

Figure 10.2 Conveyor layout for Example 10.1.

to-point accumulating conveyors, whereas LoopConv is a recirculating loop conveyor that does not allow for entity accumulation.

The conveyor specifications are as follows:

ArrConv

Type	Accumulating
Cell Size	10 feet
Velocity	25 feet per minute
Initial Status	Active
Maximum Part Size	2 feet

LoopConv

Type	Nonaccumulating
Cell Size	10 feet
Velocity	30 feet per minute
Initial Status	Active
Maximum Part Size	2 feet

ExitConv

Type	Nonaccumulating
Cell Size	10 feet
Velocity	45 feet per minute
Initial Status	Active
Maximum Part Size	2 feet

Table 10.1 Blocks/Elements Introduced in this Chapter

New Block/Element	Purpose
ACCESS block	Loads the entities onto the conveyor so they can be moved to the next station
CONVEY block	Moves the entities from the current station to the destination station using the conveyor
EXIT block	Unloads the entities from the conveyor upon arrival at the destination station
START block	Changes the status of a conveyor to active
STOP blcck	Changes the status of a conveyor to inactive
CONVEYORS element	Defines the characteristics of the conveyors that comprise the system
SEGMENTS element	Defines the paths used by the conveyors to move parts between the various stations

The distances associated with these conveyors are as follows:

Arrival Area to Drill	60 feet
Drill to Mill	70 feet
Mill to Plane	90 feet
Plane to Grinder	50 feet
Grinder to Inspect	180 feet
Inspect to Drill	250 feet
Inspect to Exit Area	100 feet

The blocks and elements introduced in this chapter for modeling conveyor systems are shown in Table 10.1.

10.2 Model Frame Conveyor Blocks

In SIMAN V, three basic steps are performed when modeling conveyors. Each step has a corresponding SIMAN V block. First, the entity is placed on the conveyor when there is enough available space (ACCESS block). Second, the entity is moved to the next station on the conveyor (CONVEY block). Third, the entity is removed from the conveyor (EXIT block). However, if an entity is to remain on the conveyor while being processed at a station, then the EXIT block should not be used.

Two other blocks that are used when modeling conveyors are START and STOP. These blocks are used to control the operational status of the conveyor. They can be used to model conveyor failures and other types of operational interruptions.

10.2.1 The ACCESS Block

The ACCESS block is used when an entity is to be placed on a conveyor. The ACCESS block must be preceded by a QUEUE block. The format for the ACCESS block is as follows:

ACCESS: *Conveyor Name, Quantity to Access*;

The operands are described as follows:

Operand	Description	Default
Conveyor Name	The name of the conveyor to be accessed [symbol name]	—
Quantity to Access	Number of consecutive cells to access [expression truncated to an integer]	1

When an entity arrives at the ACCESS block, it tries to access *Quantity to Access* empty consecutive cells of the conveyor called *Conveyor Name*. If *Quantity to Access* cells are unavailable, then the entity waits in the preceding queue until the required number of empty cells becomes available.

For a nonaccumulating conveyor, the conveyor *Conveyor Name* is disengaged when an entity accesses the conveyor. The conveyor is engaged when the accessing entity executes a CONVEY or EXIT block.

For an accumulating conveyor, the conveyor is not disengaged when an entity accesses the conveyor. Instead, the entity remains stationary at the loading point, creating a blockage, until a CONVEY or EXIT block is executed by the accessing entity.

If multiple entities try to access the same cells at the same time, SIMAN V selects the entity with the lowest numbered queue for the first ACCESS block in the model. The process continues for each ACCESS block in the model until an entity awaiting access is encountered.

An example using the ACCESS block is as follows:

```
QUEUE,
     LoopConvQ;
ACCESS:
     LoopConv, 2;
```

In this example, the conveyor *LoopConv* is nonaccumulating. The entity will wait in the queue *LoopConvQ* until two empty adjacent cells of *LoopConv* arrive at the current station. When empty, the conveyor, *LoopConv*, will stop and the entity will be placed on the conveyor.

10.2.2 The CONVEY Block

The CONVEY block is used to move an entity from the access (load) point to the destination (unload) point. The format for the CONVEY block is as follows:

CONVEY: *Conveyor Name, Destination Station*;

The operands are described as follows:

Operand	Description	Default
Conveyor Name	The name of the conveyor [symbol name]	Conveyor that the entity accessed
Destination Station	Destination station [expression truncated to an integer station number, station symbol name, or SEQ]	SEQ

After an entity has accessed the conveyor *Conveyor Name*, the entity can then be moved with the CONVEY block. Upon entering the CONVEY block, SIMAN V first evaluates the *Destination Station* value; it then sets the entity's station attribute, M, to the value of the destination. The entity then begins movement from the access point to the destination point using the conveyor specified by *Conveyor Name*. The default value of *Destination Station* is SEQ, which indicates the use of a sequence set to control the movement of an entity.

In some cases, there will be a zero delay between the time that the conveyor is accessed (i.e., loaded onto the conveyor) and the time that the entity begins to move. However, if a loading time is desired, the placement of a DELAY block between the ACCESS block and the CONVEY block can introduce the necessary time. For example:

```
QUEUE,
    BeltConv1Q;
ACCESS:
    BeltConv1, 3;
DELAY:
    3;
CONVEY:
    BeltConv1, SEQ;
```

In the above example, the entity will be delayed 3 time units after it begins accessing the conveyor and before it begins its movement.

For a nonaccumulating conveyor, the entity moves at a specified velocity from the load point to the unload point, unless the conveyor is stopped by the loading or unloading of another entity or other operational interruptions. The entity will next appear at the STATION block corresponding to the station number specified by the *Destination Station* value. The conveyor is disengaged when the entity reaches its final destination.

For an accumulating conveyor, the entity will begin moving immediately, unless there is a delay between the ACCESS block and the CONVEY block. Since an accumulating conveyor is not stopped by the loading or unloading of entities, the travel of an entity along the conveyor is only interrupted by local blockages or an accumulation point. When a blockage is cleared, the blocked entities will begin moving again. As with the nonaccumulating conveyor, the entity will reappear only when it reaches the STATION block specified by the *Destination Station* value. When the entity reaches its destination, the entity's movement will be stopped, but the conveyor will continue to move, creating a local blockage until the entity is removed from the conveyor.

An example using the CONVEY block is as follows:

```
QUEUE,
    LoopConvQ;
ACCESS:
    LoopConv, 2;
CONVEY:
    LoopConv, SEQ;
```

Once the *LoopConv* conveyor is accessed and the entity is loaded, the conveyor will restart when the CONVEY block is executed. The entity then begins its movement to the next station as specified in the SEQUENCES element.

10.2.3 The EXIT Block

The EXIT block is used to unload an entity from a conveyor. The format for the EXIT block is as follows:

EXIT: *Conveyor Name, Quantity to Release*;

The operands are described as follows:

Operand	Description	Default
Conveyor Name	The name of the conveyor to exit [symbol name]	Conveyor that the entity accessed
Quantity to Release	Number of consecutive cells to release [expression truncated to an integer]	All accessed cells

When an entity executes the EXIT block, *Quantity to Release* cells of the conveyor *Conveyor Name* are released. This means that the status of these cells is changed from occupied to empty so that they can be accessed by other entities.

On a nonaccumulating conveyor, when an entity arrives at the destination STATION block specified in the CONVEY block, the conveyor is disengaged. The EXIT block is then used to remove the entity from the conveyor and re-engage the conveyor motion. This will be an instantaneous procedure unless a DELAY block is inserted between the STATION block and the EXIT block. Such a delay may be necessary in modeling a conveyor system that requires unloading times.

With an accumulating conveyor, a local blockage is created when an entity arrives at the station block. The EXIT block removes the entity from the conveyor, thus eliminating the local blockage. As with the nonaccumulating conveyor, a DELAY block between the STATION block and the EXIT block can model unloading times.

Examples of the EXIT block are as follows:

```
STATION:
     SPlane;
EXIT;
```

Entities arriving at station *SPlane* will immediately enter the EXIT block. This will release all the cells that the entity accessed on the conveyor.

```
STATION,
    SPlane;
EXIT:
    LoopConv, 2;
```

This example releases two cells of conveyor *LoopConv* after arriving at station *SPlane*.

```
STATION,
    Paint;
DELAY:
    Unload;
EXIT:
    Overhead;
```

In this example, once the part arrives at its destination station called *Paint*, it will be delayed for a time defined by *Unload*. After this time, the part will be removed from the *Overhead* conveyor and all cells will be released. The conveyor will be re-engaged and the blockage will be cleared.

10.2.4 The START Block

The START block is used to change the status of a conveyor to **active** in the model frame. The format for the START block is as follows:

START: *Conveyor Name, Velocity;*

The operands are described as follows:

Operand	Description	Default
Conveyor Name	Name of the conveyor to activate [symbol name]	—
Velocity	Conveyor velocity [expression]	No change

The START block is used to restart a conveyor that was halted using the STOP block or to activate a conveyor that is initially idle. In addition to changing the operational status of a conveyor, the START block can be used to change the velocity of a conveyor. If the operand *Velocity* is specified, the velocity of the conveyor will be reset to this value; otherwise, the velocity will remain at its current value. The value for the current conveyor velocity is given by the SIMAN V variable *VC(Conveyor Name)*. Section 10.4 discusses further details concerning *VC(Conveyor Name)*.

Examples of the START block are as follows:

```
START:
    Roller;
```

This restarts the conveyor *Roller*.

```
START:
    Belt, 20;
```

This block will start the *Belt* conveyor with a velocity of 20 distance units per time unit.

10.2.5 The STOP Block

The STOP block is used to set the status of a conveyor to inactive. The format for the STOP block is as follows:

STOP: *Conveyor Name*;

The operand is described as follows:

Operand	**Description**	**Default**
Conveyor Name	Name of conveyor to deactivate [symbol name]	—

When an entity enters the STOP block, the conveyor *Conveyor Name* is deactivated and its movement is stopped. The only way to restart the conveyor is to use the START block. The STOP block is used in conjunction with the START block to model conveyor failures and other operational interruptions.

An example of the STOP block is as follows:

```
STOP:
    Roller;
```

This block deactivates the conveyor *Roller*.

10.3 Experiment Frame Conveyor Elements

There are two experiment frame elements that are required when using conveyors: the CONVEYORS and SEGMENTS elements. The CONVEYORS element is used to define the physical characteristics of a conveyor. The SEGMENTS element is used to define the path characteristics of the conveyor.

10.3.1 The CONVEYORS Element

The CONVEYORS element is used to define the characteristics of the conveyor. The format for the CONVEYORS element is as follows:

> CONVEYORS: *Number, Name, Segment Set ID, Velocity, Cell Size, Status, Max Cells Per Entity, Type, Accumulation Length*: repeats;

The operands are described as follows:

Operand	Description	Default
Number	Conveyor number [integer]	Sequential
Name	Conveyor name [symbol name]	—
Segment Set ID	Segment set name or number [integer or symbol name]	—
Velocity	Conveyor velocity [constant]	1.0
Cell Size	Length of each conveyor cell [integer]	1
Status	Initial status of the conveyor [Active or Inactive]	Active
Max Cells Per Entity	Maximum number of cells occupied by any entity [integer]	1
Type	Conveyor Type [Accumulating or Nonaccumulating]	Nonaccumulating
Accumulation	Accumulation length of an entity [constant, attribute name, or attribute number]	Cell size

In the model frame, conveyors are always referenced by their name. Therefore, the *Name* operand is required since the number operand will not suffice. A segment is the path between two stations. The *Segment Set ID* refers to the complete path defined in the SEGMENTS element that the conveyor follows.

The initial velocity of the conveyor is specified by the *Velocity* operand. This velocity can be changed in the model in two ways: First, by assigning a new value to the variable *VC(Conveyor Name)* as discussed in Section 10.4; or second, by specifying a different velocity with a START block.

The *Cell Size* of a conveyor is the smallest portion of a conveyor that an entity can occupy. It is defined in the same distance units given in other conveyor parameters. The *Cell Size* determines the minimum spacing of parts on the conveyor. *Max Cells Per Entity* is the maximum number of cells that a part will occupy.

The *Status* operand denotes the initial operational status of the conveyor. The conveyor will either be active or inactive. To specify active, enter A or Active. To specify inactive, enter I or Inactive. In the model frame, the operational status of the conveyor can be changed, using the START and STOP blocks.

The *Type* operand specifies whether a conveyor is accumulating or nonaccumulating. To specify an accumulating conveyor, the operand can either be A or Accum. For a nonaccumulating conveyor, enter N or Nonaccum.

The *Accumulation Length* operand is only used for an accumulating conveyor. This is the distance that a part occupies when it is accumulated. This operand can be specified as a constant if all the parts have similar accumulation lengths or as an attribute name or number (*A(k)*).

The CONVEYORS element that will be used for Example 10.1 is as follows:

```
CONVEYORS:      1, ArrConv, 1, 25, 10, A, 2, A:
                2, LoopConv, 2, 30, 10, A, 2, N:
                3, ExitConv, 3, 45, 10, A, 2, N;
```

This element defines three conveyors. The first conveyor, *ArrConv*, is an accumulating conveyor that follows segment set 1. It has an initial velocity of 25 feet per minute and has a cell size of 10 feet. Initially, the conveyor is active and the maximum number of cells that a part will use is 2. The second conveyor, *LoopConv*, is a nonaccumulating conveyor that follows segment set 2. It has an initial velocity of 30 feet per minute and has a cell size of 10 feet. Initially, the conveyor is active and the maximum number of cells that a part will use is 2. The third conveyor, *ExitConv*, is a nonaccumulating conveyor that follows segment set 3. It has an initial velocity of 45 feet per minute and has a cell size of 10 feet. Initially, the conveyor is active and the maximum number of cells that a part will use is 2.

10.3.2 The SEGMENTS Element

The SEGMENTS element defines the segments that make up the *conveyor path*. A conveyor path is described as a series of connected segments in which each segment is a directed link between two stations. The format of the SEGMENTS element is as follows:

SEGMENTS: *Identifier, Beginning Station,*
Next Station-length, ...: repeats;

The operands are described as follows:

Operand	Description	Default
Identifier	Segment set *Number* or *Name* [integer or symbol name]	Sequential
Beginning Station	Beginning station identifier [integer station number or station symbol name]	—
Next Station	Next station identifier [integer station number or station symbol name]	—
Length	Distance from the previous station [integer]	—

The segment set is defined by the operand *Identifier*. Each segment set is composed of multiple station pairs. Starting with the *Beginning Station*, the next station on the path is given by *Next Station*. The *Length* operand is the distance between these stations. The next station is then specified with the corresponding *Length*. This continues until all the stations on each conveyor path (segment set) are specified.

The *Length* operand must be specified in the same distance units used to define the conveyor in the CONVEYORS element. This length must be evenly divisible by the *Cell Size* specified in the CONVEYORS element.

If a conveyor forms a "straight line," meaning that it has unique starting and stopping points and it does not complete a circuit, then there should be no repeating of station identifiers. However, if the conveyor forms a loop, then the beginning station identifier must be the same as the last station identifier.

The SEGMENTS element that will be used for Example 10.1 is as follows:

```
SEGMENTS:    1, SArrive, SDrill-60:
             2, SDrill, SMill-70,
             SPlane-90,
             SGrinder-50,
             SInspect-180,
             SDrill-250:
             3, SInspect, SDone-100;
```

In this element, segment set 1 defines a straight line path with 60 feet between the *SArrive* station and the *SDrill* station. Segment set 2 defines a loop path from *SDrill* to *SMill* to *SGrinder* to *SInspect* and back to *SDrill*. The distance between *SDrill* and *SMill* is 70 feet, the distance between *SMill* and *SPlane* is 90 feet, and so on. Notice that *SDrill* is specified again at the end of this segment set, thus creating a looping path. Segment set 3 defines a straight line path between *SInspect* and *SDone* with a total distance of 100 feet.

10.4 SIMAN Conveyor Variables

SIMAN provides several variables related to conveyor operation. These variables can be used to gather statistical data or to obtain information while in the model. Table 10.2 lists these variables and their definitions.

In Example 10.1, we are asked to measure the utilization of each conveyor. Conveyor utilization is calculated by dividing the length of entities being conveyed by the total length of the conveyor. In SIMAN, this statistical data will be collected using the DSTATS element. In the experiment frame, the following DSTATS element will be included:

```
DSTATS:    1, LEC(ArrConv)/MLC(ArrConv), ArrConv Util.:
           2, LEC(LoopConv)/MLC(LoopConv), LoopConv Util.:
           3, LEC(ExitConv)/MLC(ExitConv), ExitConv Util.;
```

This will collect the utilization of each conveyor in the system.

Table 10.2 SIMAN Conveyor Variables

Variable	Definition
NEC(ConvName)	Number of entities currently being conveyed
NEA(ConvName)	Number of accumulated entities
LEC(ConvName)	Total length of entities currently being conveyed
CLA(ConvName)	Total length of entities currently accumulated
MLC(ConvName)	Total conveyor length
ICS(ConvName)	Conveyor status [1-Active 0-Inactive]
VC(ConvName)	Conveyor velocity

EXAMPLE 10.1 Solution

The Model Frame

The following shows the model frame for Example 8.2 modified to include the changes specified by Example 10.1. The changes, shown in italics, have been noted.

```
BEGIN;
          CREATE:                   !Create arriving jobs according to
            ED(1):                  !distribution ED(1) and save their
            MARK(ArrTime);          Arrival times in attribute ArrTime
          ASSIGN:
            JobType=ED(2):          !Define job as Type I or Type II
            NS=JobType:             !Define SEQUENCE by JobType
            M=SArrive;              Define current station as SArrive
```

Once at the station *SArrive*, the part is loaded on the conveyor *ArrConv*, using the following QUEUE and ACCESS blocks.

```
  QUEUE,
    ArrConvQ;    Wait for available conveyor space
  ACCESS:
    ArrConv, 2;  Load job onto ArrConv
```

Next, the part is conveyed to station *SDrill* as defined in the SEQUENCES element by executing the CONVEY block.

```
         CONVEY:
         ArrConv, SEQ;                 Convey by SEQUENCES
Process  STATION, SDrill-SInspect;     Stations SDrill to SInspect
```

After arriving at a processing station *(SDrill, SMill, SPlane, SGrinder,* or *SInspect)*, the part is removed from the conveyor, using the following EXIT block.

```
  EXIT;   Remove job from conveyor
```

The jobs are then processed using previously defined logic as follows:

```
QUEUE,                       !Queue number is given by M
  M,V(M), Eject;             Queue capacity is given by V(M)
SEIZE:
  Machine(M);                Capture indexed resource M
DELAY:
  ProcessTime;               Process time in SEQUENCES
RELEASE:
  Machine(M);                Free indexed resource M
BRANCH, 1, 10:               !Take one branch only
  IF, M.LE.4, Move_Job:      !If at first 4 stations continue
                             !else, when at Inspection station,
  WITH, PRedo, Rework:       !Rework with probability PRedo
  WITH, PFail, Scrap:        !Scrap with probability PFail
  ELSE, Done;                !Job is complete
```

After processing, the part is loaded on the conveyor *LoopConv* and conveyed to the next processing station as determined by the SEQUENCES element using the following QUEUE, ACCESS, and CONVEY blocks:

```
Move_Job    QUEUE,
              LoopConvQ;        Wait for available conveyor space
            ACCESS:
              LoopConv, 2;      Load job onto 2 cells of LoopConv
            CONVEY:
              LoopConv,         Convey by SEQUENCES
              SEQ;
```

Once the part has completed all its processing, it is loaded on the conveyor *ExitConv* and conveyed to the *SDone* station defined on the sequence with the following QUEUE, ACCESS, and CONVEY blocks:

```
Done        QUEUE,
              ExitConvQ;        Wait for conveyor space
            ACCESS:
              ExitConv, 2;      Load onto 2 cells of ExitConv
            CONVEY:
              ExitConv, SEQ;    Convey by SEQUENCES
            STATION,            !SDone is the station for
              SDone;            completed jobs
```

After arriving at the *SDone* station, the part is removed from the conveyor, using the following EXIT block:

```
         EXIT;                        Remove job from conveyor
         TALLY:
           Flowtime, INT(ArrTime);   Tabulate flowtime
         TALLY:
           Exit Period, BET;          Tab time between exits
         COUNT:
           JobType, 1:                !Count jobs by type and
           DISPOSE;                   destroy entity
Rework   COUNT:
           Reworks;                   Count jobs reworked
         ASSIGN:
           IS=0:                      !Restart SEQUENCES
           NEXT(Move_Job);            Move job to drilling
Scrap    COUNT:
           Fails:                     !Count jobs scrapped and
           DISPOSE;                   destroy entity
Eject    COUNT:
           Ejections:                 !Count jobs ejected and
                                      destroy entity
           DISPOSE;
```

The Experiment Frame

Exhibit 10.1 shows the corresponding experiment frame for Example 10.1. Changes to Example 8.2 include the CONVEYORS and SEGMENTS elements. In addition, the QUEUES and DSTATS elements have been modified to include the additional conveyor access queues and to collect the statistics of interest, respectively. The additions and changes are shown in italics.

Discussion of Output

Exhibit 10.2 shows the results from one replication from the output file for Example 10.1. The DSTATS variables show the utilization of each conveyor. For example, the conveyor *ArrConv* was utilized 0.1235 or 11.88% of the time. In addition, notice that the flowtime for the parts has increased from Example 8.2. This is due to the additional time required for material handling.

EXHIBIT 10.1 Experiment Frame for Example 10.1.

```
BEGIN;
PROJECT, Example 10.1, Team;
ATTRIBUTES:   ProcessTime:   !First attribute in SEQUENCES
              PRedo:         !Second attribute in SEQUENCES
              PFail:         !Third attribute in SEQUENCES
              ArrTime:
              JobType;
VARIABLES:    1, MDrill, 2:
              2, MMill, 3:
              3, MPlane, 3:
              4, MGrinder, 2:
              5, MInspect, 4;
STATIONS:     SDrill:
              SMill:
              SPlane:
              SGrinder:
              SInspect:
              SArrive:
              SDone;
RESOURCES:    Drill, 2:
              Mill, 3:
              Plane, 3:
              Grinder, 2:
              Inspector;
SETS:         MACHINE, Drill..Inspector; Machine(1) thru Machine(5)
QUEUES:       DrillQ:
              MillQ:
              PlaneQ:
              GrinderQ,
              LVF(Job Type): !Priority to JobType 1
              InspectorQ:
              ArrConvQ:      !Buffer for ArrConv requests
              LoopConvQ:     !Buffer for LoopConv requests
              ExitConvQ;     Buffer for ExitConv requests
CONVEYORS:    1, ArrConv, 1, 25, 10, A, 2, A:
              2, LoopConv, 2, 30, 10, A, 2, N:
              3, ExitConv, 3, 45, 10, A, 2, N;
SEGMENTS:     1, SArrive, SDrill-60:
              2, SDrill, SMill-70,
              SPlane-90,
              SGrinder-50,
              SInspect-180,
              SDrill-250:
              3, SInspect, SDone-100;
```

EXHIBIT 10.1 *continued.*

```
COUNTERS:     Type I Job Count:
              Type II Job Count:
              Reworks:
              Fails:
              Ejections;
SEQUENCES:    1, TypeIPart, SDrill, ProcessTime=ED(3), &
              SMill, ProcessTime=ED(4), &
              SGrinder, ProcessTime=ED(6), &
              SInspect, ProcessTime=ED(8), PRedo=.05, PFail=.05 &
              SDone:
              2, TypeIIPart, SDrill, ProcessTime=ED(3), &
              SPlane, ProcessTime=ED(5), &
              SGrinder, ProcessTime=ED(7), &
              SInspect, ProcessTime=ED(8), PRedo=.05, PFail=.05 &
              SDone;
EXPRESSIONS:  1,,EXPO(5,1):           !Interarrival time
              2,,DISC(0.7,1,1.0,2,1): !Job Type distribution
              3,,UNIF(6,9,1):         !Drilling Time
              4,,TRIA(12,16,20,1):    !Milling time
              5,,TRIA(20,26,32,1):    !Planing time
              6,,DISC(0.25,6,0.75,7,1.0,8,1):
                                      !Grinding time Type I jobs
              7,,DISC(0.1,6,0.35,7,0.65,8,0.9,9,1.0,10,1):
                                      !Grinding time Type II jobs
              8,,NORM(3.6,0.6,1);     Inspection time
TALLIES:      Flowtime:
              Exit Period;
DSTATS:       NQ(DrillQ), Drill Queue:
              NQ(MillQ), Mill Queue:
              NQ(PlaneQ), Plane Queue:
              NQ(GrinderQ), Grinder Queue:
              NQ(InspectorQ), Inspect Queue:
              NR(1)/2, Drill Utilization:
              NR(2)/3, Mill Utilization:
              NR(3)/3, Plane Utilization:
              NR(4)/2, Grinder Utilization:
              NR(5), Inspect Utilization:
              LEC(1)/MLC(1), ArrConv Util.:
              LEC(2)/MLC(2), LoopConv Util.:
              LEC(3)/MLC(3), ExitConv Util.;
REPLICATE, 5, 0, 2400;
```

EXHIBIT 10.2 **Summary report for Example 10.1.**

Summary for Replication 1 of 5

Project: Example 10.1 Run execution date: 8/24/1994
Analyst: Team Model revision date: 8/24/1994
Replication ended at time: 2400.0

TALLY VARIABLES

Identifier	Average	Variation	Minimum	Maximum	Observations
Flowtime	62.533	.22799	47.798	176.07	392
Exit Period	5.9963	.63228	1.7777	23.563	391

DISCRETE-CHANGE VARIABLES

Identifier	Average	Variation	Minimum	Maximum	Final Value
Drill Queue	.37256	1.7201	.00000	2.0000	.00000
Mill Queue	.15744	2.5637	.00000	2.0000	.00000
Plane Queue	.06513	4.1558	.00000	2.0000	.00000
Grinder Queue	.19807	2.2685	.00000	2.0000	.00000
Inspect Queue	.13647	2.5766	.00000	2.0000	.00000
Drill Utilization	.70847	.53015	.00000	1.0000	.00000
Mill Utilization	.69492	.45590	.00000	1.0000	.66667
Plane Utilization	.47580	.72288	.00000	1.0000	.66667
Grinder Utilization	.67106	.54223	.00000	1.0000	1.0000
Inspect Utilization	.65239	.72995	.00000	1.0000	.00000
ArrConv Util.	.11879	1.3680	.00000	.75000	.50000
LoopConv Util.	.07688	.49209	.00000	.18750	.12500
ExitConv Util.	.06052	1.3254	.00000	.33333	.16667

COUNTERS

Identifier	Count	Limit
Type I Job Count	271	Infinite
Type II Job Count	121	Infinite
Reworks	17	Infinite
Fails	27	Infinite
Ejections	43	Infinite

10.5 SUMMARY

In this chapter, the tools used in modeling conveyors with SIMAN V have been discussed. SIMAN V can model both accumulating and nonaccumulating conveyors. In both cases, the modeling strategy is composed of three steps: Accessing the conveyor (loading), moving the part, and exiting the conveyor (unloading). For a nonaccumulating conveyor, the entire conveyor is stopped when loading or unloading a part. On an accumulating conveyor, such as a belt, only the motion of the part is halted during loading and unloading; the conveyor continues its motion. The halted part creates a local blockage at the loading/unloading point, until it is conveyed/unloaded. The ACCESS, CONVEY, and EXIT blocks are used to control part loading, movement, and unloading, respectively. The START and STOP blocks control conveyor status. The CONVEYORS element defines all the conveyors and their characteristics that are used in the model. The SEGMENTS element defines the conveyor paths. The SIMAN variables used for specific conveyor elements are discussed in Section 10.4. These variables can be used to control, change, or collect statistics of various conveyor parameters.

10.6 EXERCISES

E10.1 Two independent conveyors deliver completed parts to a warehouse. In the warehouse, these conveyors merge placing the parts on a third conveyor to final shipping. Parts arriving on conveyor 1 follow an exponential distribution with a mean of 8 parts per minute. Parts arriving on conveyor 2 also follow an exponential distribution with a mean of 10 parts per minute. Conveyor 1 is nonaccumulating and is 15 feet long. It runs at a velocity of 15 feet per minute. Conveyor 2 is accumulating and is 8 feet long. Its velocity is 25 feet per minute. Conveyor 3 is nonaccumulating and is 10 feet long. It runs at a velocity of 30 feet per minute. Consider the parts as being disposed at the end of the last conveyor. Simulate this system for 250 minutes and find the average time that a part must wait to access conveyor 3 at the merge point. Assume that all parts are 1 foot long.

E10.2 A shipping department handles two types of packages: regular and express delivery. A 20-foot-long accumulating conveyor running at 25 feet per minute delivers the packages to the shipping department. The packages arrive at a mean rate of five per minute following an exponential distribution. Thirty percent of the packages require express delivery, the rest use regular delivery. At the end of the conveyor, regular packages are transferred to a 10-foot-long, nonaccumulating conveyor running at 20 feet per minute. Express delivery packages are transferred to a 10-foot-long accumulating conveyor running at

30 feet per minute. Consider the packages as being disposed at the ends of these last two conveyors. Regular delivery packages are 2 feet long and express delivery packages are 1 foot long. Simulate this conveyor system for 8 hours and determine the utilization of each of the conveyors.

E10.3 This assignment is a modification of Exercise 8.3. This exercise will use conveyors as the material handling device between stations. The system uses two conveyors. The first conveyor takes the incoming jobs from their arrival location to the three parallel stations (1, 2, or 3). The second conveyor takes the jobs from these stations through the remaining three stations and on to the warehouse. The distances between stations are listed as follows:

Conveyor 1	
Arrival to Station 1	20 feet
Station 1 to Station 2	10 feet
Station 2 to Station 3	10 feet

Conveyor 2	
Station 1 to Station 2	10 feet
Station 2 to Station 3	10 feet
Station 3 to Station 4	10 feet
Station 4 to Station 5	10 feet
Station 5 to Station 6	10 feet
Station 6 to warehouse	10 feet

Both conveyors are nonaccumulating and run at a velocity of 40 feet/minute. The jobs occupy 1 foot of conveyor length. Simulate the system for 3 days of continuous production and determine the following:

a. What is the utilization at each of the processing stations?

b. What is the utilization of the two conveyors?

c. What is the average number of jobs waiting for a conveyor at each station?

E10.4 An automotive assembly line uses an accumulating conveyor to move automobile frames through the engine mounting process. The engine mounting process has four areas: Preparation, Welding, Mating, and Electrical Connection. As each automobile frame arrives in each area, it is stopped and the necessary work is performed. All work is performed while the frame remains on the conveyor. Other frames entering the area will accumulate behind the frame being processed. (This problem describes the operation of a power-and-free conveyor.) The frames enter the engine mounting process according to an exponential distribution with a mean of 16 minutes. Each area has one team of workers to process the frame. The time for each area in the engine mounting process is as follows:

Preparation	Normal(13, 2.1) minutes
Welding	Uniform(12, 16) minutes
Mating	Normal(11, 1.8) minutes
Electrical Connection	Triangular(12, 13, 16) minutes

The transportation distance between processing areas is as follows:

Arrival to Preparation	140 feet
Preparation to Mount Welding	200 feet
Mount Welding to Engine Mating	250 feet
Engine Mating to Electrical Connection	180 feet

The frames are removed from the conveyor after completing the engine mounting process. The frames are 14 feet long and will accumulate with 4 feet of space between each frame. The conveyor is initially active and operates at a constant speed of 30 feet per minute. Simulate the system for 3 days of continuous production and determine the following:

a. What is the average time that a frame spends in the engine mounting process?

b. What is the utilization of each processing area?

c. How many completed frames are produced?

d. On the average, how many frames are on the power-and-free conveyor at any one time?

E10.5 A paint shop that consists of four processing areas (Cleaning, Preparation, Paint Booth 1, and Paint Booth 2) is fed by a nonaccumulating conveyor. Two colors (red and blue) are processed through this system. An average of 70% are red and 30% are blue. Red parts are cleaned, prepared, and painted in Paint Booth 1. Blue parts are cleaned, prepared, and painted in Paint Booth 2. At each processing step, the parts are removed from the conveyor and processed. Parts that are unable to be processed immediately will be placed in buffers adjacent to each processing area. After processing is completed, all parts are transferred to a 60-foot accumulating conveyor that transports them to the warehouse. Information about the processing areas (with times in minutes) is as follows:

Area	Number of Workers	Processing Time
Cleaning	1	Normal(5, .6)
Preparation	2	Normal(12, 2)
Paint Booth 1	1	Normal(18, 3)
Paint Booth 2	1	Normal(15, 2.4)

The travel distance between processing areas is as follows:

Arrival to Cleaning	15 feet
Cleaning to Preparation	20 feet
Preparation to Paint Booth 1	30 feet
Paint Booth 1 to Paint Booth 2	30 feet
Paint Booth 2 to Transfer Point	10 feet

Both parts require 3 feet of conveyor space. Assume that both conveyors are initially running with velocities of 5 feet per minute for the processing conveyor and 10 feet per minute for the warehouse conveyor. The paint shop processes 250 parts a day. Simulate the system for 5 days of production and determine the following:

a. What is the average time to complete 1 day of production?
b. On average, how many parts are waiting to be processed in each processing area?

E10.6 A 140-foot, nonaccumulating loop conveyor is used to feed jobs to seven workers that are evenly spaced around a loop. Jobs are loaded onto the conveyor 5 feet prior to worker #1. An incomplete job will continue to move around the conveyor until an idle worker is encountered. Once encountered, the job is unloaded from the conveyor, processed, and the completed job is reloaded on the conveyor. Completed jobs are unloaded from the conveyor 5 feet after worker #7. Jobs arrive exponentially with a mean of 2.2 minutes. Job processing is normally distributed with a mean of 14 minutes and a variance of 3.6 minutes^2 for all workers. The loop conveyor has a velocity of 15 feet per minute. Arriving jobs wait in a storage buffer until space on the loop conveyor becomes available. In addition, the conveyors break down according to an exponential distribution with a mean time of 9 hours. Time to repair is also exponentially distributed with a mean of 7 minutes. Simulate for 1 day of continuous production. Determine the following:

a. What is the utilization rate of the loop conveyor?
b. What is the average number of jobs on the loop conveyor?
c. How long does a typical job spend in the system?
d. What is the utilization of each worker?

E10.7 A post office uses an automatic address reader to determine envelope routing. Envelopes are read when they reach the 12-foot point on an accumulating conveyor that has a speed of 5 inches per second. Both the conveyor and reader can handle two sizes of envelopes, 6 inches and 9 inches. Envelopes arrive according to an exponential distribution with a mean of 6.5 seconds; 60% are 6 inches long and 40% are 9 inches long. The envelopes accumulate on the conveyor with zero spacing. The time for the automatic reading process is normally distributed with a mean and variance of 4 and 0.5 seconds. After the automatic reader, the envelopes continue on the conveyor for an additional 18 feet before being transferred to the sortation system. There is a 20% chance that the automatic reader will fail to read the address. If a failure occurs, the failed envelope is transferred from the automatic reader to another accumulating conveyor, which leads to a manual reading system. This conveyor is 15 feet long and has a velocity of 1.5 inches per second. The time for manual reading is uniformly distributed between 15 and 21 seconds for all envelopes. The envelopes travel another 6 feet to the sortation system after being manually read.

Upon arriving to the sortation system, the envelopes are routed to one of two possible destinations. Thirty-five percent are sent to Destination 1 while the remaining are sent to Destination 2. Envelopes traveling to Destination 1 will follow a 10-foot, nonaccumulating conveyor, while envelopes going to Destination 2 will follow an 18-foot, nonaccumulating conveyor. Both of these conveyors have a constant velocity of 7 inches per second. Assume that all conveyors are initially in operation. Simulate the system for one 8-hour shift and determine the following:

- **a.** How many envelopes are automatically read and how many are manually read?
- **b.** What is the average time spent in the entire system by all envelopes?
- **c.** How many envelopes go to each of the final destinations?
- **d.** What is the utilization of the automatic and manual readers?
- **e.** What is the utilization of the automatic reader conveyor?

11

Guided Transporters

The concept of free-path transporters was presented in Chapter 9. (Knowledge of the earlier material is a prerequisite for the material in this chapter.) Free-path transporters can move through a system without concern for delays caused by other vehicles. Guided transporters move along a fixed path and may contend with each other for space along that path.

Guided transporters move from one point to another along links that are defined by their endpoints, or intersections. The links and intersections form a network or System Map.

An *Automated Guided Vehicle* (AGV) is an example of the type of transporter described in this section. AGVs follow some path that is defined by a wire embedded in the floor or a chemical trail laid on the floor. There are other methods for guiding the vehicle, but the two mentioned are quite common. The AGVs are usually battery driven and have some form of on-board computer to control their movement and operation. One of the economic benefits of AGVs is that they are driverless.

The links that form the network are usually divided into zones. A zone can be occupied by only one AGV at a time to prevent vehicle collisions.

One of the technical problems that must be avoided when using guided vehicles is a situation known as *deadlock*. One type of deadlock occurs when two vehicles claim an intersection and neither is willing to concede that intersection to the other. In many ways guided transporters are similar to free-path transporters, even using some of the same elements and blocks. However, guided transporters are more complex than free-path transporters. This chapter provides a basic introduction to the concept of a guided transporter. For more complex systems, refer to the *SIMAN V Reference Guide* (see reference at end of chapter).

EXAMPLE 11.1 Problem Statement

Consider Example 9.1, where the forklifts traveled a fixed distance from one location to another, without any routing constraints. In this example, the forklifts will be replaced with two AGVs. Unlike the free-path forklifts, the AGVs are constrained to a fixed path. If the path is obstructed, the AGV will have to wait for it to become available or use an alternate route. The only change to the physical system of Example 9.1 is the addition of a staging area. The staging area provides space for idle vehicles to reside. A drawing of the facility layout (not to scale) is shown in Figure 11.1.

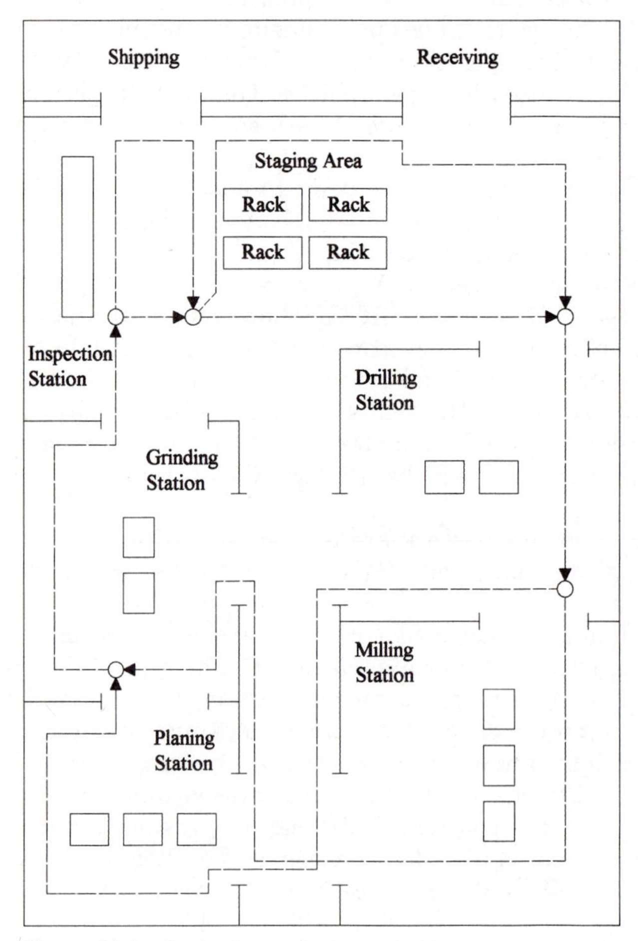

Figure 11.1 Facility layout for Example 11.1.

The dashed lines represent the AGV path and the arrows indicate direction of travel. The small circles, located at intersections where three or more paths come together, are decision points. At these locations, the AGV is directed along the shortest path to its destination. Figure 11.2 illustrates this same system in the form of a network diagram.

In Figure 11.2, the new staging area and the workstations are represented by the squares. Each square contains a station name and an integer identifier. The squares are connected by a series of arcs, and direction of travel along an arc is indicated by the arrow(s) at the end of the arc. All of the arcs are unidirectional, except for the arc between the Arrival and Drilling stations. Adjacent to each arc is the distance between the arc endpoints. Each location at which two or more arcs come together is known as an *intersection*. This means that each workstation and the staging area are intersections. The two circles are intersections that represent forks in the path. An AGV traveling through one of these intersections will choose a route and continue, without stopping. The integer values in the network symbols indicate the associated intersection number.

All processing is identical to Example 9.1. The network given in Figure 11.2 will be used to control the AGVs. Also, if there are no jobs waiting to be transported when an AGV completes a trip, the AGV will be sent to the staging area. At the start of the simulation, the AGVs are to be located in the staging area. The speed of each AGV is 120 feet

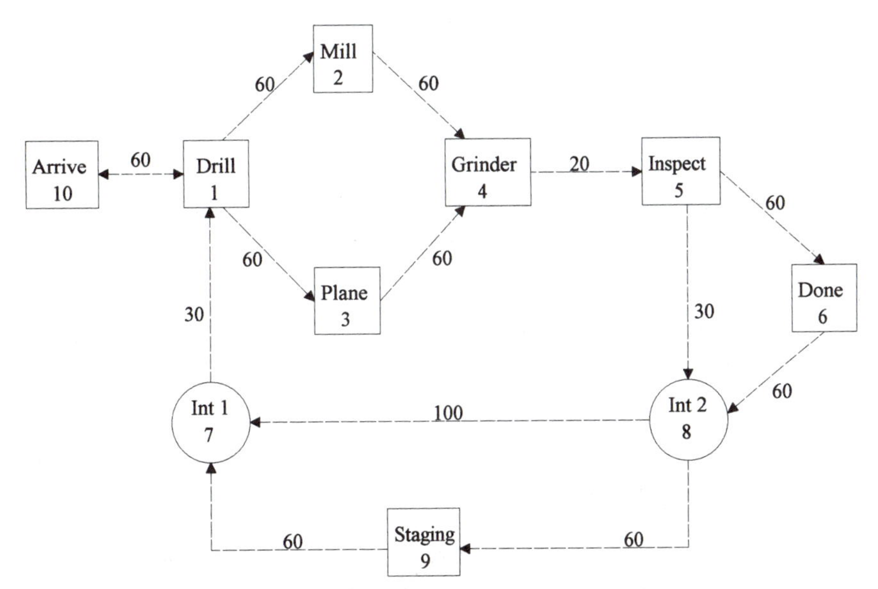

Figure 11.2 Network for Example 11.1.

Table 11.1 Elements Introduced in this Chapter

New Element	Purpose
INTERSECTIONS element	Used to define characteristics of intersections in a network
LINKS element	Used to define the links or arcs of a network
NETWORKS element	Used to define a network as a set of links

per minute. The simulation is to be run for 40 hours, and the following information is desired:

1. Average flowtime for each job
2. Average time between exits
3. Average number of jobs waiting for an AGV
4. Utilization of each of the machines
5. Total number of each type of job completed
6. Total number of jobs reworked
7. Total number of jobs that fail inspection
8. Total number of jobs ejected from queues

The new blocks and elements required to model a guided vehicle are given in Table 11.1. In addition to these new blocks, several blocks and elements introduced in the free-path transporter discussion will be revisited.

11.1 The System Map

Guided transporters move from point to point along links and through intersections defined in the experiment frame. The links can be considered as arcs and the intersections as nodes in a classical network. This network is referred to as the *System Map*. In a free-path transporter, the System Map was defined by the DISTANCES element. In a guided-path transporter, the System Map is defined by the NETWORKS element, which is defined by the INTERSECTIONS and LINKS elements.

11.2 The INTERSECTIONS Element

The INTERSECTIONS element defines the characteristics of all intersections in a guided transporter system. The format for the element is as follows:

INTERSECTIONS: *Number, Name, Travel Length, Link Selection Rule, Velocity Change Factor*: repeats;

The operands are described as follows:

Operand	Description	Default
Number	Intersection number [integer]	Sequential
Name	Intersection name [symbol name]	Blank
Travel Length	Travel length through intersection [integer]	0
Link Selection Rule	Rule used to select a link [FCFS, LCFS, LVF, HVF, CLOSEST, FARTHEST]	FCFS
Velocity Change Factor	[constant]	1

The *Number* and *Name* provide information that also is used in the System Map. *Travel Length* specifies the distance that the transporter must travel to go through the intersection. If *Travel Length* is left blank, a default value of zero is used. The *Link Selection Rule* specifies the rule used to determine which vehicle gains control of a link when multiple vehicles attempt to access the same link. The default for *Link Selection Rule* is first-come first-served (FCFS). The *Velocity Change Factor* operand controls the speed of the transporter as it passes through the intersection. A *Velocity Change Factor* less than 1 will reduce the speed of the vehicle, and a value greater than 1 indicates an accelerated speed. If *Velocity Change Factor* is defaulted, then the velocity of the transporter will remain constant as it passes through the intersection.

The following examples illustrate the use of the INTERSECTIONS element:

```
INTERSECTIONS: 10;
```

This INTERSECTIONS element defines 10 unnamed intersections with a length of zero.

```
INTERSECTIONS:  LoadingLoc, 10:
                ProcessLoc, 3:
                ReloadLoc, 12:
                UnloadLoc, 8;
```

This INTERSECTIONS element defines four intersections: *LoadingLoc, ProcessLoc, ReloadLoc,* and *UnloadLoc*. The distance through each intersection is 10, 3, 12, and 8 units, respectively. The link selection rule is first-come first-served, and the vehicle velocity remains constant as it passes through the intersection.

The available link selection rules are given in Table 11.2. The link selection rules determine which vehicle is given priority if more than one link leads into an intersection.

The *Velocity Change Factor* operand is a multiplier that is generally used to decrease the velocity of a transporter going through an intersection. Thus, a value of 0.8 in this operand multiplies thc velocity prior to entering the intersection by 0.8. An example of the INTERSECTIONS element is as follows:

```
INTERSECTIONS:  LoadLoc, 10:
                ProcessLoc, 3, CLOSEST:
                ReloadLoc, 12, , 0.8:
                UnloadLoc, 8, HVF(Card), 0.7;
```

The intersection *LoadLoc* is as described in the previous example. The intersection *ProcessLoc* is as described in the previous example, but with one change. The *Link Selection Rule* has been changed to CLOSEST. The intersection *ReloadLoc* is the same as in the previous example, except that the transporter decreases its velocity by a scale factor of 0.8 while in the intersection. The *Link Selection Rule* is defaulted to FCFS. The travel length through the intersection *UnloadLoc* is 8 distance units, the *Link Selection Rule* is based on the highest value in attribute *Card*, and the *Velocity Change Factor* is 0.7.

The following example is identical in operation to the previous example:

```
INTERSECTIONS:  1, LoadLoc, 10:
                2, ProcessLoc, 3, CLOSEST:
                3, ReloadLoc, 12, , 0.8:
                4, UnloadLoc, 8, HVF(Card), 0.7;
```

Table 11.2 Link-Selection Rules

Rule	Definition
FCFS	First-come first-served
LCFS	Last-come first-served
LVF(AttributeID)	Lowest value first based on the value of the attribute indicated
HVF(AttributeID)	Highest value first based on the value of the attribute indicated
CLOSEST	Closest to destination
FARTHEST	Farthest from destination

11.3 The LINKS Element

The LINKS element defines the characteristics of the arcs in a guided transporter network. The format for the LINKS element is as follows:

LINKS: *Number, Name, Beginning Intersection ID - Beginning Direction, Ending Intersection ID - Ending Direction, Number of Zones, Length of Zone, Link Type, Velocity Change Factor*: repeats;

The operands are described as follows:

Operand	Description	Default
Number	Link Number [integer]	Sequential
Name	Link Name [symbol name]	Blank
Beginning Intersection ID	Beginning intersection [integer intersection number or name]	—
Beginning Direction	Direction of link as it leaves the first intersection [integer]	0
Ending Intersection ID	Ending intersection [integer intersection number or name]	—
Ending Direction	Direction of link as it enters the ending intersection [integer]	Beginning direction
Number of Zones	Number of zones [integer]	1
Length of Zone	Length of each zone [integer]	0
Link Type	Link type [Unidirectional, Bidirectional, or Spur]	Unidirectional
Velocity Change Factor	[constant]	1

The link *Number* and *Name* provide a reference for the System Map and will be used by the NETWORKS element. The *Beginning Intersection ID* and *Ending Intersection ID* define the two endpoints of the intersections surrounding a link. The intersection *Number* or *Name* can be used.

The *Beginning Direction* and *Ending Direction* define the direction of the link (in degrees) as it leaves the beginning intersection and as it enters the ending intersection. The default direction for entering the ending intersection is the same as the direction for leaving the beginning intersection. The value entered should be an integer with 0 and 360 being to the right or due east. These directions are used in conjunction with the turning velocity to model a vehicle as it changes direction.

Links consist of one or more zones, as specified by the *Number of Zones* operand. Decisions are made by SIMAN V at the end of each zone, or at an intersection, concerning the continuation of travel by the guided transporter. A special case is when there is only one zone. Then, SIMAN V moves the transporter through the link as a single event. If *Number of Zones* is greater than one, then SIMAN V moves the transporter through the link one zone at a time.

The value for *Length of Zone* is the same for all zones in a link. The product of *Number of Zones* times *Length of Zone* is the total length of the link.

The *Link Type* can be *Unidirectional* (with abbreviation U), allowing the transporter to move only from *Beginning Intersection ID* toward *Ending Intersection ID*. *Unidirectional* is also the default value for this operand and is the most usual entry. *Bidirectional* (with abbreviation B) travel allows transporters to move in either direction along a link. However, only one direction of travel is possible if there are two or more transporters on a given link at the same time. *Spur* (with abbreviation S) travel allows a transporter to travel to a dead end and later, back out of the spur and rejoin the travel pattern. Only one transporter can be on a given spur at any time.

Two examples will now be given to demonstrate the use of the LINKS element.

```
LINKS:  1, , 1, 2, 4, 12:      !Intx 1-2 with 4 zones, 12 units each
        2, , 2, 3, 6, 12:      !Intx 2-3 with 6 zones, 12 units each
        3, , 3, 4, 4, 12:      !Intx 3-4 with 4 zones, 12 units each
        4, , 4, 1, 6, 12:      !Intx 4-1 with 6 zones, 12 units each
        5, , 4, 5, 1, 36, S;   Spur between Intxs 4 and 5, 36 units
```

This LINKS element defines five links forming a loop with a spur off Intersection 4. All zones have a distance of 12 units. Links 1 and 3 have four zones and Links 2 and 4 have six zones. Thus, the loop is 240 units long, and the spur is 36 units long. Travel is unidirectional on Links 1 through 4.

```
LINKS:  LoadStop, 1-0, 2, 4, 12:
        Op10Stop, 2-90, 3, 6, 12:
        Op20Stop, 3-180, 4, 4, 12:
        Op30Stop, 4-270, 1, 6, 12:
        UnloadStop, 4-90, 5, 1, 36, S;
```

This example is similar to the previous example, except the link numbers have been replaced with names, and directions have been added. Travel over the first link is to the east, or to the right. Over the second link, travel is to the north. From the third link, travel is to the west and from the fourth link, travel is to the south. Travel along the spur is to the north.

11.4 The NETWORKS Element

The NETWORKS element defines all links to be included when defining a System Map that is followed by guided transporters. SIMAN V automatically generates a shortest-path distance table between all intersections in the network using the information supplied in the INTERSECTIONS, LINKS, and NETWORKS elements. The format for the NETWORKS element is as follows:

NETWORKS: *Number, Name, Starting Link-Ending Link, ...*:
    repeats;

The operands are described as follows:

Operand	Description	Default
Number	Network *Number* [integer]	Sequential
Name	Network *Name* [symbol name]	Blank
Starting Link	Starting link in range of links to include [integer link number or link symbol name]	—
Ending Link	Ending link in range of links to include [integer link number or link symbol name]	No range

The network *Number* or the network *Name* or both can be used. If the network *Number* and following comma are omitted, then the network *Name* must be provided. If the network *Number* is given, then the network *Name* can be defaulted, but the trailing comma must be provided.

A network can consist of individual links, groups of links, or combinations of individual and groups of links. The following examples show three uses of the NETWORKS element:

NETWORKS: 1, AGVTransport, 1-5;

This NETWORKS element defines a network named *AGVTransport* that includes Links 1 through 5.

NETWORKS: 1, AGVTransport, 1-5, 11, 16-18;

This NETWORKS element defines a network named *AGVTransport* that includes Links 1 through 5, 11, and 16 through 18. The corresponding LINKS element would have to define all links appearing in this NETWORKS element.

NETWORKS: AGVEast, EastLoad-EastUnload:
AGVWest, WestLoad-WestUnload;

This element defines a network named *AGVEast* composed of links *EastLoad* through *EastUnload*. The network named *AGVWest* includes links *WestLoad* through *WestUnload*.

11.5 The STATIONS and TRANSPORTERS Elements Revisited

The STATIONS and TRANSPORTERS elements were used in Chapter 9 to model a free-path transporter. These same elements are used to model a guided vehicle system with minimal changes to the experiment frame. First, the stations used in the model must be associated with an intersection number. This includes processing stations, such as Drilling or Inspecting, and any other stations where the vehicle may stop, such as a staging or maintenance area. Second, there may be a single network, or one network divided into smaller subnetworks. Each unique network is assigned a name and each set of transporters is assigned to a network.

11.5.1 The STATIONS Element for Automatically Guided Vehicles

The STATIONS element has an additional operand that associates a station with an intersection. The format for the STATIONS element, including the intersection operand, is as follows:

STATIONS: *Number, Name, Intersection ID*: repeats;

The *Intersection ID* operand is an intersection *Name* or *Number* that provides a reference for the station. When an AGV is sent to the station, the vehicle is moved to *Intersection ID* in its current system map. This simplifies the routing of guided vehicles by allowing all movements to be based on stations.

Following is an example STATIONS element:

```
STATIONS:  Load, 1:
           Op10, 2:
           Op20, 3:
           Op30, 4:
           Unload, 5;
```

This element defines five stations. Intersections 1 through 5 are associated with the stations, so that guided transporter movements between these intersections can be performed at the stations.

11.5.2 The TRANSPORTERS Element for Automatically Guided Vehicles

The TRANSPORTERS element was introduced in Chapter 9 for free-path transporters. The operands for the TRANSPORTERS element are extended for guided-path systems. The format for the TRANSPORTERS element is given as follows:

TRANSPORTERS: *Number, Name, Number of Units, System Map Type(Map ID)-Type of Control, Velocity-Acceleration-Deceleration-Turning Velocity, Initial Position(Position ID, Zone)-Initial Status-Vehicle Size(Size Integer), ...* : repeats;

The operands are described as follows:

Operand	Description	Default
Number	Transporter *Number* [integer]	Sequential
Name	Transporter *Name* [symbol name]	—
Number of Units	Number of available units [integer]	—
System Map Type	Keyword defining the map type used for the transporter [DISTANCE or NETWORK]	DISTANCE

Operand	Description	Default
Map ID	Map ID defining the Distance Set ID or Network ID	—
Control	Type of zone control [START or S, END or E, or *k*]	START
Velocity	Velocity [constant]	1.0
Acceleration	Transporter acceleration [constant]	0.0
Deceleration	Transporter deceleration [constant]	0.0
Turning Velocity	Turning velocity factor [constant]	1.0
Initial Position	Initial Position Type [INTERSECTION, LINK, or STATION]	See below
Position ID	Station, Intersection or Link Identifier corresponding to the Initial Position Type [construct ID]	See below
Zone	Zone number if initial position is on LINK [integer]	See below
Initial Status	Initial Status [Active or Inactive]	Active
Vehicle Size	Space vehicle occupies [LENGTH or ZONE]	Zone
Size Integer	Value for initial vehicle size: number of length units or zones [integer]	1

The *Number, Name, Number of Units,* and *Initial Status* operands have the same usage as in free-path transporters. The remaining operands are required to control the complex operation of guided transporters.

The *System Map Type* and *Map ID* operands define the System Map that the transporters follow. If *System Map Type* is specified as DISTANCE, then the transporter will follow the specified distance set defined by the *Map ID* operand. This will cause the guided vehicle to operate as a free-path transporter moving between stations. A *System Map Type* of NETWORK indicates that the vehicle is guided and associates its System Map with the network *Map ID*.

The *Control* operand defines the type of zone control employed by the vehicles. Zone control determines when a vehicle controlling a zone releases that zone allowing another vehicle access to it. The three options are:

START	The transporter releases its backward-most zone as soon as it is given the next required zone.
END	The transporter waits until reaching the end of the next zone before releasing its backward-most zone.
k	The transporter releases its backward-most zone after traveling the *k* distance units through the next zone.

Velocity defines the initial velocity of the transporter. Transporter velocity may be changed in the model frame, using the VT*(Transporter ID)*. This will assign a new velocity to all of the transporters in the set with a *Name* or *Number* of *Transporter ID*. The VTU*(Transporter ID, Vehicle Number)* variable can be used to change the velocity of a specific transporter in a set. For example, if there are three vehicles called *ForkLift*, then VTU*(ForkLift, 2)* can be used with the ASSIGN block to change the velocity of the second vehicle in the *ForkLift* set. The MOVE, REQUEST, or TRANSPORT blocks may be used to alter the velocity during individual vehicle movements.

Acceleration and *Deceleration* define the acceleration and deceleration for guided transporters. They are entered as constants in (distance/time unit^2). SIMAN V applies the acceleration or deceleration when appropriate as vehicles move through the system. When a vehicle arrives at the end of a zone, a check is made to see if a stop is required. If a stop is anticipated, then deceleration will be applied. Alternately, if a vehicle is stopped or slowed, then SIMAN V may apply the appropriate acceleration to return the vehicle to normal operating velocity.

The *Turning Velocity*, turning velocity factor, is applied to slow a vehicle as it makes a turn. This operand is applied automatically if directions have been given to all links in the network.

The *Initial Position* operand defines the initial location of the transporter. Free-path transporters default the initial position to the first station specified in the DISTANCES element. Guided transporters default to the first link specified in the NETWORKS element. If all of the transporters do not fit on the first link, then an initial position other than the default must be specified for enough units such that the defaulted units can all fit on the first link. To designate an initial position other than the default location, use one of the following for the *Initial Position* operand:

Station ID	The transporter begins at the STATION block specified by *Station ID(Number* or *Name)*
INTERSECTION*(Intersection ID)*	The transporter begins at the intersection specified by *Intersection ID(Number or Name)*
LINK*(Link ID, Zone)*	The transporter begins at the zone number *Zone* (integer) on link *Link ID(Number or Name). Zone* may be defaulted to the first available zone on the link (moving from the end to the beginning of the link)
STATION*(Station ID)*	The transporter begins at the STATION block specified by *Station ID(Number or Name)*

The *Initial Status* operand indicates whether the vehicle is initially *Active* or *Inactive*. In the model frame, the HALT block may be used to switch an active transporter to inactive. The ACTIVATE block can be used to activate a currently inactive vehicle.

Vehicle Size is specified as LENGTH or ZONE. The corresponding amount of space required by the vehicle is specified with the *Size Integer* operand. If *Vehicle Size* is specified as LENGTH, then the actual area it occupies is the smallest number of zones needed to accommodate the required space on the network. Each transporter's size may be changed with the CAPTURE and RELINQUISH blocks in the model frame. For example, ZONE(2) specifies a vehicle two zones long. LENGTH(5) defines the vehicle size in distance units. If the length of a zone is also 5 units, then the transporters associated with this Vehicle Size are one zone long. If each zone is 4 units long, then two zones will be required for each transporter associated with this *Vehicle Size*.

Several examples are now given to explain the TRANSPORTERS element in the context of this chapter.

TRANSPORTERS: AGV, 4, NETWORK(AGVSystem), 50;

The four AGVs have an initial *System Map* defined by network *AGVSystem*. The velocity of the transporters is 50. Note that many of the operands are defaulted, that is, the vehicles use release-at-start zone control, the transporters have an initial position on the first link specified in the network *AGVSystem*, provided they will all fit there, and they are all active at the beginning of the simulation. In release-at-start zone control, the vehicle gains control of the next required zone before releasing its trailing zone. This concept is discussed more extensively in Section 11.5.3. They will fit if the first link has four or more zones since the default *Size* is one zone.

```
TRANSPORTERS:AGV,4,NETWORK(AGVSystem)-E,50-2-2-0.8;
```

The change from the previous example is that the vehicles use the release-at-end zone control. In release-at-end zone control, the vehicle maintains control until it reaches the end of the next zone. This concept is discussed more extensively in Section 11.5.3. Also the acceleration, deceleration, and turning velocity factors are given as the last three operands, respectively.

```
TRANSPORTERS: AGV1,4,NETWORK(AGVSystem1)-E,50-2-2-0.8:
              AGV2,2,NETWORK(AGVSystem2)-E,60-2-2-0.8;
```

Two sets of transporters are identified as *AGV1* and *AGV2*. There are four AGVs of Type 1 and two of Type 2. The first group of AGVs follows a set of links and intersections defined by the network named *AGVSystem1*, and the second group operates on the *AGVSystem2* network.

11.5.3 Controlling an Automatically Guided Vehicle

One of the most difficult problems encountered when modeling guided vehicles is a deadlock. A deadlock occurs when one vehicle requires a zone currently under the control of a second vehicle, and the second vehicle requires the zone held by the first vehicle. When this happens, SIMAN V will terminate with an error message. Another situation that may occur is when a vehicle must pass through an intersection currently occupied by an idle transporter. The traveling vehicle will stop and wait for the idle vehicle to move. If the time until a request for the idle vehicle is greater than the travel time for the waiting transporter to reach its destination, the vehicles will be prevented from any further movement. If the idle vehicle does not require a zone controlled by the waiting vehicle, SIMAN will not issue an error message and the system becomes locked. In order to avoid situations such as this, the modeler must include control logic for all possible deadlock situations.

As the complexity of a network increases, so does the opportunity for deadlocks. Multiple bidirectional links will significantly increase the possible number of deadlock situations. Before a vehicle can enter a bidirectional link, it must first control both entering zones. When this occurs, the direction of travel on that link is determined by the direction of the controlling vehicle. Multiple vehicles can travel the link simultaneously if they are all traveling the same direction. A vehicle traveling in the opposite direction will have to wait for the link to become available, or for the direction along the link to change. If the waiting vehicle controls the ending intersection of the bidirectional link and the direction along the link does not change, then the vehicle(s) already on the link cannot exit and a deadlock will occur. In order to plan for situations such as this, the modeler will first have to understand how SIMAN allocates zones to the vehicles.

The two basic types of zone control are *release-at-start* and *release-at-end*, entered as START or END (S or E) for the *Control* operand of the TRANSPORTERS element. Figure 11.3 illustrates the release-at-start method of zone control.

For simplicity, assume that the length of the vehicle and the intersections are each equivalent to one zone. The vehicle will be traveling from Intersection 1 to Intersection 2, along zones Z1 through Z5. The vehicle is initially idle at Intersection 1 (Figure 11.3A). Before the vehicle starts to move, it will take control of zone Z1, release Intersection 1, and start moving into zone Z1. As the vehicle enters zone Z1, it no longer controls Intersection 1 even though part of the vehicle may physically occupy part of the intersection (Figure 11.3B). When the vehicle reaches the end of Z1, it will attempt to gain control of zone Z2. If the zone is available, then the vehicle will take control

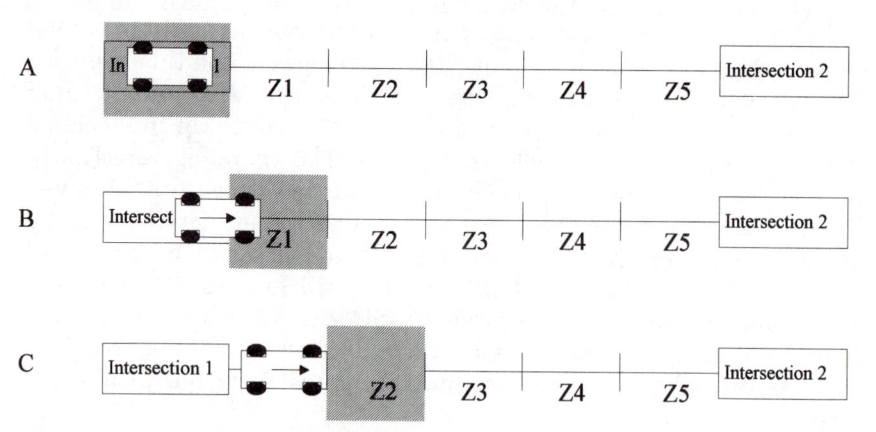

Figure 11.3 Release-at-start-zone control.

of Z2, release control of Z1, and start moving into Z2 (Figure 11.3C). This process is continued until the vehicle reaches Intersection 2. Notice that the vehicle controls the same number of zones while idle or traveling. Figure 11.4 illustrates the release-at-end method of zone control. Figure 11.4A shows the vehicle sitting idle at Intersection 1. Similar to release-at-start control, the vehicle will gain control of Z1 before starting to move into the zone (Figure 11.4B). However, as the vehicle moves into Z1, it also maintains control of Intersection 1 (Figure 11.4C). When the vehicle reaches the end of Z1, it will release Intersection 1, take control of Z2, and continue traveling. When traveling through Z2, the vehicle controls both Z2 and Z1 (Figure 11.4D). This process is continued until the vehicle reaches Intersection 2. When the vehicle reaches the end of Intersection 2, it will release zone Z5 and maintain control of Intersection 2. Thus, the vehicle requires two zones when traveling and only one when idle.

The same type of logic applies for vehicles longer or shorter than one zone. When a vehicle is longer than one zone, the minimum number of zones required to hold the vehicle will be used. For example, if a release-at-start vehicle is 6.5 units and a zone is only one unit, then the vehicle will require control of the next seven zones before it can travel. For basic modeling, vehicle lengths specified as multiples of zone lengths will provide sufficient flexibility while avoiding unnecessary complexity.

If the network contains zones or intersections of zero length, then the vehicle will continue to take control of additional zones until it

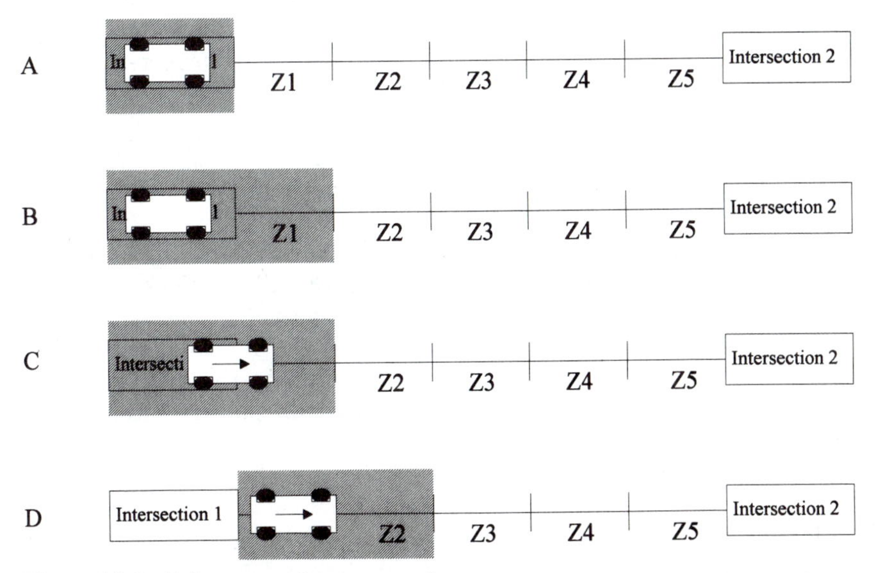

Figure 11.4 Release-at-end-zone control.

controls a segment of the network at least as long as the vehicle. For example, if Z1 of Figure 11.3A has a length of zero, then the vehicle must control zones Z1 and Z2 prior to starting movement.

SIMAN provides a wide range of methods to control guided vehicles. However, the more detailed the control is, the harder it may be to locate problems. There are some functional (but possibly less efficient) strategies to avoid control problems. For example, always send idle vehicles to a staging area. In the model frame, locate every possible point where a vehicle can become idle and insert some staging logic (see Exhibit 11.1 and associated text). Second, if there are spurs in the model, insure that they are longer than the length of the vehicle. If a spur is shorter than a visiting vehicle, then the vehicle will extend out onto the network, preventing other vehicles from passing. Finally, start simple and incrementally increase the level of complexity. For example, minimize the number of bidirectional arcs by replacing them with pairs of opposing unidirectional arcs. If possible, build the original model exclusively with unidirectional arcs. Once the original model is operating properly, start exchanging pairs of unidirectional arcs with bidirectional arcs. Continuing in this fashion, the initially simple model will evolve into a more complex model with less of a chance for having sophisticated control problems.

11.6 Guided Transporter Blocks

To this point, the elements have been described to define guided transporters and their movement systems. We now describe how empty transporters answer a request and how loaded transporters move entities through the network.

11.6.1 Allocating and Moving the Empty Transporter

The ALLOCATE, MOVE, and REQUEST blocks were discussed in Chapter 9, with particular emphasis on the REQUEST block. The ALLOCATE and MOVE blocks together are equivalent to the REQUEST block. However, more control is achieved when using the ALLOCATE and MOVE blocks. In this chapter, we will describe the differences in the REQUEST block when used with guided transporters rather than the free-path transporters of Chapter 9.

The format of the REQUEST block is given by the following:

REQUEST, *Priority, Storage ID, Alternate Path: Transporter Name, Velocity, Entity Location*;

The differences in use for guided transporters occur in the *Alternate Path* and *Entity Location*. With respect to *Alternate Path*, the difference occurs when a distance-based selection rule (SDS or LDS) is used. The *Alternate Path* operand modifies the transporter selection rule by specifying a station, link, or intersection to include in the distance calculation. The options require using the VIA keyword in one of five ways as follows:

VIA*(Station ID)*	*Station ID* is a station name or expression
VIA(STATION*(Station ID)*)	*Station ID* is a station name or expression
VIA(INTX*(Intersection ID)*)	*Intersection ID* is an intersection name or expression
VIA(LINK*(Link ID)*)	*Link ID* is a link name or expression
VIA(LINK*(Link ID, Zone)*)	*Zone* is a specific zone on link *Link ID* (*Zone* may be defaulted to 1)

The *Entity Location* tells SIMAN where the entity is located in the system when a distance-based selection rule is used. If the operand is defaulted, the entity's M value is used. The other possibilities are as follows:

Station ID	*Station ID* is a station name or expression
STATION*(Station ID)*	STATION*(Station ID)*
INTX*(Intersection ID)*	*Intersection ID* is an intersection name or expression
LINK*(Link ID, Zone)*	*Zone* is a specific zone on link *Link ID* (*Zone* may be defaulted to 1)

Several examples will be given to show the use of the REQUEST block for AGVs.

```
QUEUE,
  ToWashQ;
REQUEST:
  AGV;
```

The entity waits in queue *ToWashQ* until the transporter *AGV* becomes available. The entity remains in the REQUEST block while *AGV* moves to station M.

```
REQUEST:
  AGV;
```

If no QUEUE block precedes the REQUEST block, then entities wait in an internal queue.

```
QUEUE,
  14;
REQUEST:
  AGV(SDS), , Process;
```

The transporter unit that is closest to the station *Process* is allocated to the entity. The entities wait in queue 14 for this transporter unit to arrive.

```
QUEUE,
  A(5);
REQUEST,
  2, , VIA(INTX(4)): AGV, 30, STATION(Wash);
```

Entities wait in the queue specified by *A(5)* with request priority of 2 for the transporter *AGV*. (Assume that there is only 1 unit of *AGV* in the system.) The *AGV* moves to the entity at station *Wash* via Intersection 4. The velocity is set to 30 for the move from the *AGVs* current location to station Wash.

11.6.2 Transporting the Entity

The TRANSPORT block was discussed in Chapter 9 for use with free-path transporters. The format of the TRANSPORT block is given by the following:

TRANSPORT, *Alternate Path: Transporter Name, Destination, Velocity, Guided Trans Dest*;

The TRANSPORT block for guided transporters offers more options than for free-path transporters. These options are in the *Alternate Path* and *Guided Trans Dest* operands.

SIMAN V generates a table listing the shortest path between all locations. The *Alternate Path* operand modifies the path by specifying a specific station, link, or intersection to include in the distance calculation. The valid options require using the VIA keyword in one of five ways as follows:

VIA(*Station ID*)	*Station ID* is a station name or expression

VIA(STATION(*Station ID*))	*Station ID* is a station name or expression
VIA(INTX(*Intersection ID*))	*Intersection ID* is an intersection name or expression
VIA(LINK(*Link ID*))	*Link ID* is a link name or expression
VIA(LINK(*Link ID, Zone*))	*Zone* is a specific zone on link *Link ID*(*Zone* may be defaulted to 1)

The *Guided Trans Dest* defaults to moving the guided transporter to the intersection associated with the station of the entity controlling the transporter. *Guided Trans Dest* may also be specified as one of the following:

Station ID	*Station ID* is a station name or expression
STATION(*Station ID*)	STATION(*Station ID*)
INTX(*Intersection ID*)	*Intersection ID* is an intersection name or expression
LINK(*Link ID, Zone*)	*Zone* is a specific zone on link *Link ID* (*Zone* may be defaulted to 1)

If the destination for a guided transporter is specified as a station, or defaulted to M, then the station must have an associated intersection in the System Map.

Guided transporters may also be moved to the first intersection on the way to *Guided Trans Dest* by using the keyword FIRSTX as follows:

FIRSTX(*Station ID*)	Stop at the first intersection encountered on the way to *Station ID*
FIRSTX(STATION(*Station ID*))	Stop at the first intersection encountered on the way to *Station ID*
FIRSTX(INTX(*Intersection ID*))	Stop at the first intersection encountered on the way to *Intersection ID*
FIRSTX(LINK(*Link ID*))	Stop at the first intersection encountered on the way to *Link ID*
FIRSTX(LINK(*Link ID, Zone*))	Stop at the first intersection encountered on the way to zone *Zone* on link *Link ID*

The following example shows how the TRANSPORT block can be used:

```
TRANSPORT, VIA(LINK(2): AGV, Wash, , INTX(4);
```

A unit of the transporter AGV moves the entity to the *Wash* station. The transporter moves to intersection 4 via link 2. The velocity is unchanged.

11.7 Modeling Guided Transporter Failure

Guided-transporter failures are modeled by using the HALT and ACTIVATE blocks just like the free-path transporters discussed in Chapter 9. However, guided-path transporters remain in the network when they fail. If these vehicles fail at an intersection, then they control that intersection until they are activated. If they fail on a link, then they control that link, and so on.

If a repair is performed quickly, then it can occur at the point of failure. However, a lengthy repair can require that the vehicle be removed from the path so that other vehicles are not blocked. After the repair, the vehicle is placed back on the path. The RELINQUISH and CAPTURE blocks model these activities. Using these blocks can become rather complicated. It is suggested that you study the *SIMAN V Reference Guide: Part 1* to learn their operation if this level of accuracy in modeling is required. Another possibility is to allow vehicles to fail only at locations that will not impede the flow of traffic, such as on a spur.

EXAMPLE 11.1 Solution

The Model Frame

Exhibit 11.1 contains the entire model frame for Example 11.1. Notice that this model frame is very similar to that of Example 9.1. The major change is the addition of staging logic at the beginning of each station. When an AGV completes a transportation task, before the entity that controls the AGV will release it, a check is made to see if there are any entities awaiting transport. If there are one or more jobs waiting for the AGV (i.e., NQ*(VehicleQ)* > 0 evaluates to TRUE), then the controlling entity is redirected to a labeled block, where the AGV is released. From this point, the releasing entity will enter the QUEUE block preceding the next sequential station and the AGV will proceed to the location of the awaiting entity. If the *VehicleQ* is empty, then the controlling entity is redirected to another BRANCH block. At this BRANCH block, a duplicate entity is created, the original entity proceeds to the next sequential station, and the duplicate entity maintains control of the AGV. The entity controlling the AGV is redirected to a block labeled *StageIt*. At *StageIt*, the AGV is moved to the staging area and the duplicate entity is destroyed. The model frame for Example 11.1 is given in Exhibit 11.1. Changes and additions to Example 9.1 are shown in italics.

The Experiment Frame

Exhibit 11.2 contains the experiment frame for Example 11.1. The experiment frame is very similar to that of Example 9.1. Major changes include the omission of the DISTANCES element, addition of the INTERSECTIONS, LINKS, and NETWORKS elements, modification of the TRANSPORTERS element, and the addition of intersection identifiers to the STATIONS element components. Other changes include deleting some of the DSTATS components and changing the number of replications to one. Note that the intersection identifiers correspond to the intersection numbers in Figure 11.1.

EXHIBIT 11.1 Model Frame for Example 11.1

```
!Model Frame for Example 11.1
!
BEGIN,NO;
          CREATE: ED(1):                  !Create arriving jobs
          MARK(ArrTime);                  Save arrival times in an attribute
          ASSIGN:
            JobType=ED(2):                !Define job as Type I or II
            NS=JobType:                   !Define SEQUENCE by JobType
            M=SArrive;                    Current station is SArrive
MoveJob   QUEUE, VehicleQ;                Wait for available Vehicle
          REQUEST: Vehicle(SDS);          Request Vehicle
          TRANSPORT: Vehicle, SEQ;        Transport by SEQUENCES
Process   STATION, SDrill-SInspect;       Stations SDrill to SInspect
          BRANCH, 1:                      !Staging logic
            IF, NQ(VehicleQ) > 0, Continue:
            ELSE, Stage1;
Stage1    BRANCH, 2:
            ALWAYS, StageIt:
            ALWAYS, StartWrk;
Continue  FREE;                           Free Vehicle transporter
```

EXHIBIT 11.1 *continued.*

```
StartWrk  QUEUE, M, V(M), Eject;            !Queue no. and capacity depend on M
          SEIZE: Machine(M);                Capture indexed resource M
          DELAY: ProcessTime;               Process job - defined in SEQUENCES
          RELEASE: Machine(M);              Free indexed resource M
          BRANCH, 1, 10:                    !Take one branch only
            IF, M.LE.4, MoveJob:            !If at first 4 stations continue
            WITH, PRedo, Rework:            !otherwise, rework with probability
            WITH, PFail, Scrap:             !PRedo, scrap with probability
            ELSE, MoveJob;                  !PFail, or accept as complete
Done      STATION, SDone;                   SDone station for completed jobs
          BRANCH, 1:                        !Staging Logic
            IF, NQ(VehicleQ) > 0, Proceed:
            ELSE, Stage2;
Stage2    BRANCH, 2:
            ALWAYS, StageIt:
            ALWAYS, Leave;
Proceed   FREE;                             Free Vehicle transporter
Leave     TALLY: Flowtime, INT(ArrTime);Tabulate Flowtime
          TALLY: Exit Period, BET;          Tabulate time between exits
          COUNT: JobType, 1:                !Count jobs by type and
          DISPOSE;                          destroy entity
StageIt   MOVE: Vehicle, Staging;           !Move to the staging area and
          FREE: Vehicle;                    !free the AGV
          DISPOSE;                          !Dispose the extra entity
Rework    COUNT: Reworks;                   Count number of jobs reworked
          ASSIGN: IS=0:                     !Restart SEQUENCE
            NEXT(MoveJob);                  Move job to Drilling Station
Scrap     COUNT: Fails:                     !Count number of jobs scrapped
            DISPOSE;                        and destroy entity
Eject     COUNT: Ejections:                 !Count number of jobs ejected
            DISPOSE;                        and destroy entity
```

EXHIBIT 11.2 Model Frame for Example 11.1

```
!Experiment Frame for Example 11.1
!
BEGIN;
PROJECT, Example 11.1, Team;
ATTRIBUTES:     ProcessTime:              !First attribute in SEQUENCES
                PRedo:                    !Second attribute in SEQUENCES
                PFail:                    !Third attribute in SEQUENCES
                ArrTime:
                JobType;
```

EXHIBIT 11.2 *continued.*

VARIABLES:	1, MDrill,	2:	
	2, MMill,	3:	
	3, MPlane,	3:	
	4, MGrinder,	2:	
	5, MInspect,	4;	
STATIONS:	SDrill,	1:	
	SMill,	2:	
	SPlane,	3:	
	SGrinder,	4:	
	SInspect,	5:	
	SDone,	6:	
	Staging,	7:	
	SArrive,	8;	
RESOURCES:	Drill,	2:	
	Mill,	3:	
	Plane,	3:	
	Grinder,	2:	
	Inspector;		
SETS:	Machine, Drill..Inspector;		Machine(1) - Machine(5)
QUEUES:	DrillQ:		
	MillQ:		
	PlaneQ:		
	GrinderQ, LVF(JobType):		
	InspectorQ:		
	VehicleQ;		
INTERSECTIONS:			
	1, *DrilIntx:*		
	2, *MillIntx:*		
	3, *PlanIntx:*		
	4, *GrndIntx:*		
	5, *InspIntx:*		
	6, *DoneIntx:*		
	7, *Int1Intx:*		
	8, *Int2Intx:*		
	9, *StagIntx:*		
	10, *ArrvIntx;*		
LINKS:	1, *DriltoMill,*	1, 2, 6, 10:	
	2, *DriltoPlan,*	1, 3, 6, 10:	
	3, *MilltoGrnd,*	2, 4, 6, 10:	
	4, *PlantoGrnd,*	3, 4, 6, 10:	
	5, *GrndtoInsp,*	4, 5, 2, 10:	
	6, *InsptoInt2,*	5, 8, 3, 10:	
	7, *Int2toInt1,*	8, 7, 10, 10:	
	8, *Int2toStag,*	8, 9, 6, 10:	
	9, *StagtoInt1,*	9, 7, 6, 10:	
	10, *Int1toDril,*	7, 1, 3, 10:	
	11, *InsptoDone,*	5, 6, 6, 10:	
	12, *DonetoInt2,*	6, 8, 6, 10:	
	13, *DriltoArrv,*	1, 10, 6, 10, Spur;	

EXHIBIT 11.2 *continued.*

```
NETWORKS:         Handler Sys, 1-13;
TRANSPORTERS:     Vehicle, 2, NETWORK(Handler Sys), 120, LINK(8)-A;
COUNTERS:         Type I Job Count:
                  Type II Job Count:
                  Reworks:
                  Fails:
                  Ejections;
SEQUENCES:        1, TypeIPart, SDrill, ProcessTime=ED(3), &
                   SMill, ProcessTime=ED(4), &
                   SGrinder, ProcessTime=ED(6) ,&
                   SInspect, ProcessTime=ED(8),PRedo=.05,PFail=.05 &
                   SDone:
                   2, TypeIIPart,SDrill, ProcessTime=ED(3), &
                   SPlane, ProcessTime=ED(5), &
                   SGrinder, ProcessTime=ED(7), &
                   SInspect, ProcessTime=ED(8),PRedo=.05,PFail=.05 &
                   SDone;
EXPRESSIONS:      1, , EXPO(5, 1):              !Interarrival time
                  2, , DISC(0.7, 1, 1.0, 2, 1): !Job Type distribution
                  3, , UNIF(6, 9, 1):           !Drilling time
                  4, , TRIA(12, 16, 20, 1):     !Milling time
                  5, , TRIA(20, 26, 32, 1):     !Planing time
                  6, , DISC(0.25, 6, 0.75, 7, 1.0, 8, 1):
                                                !Grinding times for Type 1 jobs
                  7, , DISC(0.1, 6, 0.35, 7, 0.65, 8, 0.9, 9, 1.0, 10, 1):
                                                !Grinding times for Type 2 jobs
                  8, , NORM(3.6, 0.6,1);        Inspection time
TALLIES:          Flowtime:
                  Exit Period;
DSTATS:           NQ(VehicleQ), Vehicle Queue
                  NR(1)/2, Drill Utilization:
                  NR(2)/3, Mill Utilization:
                  NR(3)/3, Plane Utilization:
                  NR(4)/2, Grinder Utilization:
                  NR(5), Inspect Utilization;
REPLICATE, 1, 0, 2400;
```

EXHIBIT 11.3 **Summary Report for Example 11.1**

```
                 Summary for Replication 1 of 1

Project: Example 11.1       Run execution date :  8/24/1994
Analyst: Team               Model revision date:  8/24/1994
Replication ended at time: 2400.0
```

EXHIBIT 11.3 *continued*

TALLY VARIABLES

Identifier	Average	Variation	Minimum	Maximum	Observations
Flowtime	58.114	.24103	38.952	149.15	377
Exit Period	6.2705	.71865	.75046	33.466	376

DISCRETE-CHANGE VARIABLES

Identifier	Average	Variation	Minimum	Maximum	Final Value
Vehicle Queue	1.0097	1.2967	.00000	8.0000	4.0000
Drill Utilization	.68705	.55504	.00000	1.0000	.50000
Mill Utilization	.69139	.46970	.00000	1.0000	.66667
Plane Utilization	.43823	.78849	.00000	1.0000	1.0000
Grinder Utilization	.64184	.57544	.00000	1.0000	.50000
Inspect Utilization	.63219	.76276	.00000	1.0000	1.0000

COUNTERS

Identifier	Count	Limit
Type I Job Count	273	Infinite
Type II Job Count	104	Infinite
Reworks	17	Infinite
Fails	27	Infinite
Ejections	39	Infinite

Simulation run complete.

Discussion of Output

The solution to Example 11.1 is given in Exhibit 11.3. The information required is the same as that in Example 9.1, except the data is provided concerning the vehicle queue. The average value is 1.0097 and the queue reaches a maximum of 8.

11.8 SUMMARY

This chapter presented the basics for guided transporters. An example problem was stated, and then the INTERSECTIONS, LINKS, and NETWORKS elements were introduced. The STATIONS and TRANSPORTERS elements were discussed in terms of the additional operands that are used in guided-path transporters. We then described how the empty transporter is called by the REQUEST block and how the entity is delivered to its destination with the TRANSPORT block. Modeling transporter failures was introduced, and a solution to the example problem was given.

11.9 REFERENCE

SIMAN V Reference Guide (1994), Systems Modeling Corporation, Sewickley, Penn.

11.10 EXERCISES

E11.1 Work Example 3.1 with the following changes. The jobs are transported between work areas with an AGV. The distance between each work area is 20 meters. The only change to the system is the addition of shipping and receiving areas. Upon arrival, the job is transported from receiving to the drilling area. When a job is finished, it is transported from the grinding area to the shipping area. The network will consist of five nodes (Receiving, Drilling, Milling, Grinding, and Shipping), each 20 meters apart. At the start of the simulation, the AGV is to be located at the receiving area. The AGV velocity is 80 meters per minute. Model this system according to the following:

- **a.** Use all bidirectional links, with one link connecting each node.
- **b.** Use all unidirectional links, forming a closed loop. The distance from the shipping area to the receiving area is 100 meters. Compare the results with part a.

E11.2 The SIMAN NT*(Transporter ID)* variable returns the number of busy transporters identified by *Transporter ID*. This variable, divided by the number of transporters in the set, could be used as a DSTAT entry to monitor the utilization of the transporters. However, when a guided transporter is being sent to a staging area, the NT variable recognizes the transporter as being in use. This means that simply including NT*(Transporter ID)*/(Number of Transporters) in the DSTATS element will produce erroneously high utilization rates for the guided vehicles.

- **a.** With this in mind, discuss a possible method for more accurately monitoring the utilization of a guided vehicle.

b. Make necessary changes to Example 11.1 to illustrate your solution to part a. How does this solution compare to the reults of NT*(Transporter ID)*/(Number of Transporters)?

E11.3 Use the base system from Example 11.1 for each of the following:

a. An additional link is to be added between the Grinder and the Staging area. The link is unidirectional and 50 meters long. Discuss the effects of this additional link.

b. Due to quality control problems, an additional inspection station has been added to the process. Type I jobs are inspected immediately after being drilled. Type II jobs are inspected immediately prior to being sent to the Grinder. There is a 95% success rate at the new inspection station with 2% of the jobs being rejected and the remaining 3% reworked. The inspection times are normally distributed with a mean of 3 minutes and a standard deviation of 30 seconds. The original inspection station has observed a decrease in rejections and reworks down to 4% and 3%, respectively. Due to higher confidence, the mean inspection times at the original inspection station have decreased to 2.5 minutes, with a standard deviation of 24 seconds. The new inspection station is located such that the distance from any of the four processing stations is 20 meters. Compare this system to the original system.

E11.4 Three types of items arrive according to an exponential distribution with a mean of 30 minutes. There is a 1/3 probability that an arriving item is a specific type. The items are sent to Inspection area 1 prior to being accepted for use. Inspection times are normally distributed with a mean of 12 minutes and a standard deviation of 2 minutes. Observations have indicated that only 2% of the incoming items are rejected. These rejected items are returned to the Shipping/Receiving area, whereas the rest are sent to Processing area 1. Processing times at Processing area 1 are normally distributed with a mean and standard deviation of 20 and 3.2 minutes, respectively. The items are then sent to the second processing area for painting. The painting process is normally distributed with a mean of 27 minutes and a standard deviation of 4.3 minutes. Type I and Type II items are then sent to the second inspection station, whereas Type III items are returned to Processing area 1 for additional work. After the additional work, Type III items are also sent to the second inspection area. Inspection times at area 2 are uniformly distributed between 5 and 35 minutes. The success rate at the second inspection station is 90%, with 2% being rejected, 3% returned to Processing area 1, and 5% returned to Processing area 2. Rejected and accepted items are taken to the Shipping/Receiving area for disposal or shipping. Type III items that are sent back to Processing area 1 do not have to go through Processing area 2 again. In other words, when reworked Type III items are finished at Processing area

1, they are then sent back to Inspection area 2. However, Type III items sent back to Processing area 2 must proceed to Processing area 1 prior to returning to Inspection area 2. Type I and Type II rework jobs proceed normally through the system. A schematic representation of this system is illustrated in Figure 11.5.

There are no buffer space limitations at resource locations or for items waiting to be transported. The two processing areas are each 50 meters long. There are two AGVs, each with a speed of 15 meters per minute. Use random number stream 1 for all distributions. Simulate this system for 160 hours and answer the following questions:

- **a.** How many items are waiting to be transported?
- **b.** What is the utilization of each of the four resources?
- **c.** How many items are waiting at the two processing stations?
- **d.** How many of each type of item is completed?
- **e.** How many items are sent back to each processing area?
- **f.** How many items are rejected at the inspection stations?

E11.5 Items arrive according to an exponential distribution with a mean of 10 minutes. The items are processed according to the system shown in Figure 11.6. Processing times are as follows:

Station ID	Processing Time Distribution
1	Normal (8.0, 1.2)
2	Normal (30.0, 5.8)
3	Uniform (8.0, 12.0)
4	Triangular (10.0, 14.0, 16.0)
5	Normal (14.0, 2.1)
6	Uniform (7.0, 9.0)

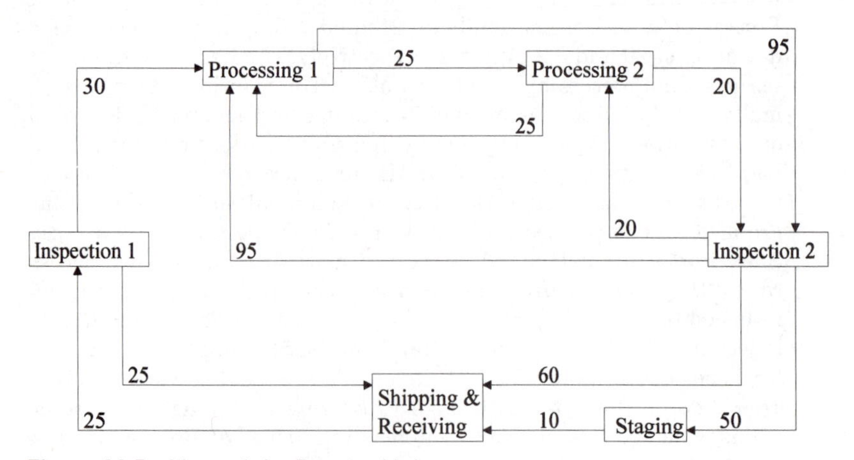

Figure 11.5 Network for Exercise 11.4.

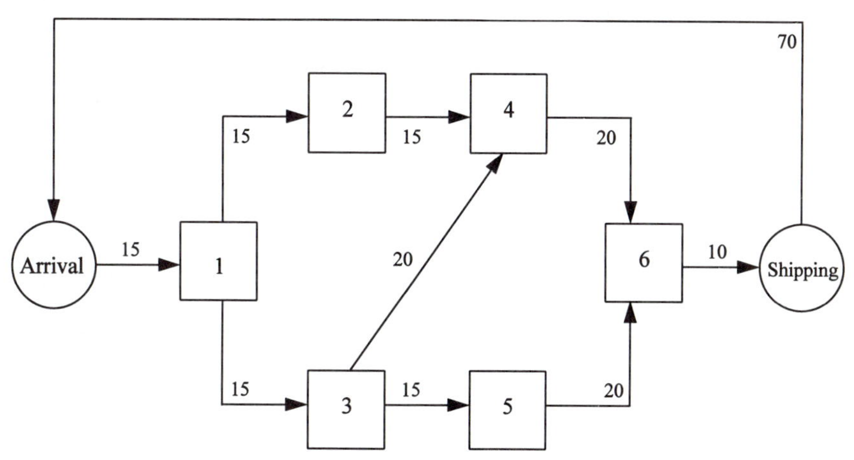

Figure 11.6 Network for Exercise 11.5.

Upon arrival, items are transported to area 1 for cleaning, inspecting, and sorting. Twenty-five percent are then sent to area 2, and the rest are sent to area 3. After processing at area 2, items are transported to area 4. One third of the items at area 3 are also sent to area 4, whereas the remainder are sent to area 5. Items proceed from areas 4 and 5 to area 6 for inspection. Ninety-five percent of the inspected items are sent to shipping, and the rest are transported back to area 1 for rework. Distances between workstations are located along the arcs.

There are no buffer space limitations at resource locations or for items waiting to be transported. There are two AGVs, each with a speed of 40 meters per minute. Simulate this system for 40 hours and answer the following questions:

- **a.** How many items are waiting to be transported?
- **b.** What is the utilization of the resources?
- **c.** What is the utilization of the transporters?
- **d.** How many jobs are completed?
- **e.** How many items are sent back to be reworked?

12

Advanced Concepts

This chapter introduces blocks, elements, and modifiers that facilitate the simulation of more complex systems. In particular, this chapter covers different queue management concepts and methods of handling groups of entities. The concepts discussed are useful when modeling systems with parallel queues, accumulation points with periodic releases, temporary or permanent grouping of entities, and other similar situations.

The blocks, elements, and modifiers introduced in this chapter are not integrated into one of the previous example problems. Instead, the statements are discussed in detail and illustrated with multiple code fragments.

12.1 Advanced Selection Concepts

In some situations, the modeler will be required to simulate sophisticated queuing systems. This may include multiple parallel queues containing entities competing for the same resource or a resource supplying multiple queues. The system may require that entities be placed in a fixed position of a queue, or certain entities must be located and removed from a queue. These and other queue management strategies are discussed in the following section.

12.1.1 The QPICK Block and DETACH Modifier

The QPICK block is used to model multiple parallel queues preceding a resource. In this situation, the resource is downstream from the queues and the QPICK block provides the mechanism required to determine from which queue the entity will be taken. When the downstream resource becomes available, the QPICK block removes an entity from one of the preceding QUEUE blocks and sends it to the following hold-type (ACCESS, ALLOCATE, PREEMPT, REQUEST, SEIZE, or SELECT) block.

The QPICK block is preceded in the model frame by two or more QUEUE blocks. Each associated QUEUE block must have the DETACH modifier. When an entity enters a QUEUE block with the DETACH modifier, the modifier causes the entity to stay there until it is removed by another block. Without the DETACH modifier, the arriving entity would flow through the first QUEUE block to the second QUEUE block and so on until it reaches the resource or enters a non-empty queue. The format for the QPICK block is as follows:

QPICK, *Queue Selection Rule*: *Queue Label,* repeats;

The operands are described as follows:

Operand	**Description**	**Default**
Queue Selection Rule	Queue Selection Rule [CYC, ER*(RuleID)*, LNQ, LRC, POR, RAN, SNQ, SRC, UR*(k)*]	POR
Queue Label	The block label of each preceding detached QUEUE block [block label]	—

Note: There are no modifiers permitted for use with the QPICK block.

When the downstream resource becomes available, the QPICK block will select an entity from one of the preceding labeled QUEUE blocks based on the *Queue Selection Rule*, which may include one of the following: CYClic, RANdom, Preferred Order Rule, Largest Number in Queue, Smallest Number in Queue, Largest Remaining Capacity, Smallest Remaining Capacity, user-coded User Rule number *k*, and Experimental Rule number or name *Identifier* as defined by the Experiment frame RULES element. The RULES element is discussed in Section 12.1.7.

Each *Queue Label* operand identifies an immediately preceding QUEUE block associated with the QPICK block. All QUEUE blocks referenced by a *Queue Label* operand must have a symbolic name or integer value specified for the *Queue ID* operand in the Experiment frame QUEUES element. As previously mentioned, the associated QUEUE blocks must also have the DETACH modifier.

The Preferred Order Rule (POR) is used to break ties or to select another queue if the specified queue is empty. The following segment of code illustrates the use of the QPICK block:

```
Truck   QUEUE,    ReceiveQ:  DETACH;
Pallet  QUEUE,    StorageQ:  DETACH;
Shelf   QUEUE,    BenchQ:    DETACH;
        QPICK, LNQ:
                  Truck:
                  Pallet:
                  Shelf;
        SEIZE:    Worker;
```

In this example, when a unit of the resource *Worker* becomes available, the QPICK block evaluates the number of entities in the queues with labels *Truck, Pallet, and Shelf.* An entity is taken from the queue with the largest number of entities. For example, if NQ*(StorageQ)*> NQ*(BenchQ)* > NQ*(ReceiveQ)*, then an entity is taken from the queue named *StorageQ*.

Following is an example of two queues supplying one conveyor:

```
Mold    QUEUE,    MoldQ:        DETACH;
Insp    QUEUE,    InspectQ:     DETACH;
        QPICK, LNQ:
                  Mold:
                  Insp;
        ACCESS:   PartConvey,   PartSize;
```

This example illustrates the use of the QPICK block in loading a conveyor from two different locations. The QPICK block uses the Largest Number in Queue rule for selecting entities from two queues. When *PartSize* consecutive cells are available on *PartConvey* at the entity's station location, then an entity is taken from the queue with the largest number of waiting entities and is placed on the conveyor. Conveyor concepts are covered in Chapter 10.

12.1.2 The PICKQ Block

The PICKQ block operates in a similar way to the QPICK block. The PICKQ block sends an arriving entity to one of two or more downstream queues. Similar to the BRANCH block, the PICKQ block can be used to send entities to queues serving as entry points for different resources. In this sense, the PICKQ block is used to model parallel resources. The format for the PICKQ block is as follows:

PICKQ, *Queue Selection Rule, Balk Label*: *Queue Label*: repeats;

The operands are described as follows:

Operand	Description	Default
Queue Selection Rule	Queue Selection Rule[CYC, ER*(RuleID)*, LNQ, LRC, POR, RAN, SNQ, SRC, UR*(k)*]	POR
Balk Label	The block label of the destination block for balking entities	Dispose of the entity
Queue Label	The labels of following QUEUE blocks	—

Note: There are no modifiers permitted for use with the PICKQ block.

The destination queue for the arriving entity is determined by the specified *Queue Selection Rule*, which may include one of the following: CYClic, RANdom, Preferred Order Rule, Largest Number in Queue, Smallest Number in Queue, Largest Remaining Capacity, Smallest Remaining Capacity, user coded User Rule *k*, and Experimental Rule *k* as defined in the Experiment frame RULES element.

An entity arriving to find all associated queues full will be sent to the block labeled *Balk Label*. For the same situation, if *Balk Label* is omitted, the arriving entity will be destroyed. Note that if at least one of the associated queues has infinite capacity, the arriving entity will never be sent to *Balk Label* or destroyed.

The *Queue Label* operands are the alphanumeric block labels of all QUEUE blocks associated with the PICKQ block. The PICKQ block uses the Preferred Order Rule to break any ties. If the specified queue selection rule results in a queue currently at capacity, the POR rule is used to select another queue.

Following is an example of the PICKQ block:

```
         PICKQ, LRC,  Recycle:
                      Grind1:
                      Grind2;
Grind1   QUEUE,       GrindQue1, 5;
         SEIZE:       Grinder1;
           :
Grind2   QUEUE,       GrindQue2, 9;
         SEIZE:       Grinder2;
           :
Recycle  COUNT:
         NumBalks:
         DISPOSE;
```

In this example, an arriving entity will be sent to the queue with the largest remaining capacity. Note: *GrindQue1* and *GrindQue2* are both finite capacity queues. If both of these queues are full, the arriving entity will be sent to the block labeled *Recycle* where it is counted and disposed.

12.1.3 The INSERT and REMOVE Blocks

The INSERT and REMOVE blocks give the modeler the ability to manipulate individual entities in, or arriving to a queue. The INSERT block is used to place an arriving entity at a specified position in the queue. The REMOVE block takes a specific entity from the queue and sends it to another location. The format for the INSERT block is as follows:

INSERT: *Queue Label, Rank to Place Entity*;

Operand	Description	Default
Queue Label	Label of the QUEUE block into which the entity is to be inserted	—
Rank to Place Entity	Position in the queue to place the entity [integer expression]	Queue Ranking Rule

Note: MARK is the only modifier allowed for use with the INSERT block.

Queue Label is the label of the target QUEUE block into which the entity will be inserted. If the target QUEUE block does not immediately follow the INSERT block, entities entering the INSERT block will never visit the following block.

The *Rank to Place Entity* operand specifies the position in the queue to place the entity. If defaulted, the ranking rule specified in the experiment frame QUEUES element is used to position the inserted entity. The *Rank to Place Entity* specified must be between 1 and the number in queue plus 1; in other words, in the range [1 .. NQ(*Queue ID*) + 1], where *Queue ID* is the queue number or name.

If the target queue uses HVF or LVF (High or Low Value First) ranking rules, later arrivals to the queue may be positioned incorrectly. This is best illustrated with an example in which the experiment frame contains the following element:

QUEUES: Queue1, HVF(PartType);

The associated model frame has the following block:

INSERT: Queue1, 2;

The QUEUES element specifies that arriving entities are to be ordered based on the value of an attribute named *PartType*. On the other hand, an entity arriving to the INSERT block will be placed in the second position of *Queue1*, regardless of the value of *PartType*. For this example, assume that at some point during the simulation *Queue1* contains five entities with the following *PartType* values:

2-5-8-13-23

The entities are ranked highest value first based on *PartType*. The entity at rank 1 has *PartType* value 23 and the last entity has a *PartType* value of 2. When an entity enters the INSERT block, it is unconditionally placed in the second position in the queue. For example, if an entity with a *PartType* value of 10 enters the INSERT block, it is positioned at rank 2 and the *PartType* order of entities in the queue becomes:

2-5-8-13-10-23

Notice that the queue is no longer ordered HVF*(PartType)* and that this could affect future arrivals to the queue. If the next entity enters the queue through a QUEUE block, SIMAN will scan the queue from back to front until the position for the new arrival is found. For example, if an entity with a *PartType* value of 11 enters *Queue1* through a QUEUE block, it will be placed before the entity with *PartType* value 13 and the new order will be:

2-5-8-11-13-10-23

The order of a queue that employs LVF ranking can be affected in a similar manner.

The INSERT block can be used to ensure that a preempted entity will be guaranteed the next seizure of a resource. The PREEMPT block is discussed in Chapter 13. Briefly, a preempted entity is an entity that has control of a resource and is removed before the scheduled completion time. This could be due to a simulated breakdown, a higher priority job, and so on. Usually, we would want the interrupted entity to resume processing as soon as possible. This can be done by inserting the preempted entity in the first position of the queue.

The SEARCH block, discussed in the next section, can be used to search the queue to find the correct position to insert an entity into a queue.

The REMOVE block removes the entity at a specified location in a queue and sends it to another location in the model frame. The format for the REMOVE block is as follows:

REMOVE: *Rank of Entity, Queue ID, Removed Entity Destination*;

The operands for the REMOVE block are described as follows:

Operand	Description	Default
Rank of Entity	Rank of the entity to be removed [keyword NQ or integer expression]	—
Queue ID	Queue from which the entity is to be removed [queue name or number]	—
Removed Entity Destination	Label of block where the removed entity is to be sent	—

The *Rank of Entity* is the position of the entity to be removed from the queue. If a *Rank of Entity* greater than the number of entities in the queue is used, a runtime error will occur. Therefore, the largest possible value for *Rank of Entity* is NQ*(Queue ID)*.

Queue ID is the integer queue number or symbolic queue name as defined in the experiment frame QUEUES element. A runtime error will occur if the queue *Queue ID* is not defined. The removed entity will be sent to the block labeled *Removed Entity Destination*.

Following are some examples of the REMOVE block:

REMOVE: 1, 3, Que2;

This example removes the first entity from queue number 3 and sends it to a block labeled *Que2*.

REMOVE: NQ(MachQue1), MachQue1, MachQ2;

In this example, the last entity is removed from the queue *MachQue1* and sent to the block labeled *MachQ2*.

Note: Both these examples assume that there is at least one entity in each queue. A logical construct, such as IF - ELSE, should be used to ensure that an entity actually resides in the queue before allowing an entity to enter the REMOVE block. Following is an example that will prevent any runtime errors:

```
IF:         NQ(MachQue1) > NQ(MachQue2);
     REMOVE: NQ(MachQue1), MachQue1, MachQ2;
ELSEIF:     NQ(MachQue2) > NQ(MachQue1);
     REMOVE: NQ(MachQue2), MachQue2, MachQ1;
ENDIF;
```

This example illustrates one possible method of equalizing queue lengths. When an entity arrives to the IF block, the size of *MachQue1* is compared to the size of *MachQue2*. If the first is larger than the second, then the last entity is removed from *MachQue1* and sent to *MachQue2*. The ELSEIF block checks the opposite condition. Note: An error will not occur if both queues are empty, because all conditions will evaluate to FALSE.

An entity arriving to the REMOVE block simply triggers the action and proceeds to the next block.

Block modifiers on the REMOVE block apply to the arriving entity, not the entity removed from a queue. For example:

```
Check   IF:      NQ(MachQue1) > NQ(MachQue2) .AND.
                 NQ(MachQue1) > OptFlowSize;
          REMOVE: NQ(MachQue1), MachQue1, MachQ2:
                 NEXT(Check);
        ENDIF;
        DISPOSE;
```

When an entity arrives to the IF block, the number of entities in *MachQue1* is compared to the number in *MachQue2* and to the value of a variable named *OptFlowSize*. If both components of the IF block condition evaluate to TRUE, then the last entity in *MachQue1* is removed and sent to a block labeled *MachQ2*. The NEXT modifier then sends the arriving entity back to the IF block labeled *Check*. This sequence will continue until at least one of the components of the IF block condition is FALSE, at which time the arriving entity exits the IF block and is disposed.

12.1.4 The SEARCH Block

The SEARCH block examines a group or queue for an entity that satisfies a specified condition. The position of the first entity satisfying the condition is returned in the global system variable J. If the condition is not satisfied, J is returned with a value of zero. The format of the SEARCH block is as follows:

SEARCH, *Search Item, Starting Index, Ending Index*: *Search Condition*;

The operands are described as follows:

Operand	**Description**	**Default**
Search Item	Queue or group to be searched [keyword GROUP, expression truncated to integer queue number, or a queue name]	Search current entity's group
Starting Index	Starting index rank for the search [integer expression]	1
Ending Index	Ending index rank for the search [integer expression]	Last entity
Search Condition	Search condition [condition, MIN(*Expression*) or MAX(*Expression*), and *Expression* should contain one or more attributes]	—

The *Search Item* operand specifies the type of search to be completed. An integer queue number, expression truncated to a queue number, or a queue name causes the SEARCH block to search a queue, whereas a blank or the keyword GROUP will search the arriving entity's group.

Starting Index and *Ending Index* specify the start and endpoints of the search. If *Starting Index* is greater than *Ending Index*, then a backward search is executed. When defaulted, *Starting Index* is always set to 1 and *Ending Index* is set to NG or NQ(*Queue ID*), which is determined by *Search Item*.

The *Search Condition* operand includes standard logical operators as well as the MIN(*Expression*) and MAX(*Expression*) functions. If the MIN or MAX function is used, *Expression* should contain at least one attribute.

Upon completion of the search, J is returned with the rank of the satisfying entity or zero if the condition is not satisfied. If the MIN or MAX functions are used, the entire group or queue is searched until the satisfying entity is located. Following are some examples of the SEARCH block:

```
ASSIGN:    Position = A(3);
SEARCH,    MachQue1: A(3) < Position;
IF:        J > 0;
  INSERT:  MachQue1, J;
ENDIF;
QUEUE,     MachQue1;
```

In this example, the variable *Position* is assigned the value of the third attribute of the arriving entity. This is done because attributes referenced in the SEARCH block *Search Condition* operand refer to entities in the queue, not the arriving entity. The search is conducted on *MachQue1* over the default range (1 to NQ(*MachQue1*)). When an A(3) less than *Position* is found, the location of the satisfying entity is stored in the global system variable J and the search terminates. The arriving entity then enters the IF block where the value of J is tested against zero. If J is greater than zero (a satisfying entity was found), the arriving entity is placed immediately in front of the satisfying entity in *MachQue1*. Otherwise, the arriving entity goes to the block following the ENDIF block, which places the entity in the back of the queue. Note: If *MachQue1* had been defined in the experiment frame QUEUES element with a LVF(A(3)) ranking rule, then the model frame block QUEUE, *MachQue1*; would produce the same results. Following is another example of the SEARCH block:

```
SEARCH,    WIPQue: A(4) > ExpireDate;
IF:        J > 0;
  REMOVE:  J, WIPQue, Trash;
  COUNT:   Spoiled;
ENDIF;
DISPOSE;
```

In this example, the *WIPQue* is searched from beginning to end for an entity with attribute 4 greater than the value of *ExpireDate*. If a satisfying entity is found, it is removed from the queue and sent to a block labeled *Trash*. The arriving entity proceeds to the next block where a counter is incremented and is then disposed. If there are no entities in the queue satisfying the condition, J will be zero, so the arriving entity is just sent to the DISPOSE block following the ENDIF block.

12.1.5 The SCAN Block

The SCAN block holds entities in the preceding queue block until a condition is satisfied. A QUEUE block must precede the SCAN block. The SCAN block only releases one entity at a time, and this is always the first entity in the associated queue. The format for the SCAN block is as follows:

SCAN: *Condition*;

The operand is described as follows:

Operand	Description	Default
Condition	Scan condition [condition or expression]	—

The *Condition* includes all standard logical operators and numerical expressions. Any nonzero expression is interpreted as TRUE.

If the queue preceding the SCAN block is empty, then the arriving entity will proceed directly to the SCAN block. If *Condition* is TRUE, the entity will continue; otherwise, the entity is returned to the queue. If the associated queue is not empty, then the arriving entity is placed according to the queue ranking rule specified in the experiment frame.

When *Condition* evaluates to TRUE, the first entity is processed until it encounters a time delay where it is scheduled as the next event to take place. If *Condition* is still TRUE, the next entity is processed. This sequence continues until *Condition* becomes FALSE or the associated queue is emptied.

If there are multiple queues preceding one SCAN block, when *Condition* is TRUE, then the first entity is released from the lowest-numbered queue containing entities.

Multiple SCAN blocks in the same model frame are checked from top to bottom. Since each entity passing through a SCAN block is placed first on the event list, entities waiting at later SCAN blocks will be placed in front of entities at previous SCAN blocks.

Following are some examples of the SCAN block:

```
QUEUE,    Staging;
SCAN:     ForkStatus == Empty;
```

In this example, entities are held in a queue named *Staging* until the *ForkStatus* variable is equal to the value of *Empty*. When the condition is TRUE, the first entity in *Staging* goes to the block following the SCAN block.

```
QUEUE,     A(2) + Location;
SCAN:      NQ(A(2) + Location) > Capacity;
```

In this example, entities are placed in queues according to the value of the sum of the second attribute and a variable named *Location*. For example, if *Location* is set to 10, the first four entities have A(2) values of 1, 2, 3, and 4, and the condition is FALSE, then the entities will be placed in queues 11, 12, 13, and 14. For each time advance, the SCAN block will evaluate the condition for each of the queues. If the number of entities in queue 11 is greater than *Capacity*, then the first entity in queue 11 is released, processed until it encounters a time delay, and placed first on the event list. The SCAN block then evaluates queue 12. If the number in queue 12 is greater than *Capacity*, the first entity in that queue is released, processed until it encounters a time delay, and placed first on the event list. Note: The entity from queue 12 is now first on the event list and the entity released from queue 11 is second.

Since SCAN blocks are evaluated after each time advance, the use of many SCAN blocks will significantly increase the execution time of a simulation run.

12.1.6 The FINDJ Block

The FINDJ block searches within a specified range for an integer value of the global variable J that satisfies a condition. Typically, the FINDJ block is used as a decision function in the routing of arriving entities. The format for the FINDJ block is as follows:

FINDJ, *Start of Range*, *End of Range*: *Search Condition*;

The operands are described as follows:

Operand	Description	Default
Start of Range	Start limit of range to search [integer expression]	—
End of Range	End limit of range to search [integer expression]	—
Search Condition	Search condition containing the index J [condition, *Expression*, MIN(*Expression*) or MAX(*Expression*)]	—

Start of Range is the integer starting value of J and *End of Range* is the last possible value of J. The FINDJ block searches from *Start of Range* to *End of Range* for a value of J that satisfies *Search Condition*. A *Start of Range* value greater than *End of Range* causes a backward search. If *Search Condition* is satisfied prior to reaching *End of Range*, the value of J is retained and the block is exited. If *Search Condition* is not satisfied, J is set to zero. If a MIN or MAX *Search Condition* is used, the entire range from *Start of Range* to *End of Range* is searched and FINDJ will always return a nonzero value of J. In the instance of a tie, the first satisfying value of J is retained. Following are some examples of the FINDJ block:

```
FINDJ, 1, NumCounters: (NC(J)) > Limit;
```

This FINDJ block searches from counter 1 through *NumCounters* for the first counter with a value greater than the value of *Limit*.

```
FINDJ, 7, 13: MIN(NR(J) / MR(J));
```

This example returns J that corresponds to the resource with the minimum average number of units in service. Note: NR(J) / MR(J) is the number of busy units of resource J divided by the total number of units of resource J.

```
           FINDJ,      First, Last: MAX(NQ(J));
           ASSIGN:     Longest = J;
           FINDJ,      First, Last: MIN(NQ(J));
           ASSIGN:     Shortest = J;
           IF:         Longest <> Shortest;
             REMOVE:   NQ(Longest), Longest, Shift;
           ENDIF;
           DISPOSE;
     Shift QUEUE,      Shortest;
```

In this example, two searches are conducted on the queues in the range *First* to *Last*. The first search returns the queue number with the largest number of waiting entities. The queue number is then assigned to a variable name *Longest*. The second FINDJ block then searches for the queue with the smallest number of entities waiting. The queue number satisfying this search is assigned to a variable named *Shortest*. If *Longest* is not equal to *Shortest*, then there are at least two queues of unequal length. In this case, the last entity in queue number *Longest* is removed and sent to the block labeled *Shift*. At this block, the removed entity is placed in queue number *Shortest*. The arriving entity that traveled through the FINDJ blocks, the IF block, and the REMOVE block is disposed after the ENDIF block. If *Longest* is equal to *Shortest* (i.e., all queues are empty or have the same number of waiting entities), then the arriving entity is disposed and no shuffling takes place.

12.1.7 The RULES element

The RULES element specifies selection rules to be used by the ALLOCATE, PICKQ, QPICK, REQUEST, and SELECT blocks. The SELECT block is discussed in Section 12.2.3. The RULES element is particularly useful when the decision logic for a model has not been defined, or is expected to change frequently. If this is the case, the modeler can change the RULES element instead of editing all associated blocks in the model frame. The format for the RULES element is as follows:

RULES: *Identifier, Selection Rule*: repeats;

The operands are described as follows:

Operand	Description	Default
Identifier	Rule number or name [integer or symbol name, or range]	—
Selection Rule	Queue, Resource, or Transporter Selection Rule	—

The *Identifier* operand is the integer number, inclusive range, or symbol name associated with the *Selection Rule* operand. A specific rule may be referenced in the model frame with ER(*Identifier*).

Following is an example RULES element:

```
RULES:  JobTrn, SDS:
        QRule, SNQ:
        3-7, CYC:
        8, RAN;
```

This RULES element defines the application of selection rules in eight instances that may be referenced in the model frame. The first selection rule has Identifier *JobTrn* and specifies the *Smallest Distance to Station* rule. In the model frame, this rule may be referenced with ER(*JobTrn*) or ER(1). The second entry, *QRule*, specifies the *Smallest Number in Queue* decision rule. This rule may be referenced in the model frame with ER(*QRule*) or ER(2). The third entry has an *Identifier* range from

3 to 7 and specifies the CYClic selection rule. Model frame blocks may reference this rule with ER(3) through ER(7). The last entry has *Identifier* 8 and specifies the RANdom selection rule. A model frame block may only reference this rule with ER(8).

12.2 Advanced Flow Control Blocks

12.2.1 The WAIT and SIGNAL Blocks

The WAIT and SIGNAL blocks are used to accumulate entities and then release them in batches. The WAIT block acts as a gate that impedes entity flow and causes the entities to accumulate in a preceding queue. The SIGNAL block sends a code to WAIT blocks that will allow entities with a matching code to be released. The format for the WAIT block is as follows:

WAIT: *Signal Code*, *Number to Release*;

The operands are described as follows:

Operand	Description	Default
Signal Code	Signal code [expression truncated to an integer signal code]	—
Number to Release	Maximum number of entities to release when the signal code is received [expression truncated to an integer]	—

When an entity arrives at the WAIT block, it is assigned the current value of *Signal Code*. Since the *Signal Code* can be specified as an expression, arriving entities may be assigned different *Signal Code* values. This is useful when simulating a system that handles families of jobs. When the WAIT block receives a signal, a maximum of *Number to Release* entities with a *Signal Code* equal to the received signal are released from the preceding queue.

A WAIT block must be preceded by a QUEUE block. The number of entities released by a WAIT block can be controlled with the *Number to Release* operand or by a SIGNAL block. If there is no limit set by the WAIT block *Number to Release* operand or by the broadcasting SIGNAL block, all waiting entities with a *Signal Code* value equal to the broadcast signal will be released.

Following are some examples of the WAIT block. For these examples, it is assumed that the number of entities that can be released from a WAIT block is limited only by the *Number to Release* operand.

```
QUEUE,      1;
WAIT:       10;
```

In this example, arriving entities are assigned a *Signal Code* value of 10 and placed in queue number 1. The *Number to Release* operand has been defaulted to infinite. All entities residing in queue 1 will be released when the WAIT block receives a signal of 10.

```
QUEUE,      DrillQ;
WAIT:       1, Bundle;
```

In this example, arriving entities are assigned a *Signal Code* of 1 and placed in a queue named *DrillQ*. When the WAIT block receives the appropriate signal, a maximum of *Bundle* entities may be released. If there are more than *Bundle* entities in the queue, the remaining entities will wait for the next signal.

```
QUEUE,      DrillQ;
WAIT,       WorkerID, Pace(WorkerID);
```

In this example, arriving entities are placed in a queue named *DrillQ* and assigned a *Signal Code* equal to the current value of the *WorkerID* variable. *WorkerID* also serves as an index for a one-dimensional array named *Pace*, which would be defined in the experiment frame VARIABLES element. The value returned by *Pace(WorkerID)* determines the maximum number of entities that may be released by this WAIT block. For example, if the current value of *WorkerID* is 2, and *Pace* is defined in the experiment frame as:

```
VARIABLES: Pace(5), 4, 9, 6, 3, 8;
```

then an arriving entity is assigned a *Signal Code* of 2. When a signal of 2 is broadcast, if *WorkerID* is still equal to 2, a maximum of nine entities can be released by the WAIT block. Note: the signal does not affect the value of *WorkerID*. If *WorkerID* is equal to 3 and a signal of 2 is received, a maximum of six entities with a *Signal Code* value of 2 can be released.

The SIGNAL block sends a signal to all WAIT blocks in the model frame. The SIGNAL block also specifies the total number of entities that can be released from all WAIT blocks. The format for the SIGNAL block is as follows:

SIGNAL: *Signal Code, Release Limit*;

The operands are described as follows:

Operand	Description	Default
Signal Code	Signal code [integer expression]	—
Release Limit	Total number of entities to release [integer expression]	Infinite

Signal Code is the value of the signal sent to all of the model frame WAIT blocks. *Release Limit* is the maximum number of entities in the entire model frame that may be released with the current *Signal Code*. Note: *Release Limit* constrains the maximum total number of entities that may be released, whereas the WAIT block *Number to Release* operand limits the maximum number of entities that may be released from one WAIT block.

Following are some examples of the SIGNAL block:

SIGNAL: 10;

When an entity arrives to this SIGNAL block, a signal of 10 is sent to all WAIT blocks in the model frame. Since *Release Limit* is defaulted to infinite, either all waiting entities with a *Signal Code* of 10 will be released or individual WAIT blocks will control the number of entities released. The entity arriving to the SIGNAL block continues processing after the waiting entities have been released from the WAIT blocks.

SIGNAL: 1, 2 * Bundle;

This SIGNAL block broadcasts a signal of 1. The maximum number of entities that may be released is specified as two times the value of *Bundle*. If the model frame contains multiple WAIT blocks, each with a *Number to Release* equal to *Bundle* and each with at least *Bundle* entities waiting, then only the first two WAIT blocks will be able to release entities. Entities in excess of two times *Bundle* will have to wait for the next signal.

SIGNAL: WorkerID, 20;

In this example, the SIGNAL block sends a signal equal to the current value of *WorkerID*. A maximum of 20 entities with a *Signal Code* equal to *WorkerID* may be released in the model frame.

12.2.2 The MATCH Block

The MATCH block synchronizes the advance of entities from multiple preceding detached queues. The entities may have an attribute value used to "match" them with other entities, or the MATCH block may be used to block entity flow until there is at least one entity in each associated queue. The format for the MATCH block is as follows:

MATCH, *Match Attribute*: *Queue Label, Destination Label*:
repeats;

The operands are described as follows:

Operand	Description	Default
Match Attribute	Match attribute [integer attribute number, A(*k*), or attribute symbol name]	No Attribute
Queue Label	Label of the detached QUEUE block where the entity resides	—
Destination Label	Destination label for the entity	Dispose the entity

Note: No modifiers are permitted for use with the MATCH block. All QUEUE blocks associated with a MATCH block must be labeled and have the DETACH modifier. The *Queue ID* operand of the associated QUEUE blocks must be specified as an integer queue number or the queue name. An expression type *Queue ID* will result in a linker error.

The *Match Attribute* operand is the attribute name or number that is used to match an entity with entities in other queues. When each associated queue has one or more entities with equal *Match Attribute* values, the first entity with this *Match Attribute* value is released from each queue. If *Match Attribute* is defaulted, the MATCH block will release the first entity from each associated queue when all of the queues have one or more residing entities.

Queue Label operand values are the QUEUE block labels corresponding to the detached queues where entities wait for a match. These QUEUE blocks may not be referenced by any other MATCH or QPICK blocks.

The *Destination Label* operand is the label of the block where the entity released from *Queue Label* will be sent. If *Destination Label* is defaulted, the leaving entity is destroyed. Following are some examples of the MATCH block:

```
Engine   QUEUE,  EngineQ:  DETACH;
Tranny   QUEUE,  TransQ:   DETACH;
         MATCH:
                 Engine, Asmbly:
                 Tranny;
```

In this example, as soon as there is at least one entity in *EngineQ* and *TransQ*, the first entity in *EngineQ* is sent to a block labeled *Asmbly* and the first entity in *TransQ* is destroyed. A possible application of this code fragment is in the manufacture of automobiles where assembled engines and transmissions are carried forward in the process.

```
MthrBrd   QUEUE,   BoardQ:  DETACH;
DskDrv    QUEUE,   DiskQ:   DETACH;
Monitor   QUEUE,   7:       DETACH;
          MATCH,   PCType:
                   MthrBrd, Asmbly:
                   DskDrv:
                   Monitor, Package;
```

In this example, entities are matched based on the value of the attribute named *PCType*. A match occurs when each queue has an entity with the same value for *PCType*. When there is a match, the first entity in *BoardQ* with a matching *PCType* is sent to the block labeled *Asmbly*, the first entity in *DiskQ* with a matching *PCType* is destroyed, and the first matching entity in queue number 7 is sent to a block labeled *Package*. Similar to the first example, the entity from *BoardQ* is probably treated as a single assembly from this point forward.

12.2.3 The SELECT Block and Keyword

The SELECT block is the resource version of the PICKQ block. An entity arriving to the SELECT block chooses one resource from a group of resources based on the specified resource selection rule. Following the SELECT block are labeled SEIZE blocks from which the arriving entity will select. In most cases, the SELECT keyword may be used with a SEIZE block instead of the SELECT block. Examples of both cases will be given. The format of the SELECT block is as follows:

SELECT, *Resource Selection Rule*: *Seize Label*: repeats;

The operands are described as follows:

Operand	Description	Default
Resource Selection	Resource Selection Rule [CYC, ER(*RuleID*), LNB, LRC, POR, RAN, SNB, SRC, UR(*k*)]	POR Rule
Seize Label	The block label of each following SEIZE block	—

Note: No modifiers are permitted for use with the SELECT block. The SELECT block must be preceded by a QUEUE or QPICK block and followed by two or more labeled SEIZE blocks.

The *Resource Selection Rule* operand specifies which resource selection rule will be used. Among these are CYClic, Largest or Smallest Number Busy, Largest or Smallest Remaining Capacity, RANdom, Preferred Order Rule, user coded User Rule number k, and Experimental Rule (*RuleID*) from the experiment frame RULES element. *Seize Label* refers to the label of an associated SEIZE block.

Following are some examples of the SELECT block:

```
        QUEUE,   DrillQ;
        SELECT:
                 Drill1:
                 Drill2;
Drill1  SEIZE:   NewDrill;
        DELAY:   UNIF(1, 6);
          :
Drill2  SEIZE:   OldDrill;
        DELAY:   UNIF(4, 7);
          :
```

In this example, entities are stored in a queue name *DrillQ* and choose from two resources. The default Preferred Order Rule is applied, so the block labeled *Drill1* is always tried first. If *NewDrill* is busy, the block labeled *Drill2* is tried. If both are busy, the entity is returned to *DrillQ* to wait for one of the resources to become available. The selection portion of this example can also be modeled with the SELECT keyword. If the experiment frame contained the following:

```
RESOURCES: Drill1: Drill2;
SETS:      Drills, Drill1, Drill2;
```

Then the model frame could use:

```
QUEUE,     DrillQ;
SEIZE:     SELECT(Drills, POR, DrillID);
```

to achieve the same selection results as the previous example. The SELECT keyword selects from the set using the same selection rule as before and stores the number of the seized resource in an attribute named *DrillID*. A subsequent RELEASE: *Drills(DrillID)* block will insure that the correct resource is released. Not shown here are the blocks required to duplicate the time delays. An IF block could be used to determine which drill was seized and assign appropriate parameter values for the uniform distribution. Following is a QPICK block and SELECT block combination:

```
Truck   QUEUE, ReceiveQ:   DETACH;
Pallet  QUEUE, StorageQ:   DETACH;
Shelf   QUEUE, BenchQ:     DETACH;
        QPICK, SNQ:
          Truck:
          Pallet:
          Shelf;
        SELECT, LRC:
          Work1:
          Work2:
          Work3;
Work1   SEIZE: Joe:        NEXT(Process);
Work2   SEIZE: Mary:       NEXT(Process);
Work3   SEIZE: Bill:       NEXT(Process);
```

In this example, the QPICK block selects entities from three queues based on the *Smallest Number in Queue* rule. The SELECT block uses the *Largest Remaining Capacity* resource selection rule for resource allocation. When an entity successfully seizes a unit of one of the resources, it is then sent to the block labeled *Process*. This example could also be modeled using the SELECT keyword with the SEIZE block. If the experiment frame contained the following:

```
RESOURCES:   Joe: Mary: Bill;
SETS:        Worker, Joe..Bill;
```

then the model frame could contain:

```
Truck   QUEUE, ReceiveQ:   DETACH;
Pallet  QUEUE, StorageQ:   DETACH;
Shelf   QUEUE, BenchQ:     DETACH;
        QPICK, SNQ:
          Truck:
          Pallet:
          Shelf;
        SEIZE: SELECT(Worker, LRC, EmpID):
          NEXT(Process);
```

which provides the same results as the preceding example. The resource number is stored in the entity attribute named *EmpID* for later release. Note: For an *EmpID* of 2, *Worker*(2) and *Worker(EmpID)* both refer to the resource named *Mary*.

12.3 Grouping Entities

Sometimes it is necessary to treat a collection of entities as a single object. This is common in simulations that model the assembly of a product. A good example is automotive production. A simulation may model production of automotive subassemblies and the assembly of these subassemblies into an automobile. When the parts are assembled, the vehicle becomes the primary concern and the subassemblies are no longer significant. In this case, the set of contributing entities can be permanently combined to form one entity. In other situations, entities may be placed in a temporary group, which can be processed as a single entity or separated into component entities.

12.3.1 The GROUP Block

The GROUP block temporarily combines entities into a new representative entity. Once the group is formed, the representative entity can be treated like any other entity in the model. The SEARCH block can be used to evaluate the attributes of entities in a group. Since the group is temporary, it can be split into individual entities. If there is not a QUEUE block preceding the GROUP block, arriving entities are stored in an internal queue. The format for the GROUP block is

GROUP, *Match Expression*: *Quantity to Group, Save Criterion*;

The operands are described as follows:

Operand	Description	Default
Match Expression	The expression used to group matching entities [expression]	No matching expression
Quantity to Group	The number of entities to temporarily group together [integer expression]	—
Save Criterion	Save criterion [FIRST, LAST, PRODUCT, SUM]	LAST

The *Match Expression* operand is a comparison value used to combine entities with similar characteristics. If *Match Expression* is specified, it is usually an entity attribute or expression containing one or more attributes. *Quantity to Group* specifies the number of entities to be put in one group.

Save Criterion determines how the representative entity will be assigned attribute values. The FIRST *Save Criterion* will assign the attribute values of the first entity to enter the group to the representative entity. The SUM *Save Criterion* assigns the sum of component entity attribute A(k) to the representative entity attributes A(k). For example, A(1) of the representative entity will be the sum of all component entity A(1) values. LAST and PRODUCT are handled in a similar manner.

If *Save Criterion* is specified as FIRST, SUM, or PRODUCT, then the representative entity STATIONS and SEQUENCES attributes (M, NS, and IS) are the same as the first entity placed in the group; otherwise, these values are determined by the last entity placed in the group.

If *Match Expression* is not specified, then the first *Quantity to Group* entities to arrive to the GROUP block form a group; otherwise, the first *Quantity to Group* entities to arrive to the GROUP block with the same *Match Expression* value will form a group. The component entities of a group are stored internally for later retrieval. Each new group causes the creation of a representative entity, which adds one to the total number of entities in the system.

A representative entity should not be disposed until it has been separated into individual entities. If this is done, the component entities remain in the system and cannot be accessed. If duplicate representative entities are created, group size changes to the original or duplicate representative entities will apply to the original and other duplicates. If the group size of one of the representative entities for a particular group is reduced to zero, all other associated representative entities will have a group size of zero. Following are some examples of the GROUP block:

```
QUEUE,    PalletQ;
GROUP:    25;
```

In this example, entities accumulate in a queue named *PalletQ*. Since *Match Expression* is defaulted, every 25 entities to arrive to the queue form a group. The representative entities created by this GROUP block are assigned the attribute values of the last entity placed in each group since *Save Criterion* defaults to LAST.

```
QUEUE,    PalletQ;
GROUP,    PartType: Pallet(PartType);
```

In this example, entities are grouped according to the value of an attribute named *PartType*. The number of entities in a group is determined by the entry in the one-dimensional array called *PartSize*, which would be defined in the experiment frame VARIABLES element. Once again, the representative entity attributes are determined by the last entity to be placed in the group.

```
QUEUE,  WipQ;
GROUP,  ShipToID: QtyOrdered, SUM;
ROUTE:  A(4), Unload;
```

In this example, entities are grouped according to the value of an attribute named *ShipToID*. The number of entities in each group depends on the value of the variable *QtyOrdered*. The representative entity attributes are the sums of individual component entity attribute values. The M, NS, and IS representative entity values are the same as the first entity placed in each group. Once a group is formed, the representative entity is transported to the station named *Unload* by the ROUTE block. The duration of the transport is specified as the fourth attribute of the representative entity, which is the sum of the fourth attribute of all component entities in the group.

12.3.2 The SPLIT Block

The SPLIT block separates a temporary group formed by the GROUP block and destroys the associated representative entity. The entities in the group can be retrieved with original attribute values or assigned the attribute values of the representative entity. The format of the SPLIT block is as follows:

SPLIT: *Attributes*;

The operand is described as follows:

Operand	Description	Default
Attributes	List of entity attributes that are assigned to members of the temporary group [M, NS, IS, *, A(*), *AttributeID*]	No attributes assigned

The *Attributes* operand values identify leaving entity attributes that will be assigned the value of the corresponding representative entity attributes. Following are some examples of the SPLIT block:

```
SPLIT;
```

When a representative entity arrives to this SPLIT block, all of the component entities are restored unchanged and the representative entity is destroyed. The component entities proceed from the SPLIT block in the same order that they were placed in the group.

SPLIT: A(4), TimeOfArr;

This SPLIT block assigns the value of the representative entity attributes A(4) and *TimeOfArr* to component entities recovered from the group. Other component entity attributes remain unchanged, and the representative entity is destroyed.

SPLIT: * ;

This SPLIT block assigns the current value of all representative entity attributes to the recovered entities and the representative entity is destroyed.

12.3.3 The PICKUP and DROPOFF Blocks

The PICKUP block provides the ability to add more entities to a previously created group. The DROPOFF block is used to remove a specified number of entities from a group without disturbing the remaining component entities. These blocks are very useful in modeling material handling systems. The format for the PICKUP block is as follows:

PICKUP: *Queue ID, Starting Rank, Quantity to Pickup*;

The operands are described as follows:

Operand	Description	Default
Queue ID	Source queue from which entities are picked [expression truncated to an integer queue number, or queue symbol name]	—
Starting Rank	Starting rank of the entities in *Queue ID* [expression truncated to an integer]	1
Quantity to Pickup	Number of entities to pick [expression truncated to an integer]	1

The *Quantity to Pickup* of entities removed from the queue *Queue ID* are added to the end of the current entity's group. These entities may be removed from the group using a DROPOFF or SPLIT block. A runtime error occurs if there are less than *Starting Rank* + *Quantity to Pickup* entities. A BRANCH block is suggested to avoid this problem. An entity entering a PICKUP block becomes the group representative entity. This representative entity should not be disposed until all group members have been removed or a runtime warning will be issued. (An entity entering a GROUP block becomes a group member, rather than a rcpresentative.)

Following are some examples of the PICKUP block:

PICKUP: DrillQ;

The entity entering this block picks up the first entity in the queue *DrillQ* and adds the entity to the end of the group.

PICKUP: DrillQ, 2, Extra;

This block takes the quantity given by the named attribute or variable *Extra* from the queue named *DrillQ* starting with rank 2 and adds them to the end of the group.

The format for the DROPOFF block is as follows:

DROPOFF, *Rank of Entity, Quantity to Dropoff*: *Dropoff Location, Attributes*;

The operands are described as follows:

Operand	**Description**	**Default**
Rank of Entity	Starting rank of entities to drop off [expression truncated to an integer]	1
Quantity to Dropoff	Number of entities to drop off [expression truncated to an integer]	1
Dropoff Location	Label of the block where dropped entities are sent [block label]	—
Attributes	List of representative entity attributes that are assigned to the entities being dropped [*, A(*), M, NS, IS, *AttributeID*]	No attribute assignment

GROUP and PICKUP blocks create a single entity to represent a group of entities. The DROPOFF block removes *Quantity to Dropoff* entities from the current entity group beginning with the entity *Rank of Entity* and sends them to the block at *Dropoff Location*. The representative entity can have different attribute values than the entities comprising the group. Attribute values can be assigned to the entities leaving the group with the *Attributes* operand. To assign individual attributes, use *AttributeID* (attribute symbol name or A(k)), M, NS, or IS. The attribute variable with an asterisk [A(*)] assigns all general purpose attributes of the representative entity to exiting entities and a standalone asterisk (*) will assign all attributes including M, NS, and IS to the exiting entities.

The variable NG evaluates the group size for the current entity. This can be used to drop off all entities in a group and have the representative entity continue to another block location. Following are some examples of the DROPOFF block:

```
DROPOFF:     LineA;
```

The first entity in the group is sent to the block labeled *LineA*. No attributes are assigned to the departing entity. The representative entity continues to the next block.

```
ASSIGN:      Color = 3;
DROPOFF,     2, X(6):
             PaintShop, Color;
```

Prior to the DROPOFF block, the attribute *Color* is assigned the value 3. The DROPOFF block sends X(6) entities, beginning with the second to the block label *PaintShop*. Each entity leaving the group has its attribute named *Color* set to 3. The other attributes, if any, are not affected.

```
DROPOFF,     1, NG:
             ProcessB, M, A(2):
             DISPOSE;
```

All entities in the group are sent to the block labeled *ProcessB*. The values of M and A(2) from the representative entity are assigned to the departing entities. The representative entity is disposed.

```
DROPOFF,     1, NG:
             ProcessB, *:
             DISPOSE;
```

This is the same as the previous example, except the value of all the representative entity's attributes are assigned to the departing entity's attributes.

12.3.4 The COMBINE Block

The COMBINE block permanently combines a specified number of entities into one representative entity. Unlike the GROUP block, entities grouped with the COMBINE block cannot be recovered. Using the COMBINE block to group entities will effectively reduce the total number of entities in the system. If the COMBINE block is not preceded by a QUEUE block, entities arriving to the COMBINE block will be stored in an internal queue. The format for the COMBINE block is as follows:

> COMBINE, *Match Expression*: *Quantity to Combine, Save Criterion*;

The operands are described as follows:

Operand	Description	Default
Match Expression	Expression used to combine matching entities [expression]	No matching expression
Quantity to Combine	Number of entities to permanently combine [integer expression]	—
Save Criterion	Save criterion [FIRST, LAST, PRODUCT, SUM]	LAST

The *Match Expression* operand is a comparison value used to combine entities with similar characteristics. If *Match Expression* is specified, it is usually an entity attribute or expression containing one or more attributes.

Quantity to Combine specifies the number of entities to be permanently combined into one representative entity. If *Quantity to Combine* is an expression, then each arriving entity causes the expression to be evaluated. The new value of *Quantity to Combine* will then be applied to all entities waiting to be combined.

Save Criterion determines how the representative entity will be assigned attribute values. The LAST *Save Criterion* will assign the attribute values of the last entity in the queue to be combined with the representative entity. The PRODUCT *Save Criterion* assigns the product of component entity attributes A(k) to the representative entity attribute A(k). For example, A(1) of the representative entity will be the product of all component entity A(1) values. FIRST and SUM are handled in a similar manner.

If *Save Criterion* is specified as FIRST, SUM, or PRODUCT, then the representative entity STATIONS and SEQUENCES attributes M, NS, and IS are the same as the first entity taken from the queue; otherwise, these values are determined by the last entity in the queue to be combined.

If *Match Expression* is not specified, the COMBINE block creates a representative entity for each group of *Quantity to Combine* entities to accumulate in the queue and then destroys the component entities. Otherwise, the first *Quantity to Combine* entities to arrive to the queue with the same value for *Match Expression* will be combined to form a representative entity and the component entities will be destroyed.

Unlike the MATCH block, a QUEUE block preceding a COMBINE block may have an expression for the *Queue ID* operand. This allows the entities to accumulate in multiple queues and is helpful when using a station type model.

Following are some examples of the COMBINE block:

```
QUEUE,    1;
COMBINE: 25;
```

Each time 25 entities accumulate in queue 1, a representative entity is created with the same attribute values of the last entity to enter the queue. The 25 waiting entities are then destroyed, and the representative entity proceeds to the next block.

```
QUEUE,    DrillQ;
COMBINE, JobType: 4, SUM;
```

In this example, entities accumulate in a queue named *DrillQ*. When four entities with the same value of the attribute *JobType* are present, a representative entity will be created. The representative entity attribute values will be the sum of the four component entity attributes. The M, NS, and IS values of the representative entity are assigned the values of the first entity taken from the queue. The four matching entities are then destroyed and the representative entity proceeds to the next block.

```
QUEUE,    Location(M);
COMBINE, M: 7, PRODUCT;
```

In this example, the queue number is returned from a one-dimensional array named *Location* based on the value of the station attribute M.

The COMBINE block combines groups of seven entities that have the same M value. The representative entity created will have M, N, and IS values of the first entity taken from the queue. Remaining attribute values will be the product of component entity attributes. For example, if the value of *Location*(3) is 9, then all entities with an M value of 3 will be placed in queue 9. When queue 9 contains seven entities with M equal to 3, a representative entity is created and the seven component entities are destroyed.

12.4 SUMMARY

This chapter introduces important blocks and elements for simulating complex systems. For modeling systems with parallel queues, the QPICK and PICKQ blocks provide the capability to selectively choose entities from preceding queues and to specify the destination queue for exiting entities. The INSERT and REMOVE blocks can be used to manipulate queues one entity at a time. The INSERT block is used to place an entity in a specific position in a queue, and the REMOVE block takes an entity at a specific position out of a queue. The SEARCH block returns the position of the first entity in a group or queue that satisfies a conditional expression. The SEARCH is commonly used in conjunction with INSERT and REMOVE to manage queues on an entity level. The SCAN block is used to block entity flow, causing the entities to accumulate in the preceding queue, until a logical expression is evaluated to be TRUE or a mathematical expression returns a nonzero value. The WAIT and SIGNAL blocks are used to accumulate entities and then to release them in batches. The WAIT block causes the entities to accumulate until a releasing signal is received from a SIGNAL block. The FINDJ block is also used to manage entity flow. The FINDJ block returns the value of the global variable J that satisfies an expression. The value of J can then be used to make processing decisions in the model frame. The MATCH block is used to synchronize the processing of entities as they progress through the model frame. The SELECT block is used to select one resource from multiple parallel resources. The GROUP and SPLIT blocks are used to temporarily combine entities into a representative entity and later break the representative entity down into component entities. The PICKUP and DROPOFF blocks are used to add or remove entities to a temporary group formed with the GROUP block. The last block introduced was the COMBINE block, which combines entities into a permanent representative entity. Once the permanent representative entity has been created, the component entities are destroyed.

12.5 EXERCISES

E12.1 A production process has six machines. Machines 1, 2, and 3 are in parallel followed by Machines 4 and 5, also in parallel, and finally, Machine 6. Jobs arrive to be processed according to an exponential distribution with a mean interarrival time of 10 minutes. Upon arrival, jobs select one of Machines 1, 2, or 3 based on the smallest number in queue. The times spent in Machines 1, 2, and 3 are exponentially distributed with means 22, 26, and 36 minutes, respectively. After processing on one of Machines 1, 2, and 3, the jobs are split into two parts, A and B. Part A is processed on Machine 4 in a uniformly distributed time that ranges from 2 to 16 minutes. Part B is processed on Machine 5 in a time of 6 minutes with probability 0.6, a time of 9 minutes with probability 0.3, or 20 minutes with probability 0.1. After processing on Machines 4 and 5, matching pairs are reassembled and sent to be inspected at Machine 6. Some 3% of the jobs fail inspection and must go through Machines 4 and 5 again, and then reinspected. Inspection takes 4 to 14 minutes, uniformly distributed.

a. Simulate this system for 24 hours and collect statistics on each queue, the number of reworked parts, and resource utilization.

b. A spotter comes to the inspection machine every hour on the hour and removes the last job (if any) from the queue for destructive testing purposes. Simulate for 24 hours and collect statistics on all queues, resources, number reworked, and number removed.

E12.2 Repeat E12.1 (a and b) with the following change to the system. There is only one queue serving the first three machines. There are still seperate queues for machines 4, 5, and 6.

E12.3 Jobs arrive at a shop to be processed according to an exponential distribution whose mean is 20 minutes. There are two identical machines, A and B, which can process the jobs. A queue forms in front of each machine. The arriving jobs are placed in the shortest queue. The processing time on each machine is normally distributed with a mean of 32 minutes and a standard deviation of 4 minutes. The shop opens at 8:00 A.M. Machine A is serviced from 11:00 A.M. to Noon each day. Machine B is serviced from 1:00 P.M. to 2:00 P.M. each day. After completing the machining, the jobs are inspected, with 85% passing inspection. Inspection times are uniformly distributed between 15 and 20 minutes. Sixty percent of those failing inspection must be reprocessed on the same machine as previously used. The other jobs that fail to pass inspection are scrapped.

Simulate the operations of the system for 1 day. A day ends after all jobs arriving by 4:30 P.M. have been processed. Collect statistics on all queues and resources. Note: At simulation completion, all resources and their respective queues should be empty.

E12.4 Three different types of candy arrive to a packaging area according to an exponential distribution with a mean of 1.5 minutes. Twenty-one percent of the arriving candies are type A, 42 percent are type B, and the rest are type C. As the candies arrive, they are placed in separate queues. Each queue is equipped with an automatic sorter that combines families of candy. There are two equally likely combinations for each type of candy. Type A candy may be combined in groups of one or three pieces, type B may be combined into groups of two or six pieces, and type C may be combined into groups of two or four pieces. These packages are then automatically placed in separate queues to await further processing.

There are two workstations that perform final packing for shipment. When each queue from the sorters has at least one package of candy, one package of each type of candy is taken from their respective queues and packed for shipping. The two workstations are used cyclically. The worker at the first station requires a normally distributed packing time with a mean of 12 minutes and a standard deviation of 1.9 minutes. The worker at the second station also operates according to a normally distributed time with a mean and standard deviation of 18 and 2.8 minutes, respectively.

Simulate this system for 4800 minutes. Maintain a count of each type of candy that arrives and the total number of packages produced. Also produce statistics on all resources and queues.

E12.5 Jobs arrive to a machine shop according to an exponential distribution with a mean of 8 minutes. The jobs are placed in a queue until one of three identical machines becomes available. Processing times for the machines are normally distributed with a mean and standard deviation of 21.5 and 3.4, respectively. When processing is completed, the job is placed in one of two queues, based on the least number in queue. From these two queues, the jobs proceed to one of four identical workstations. Processing times at the workstations are uniformly distributed between 20.2 and 37.5 minutes. From the workstations, the jobs are sent to be inspected. Inspection times are normally distributed with a mean and standard deviation of 7.3 and 1.2 minutes, respectively. Records indicate that 4% of inspected jobs are rejected. Of this 4%, 1% is sent back to the first three machines and 3% is sent back to the four workstations. The rejected jobs are considered highest priority and therefore are placed at the front of any queues encountered during rework.

Simulate this system for four 40-hour weeks. Record the number of jobs completed, the number sent to the three machines, and the number sent back to the workstations. Also, collect statistics on each resource and queue. Use the Largest Remaining Capacity Rule for selecting among resources.

13

Additional Concepts

This chapter presents additional blocks and elements that can be useful in constructing more advanced models and experiments. These blocks and elements are described with code fragments, but are not integrated into an example problem being solved. If the reader has conscientiously completed a selection of the exercises found in the text, using the concepts in this chapter can be aided by completing the exercises at the end of the chapter.

13.1 Naming

The NICKNAMES element substitutes user-defined symbol names *Nickname* for a SIMAN attribute or system variable, a previously defined symbol name or a constant, *Name or Constant*. More than one name can be given to the same item. User-defined nicknames may be used in the experiment only after being defined in the NICKNAMES element. The format is as follows:

NICKNAMES: *Nickname, Name or Constant*: repeats;

The operands are described as follows:

Operand	Description	Default
Nickname	User-defined symbol name	—
Name or Constant	SIMAN variable or attribute name, user-defined symbol, or constant	—

Two examples of the use of the NICKNAMES element are:

```
NICKNAMES:
        MachineCenter, M:
        PartType, NS:
        ProcessStep, IS;
```

This NICKNAMES element defines the symbol name *MachineCenter* for station attribute M, *PartType* for sequence set attribute NS, and *ProcessStep* for sequence index attribute IS. In the model frame, the names given can be substituted anywhere for the SIMAN names M, NS, and IS.

```
ATTRIBUTES:
        DrillTime:
        MillTime;
VARIABLES:
        Inventory;
:
NICKNAMES:
        OP20Time, MillTime:
        WorkInProcess, V;
```

Individual elements of an array may not be nicknamed unless they have been defined with a unique name. In this example, the SIMAN attribute A(2) can be given a nickname since it was defined in the ATTRIBUTES element with a unique name, *MillTime*. Note: *MillTime* is specified as the *Name or Constant* in the NICKNAMES element. Similarly, the SIMAN variable V(1) was defined as *Inventory*. Now, A(2), *MillTime*, and *OP20Time* can be used interchangeably in the model or experiment as can V(1), *Inventory*, and *WorkInProcess*.

13.2 Setting Initial Values

We have previously discussed the ASSIGN block, ATTRIBUTES, and VARIABLES elements that assign values. Values may be assigned also, at the time of creation, within the experiment frame, using the INITIALIZE element and the ARRIVALS element.

13.2.1 The INITIALIZE Element

The INITIALIZE element specifies initial values for SIMAN user assignable variables and special-purpose attributes including the integer global variable J, elements of the X(k) variable array, and attributes M, NS, and IS. If real numbers are entered for J, M, NS, or IS, they will be truncated to integers. Values specified by the INITIALIZE element for attributes M, NS, and IS are applied to all entities created during the simulation run. The format for the INITIALIZE element is as follows:

INITIALIZE: *Variable* = *Value*: repeats;

The operands are described as follows:

Operand	Description	Default
Variable	Variable or integer attribute to be initialized [IS, J, M, NS, or X(*k*), *k* an integer]	—
Value	Value to be assigned to the variable [constant, (integer for IS, J, M, or NS)]	—

An example using the INITIALIZE element is as follows:

```
INITIALIZE:
    X(12) = 17.35:
    J = 4:
    M = 3;
```

This element initializes X(12) to 17.35 and integer global variable J to 4. All entities entering the simulation will have their station attribute, M, set to 3.

13.2.2 The ARRIVALS Element

The ARRIVALS element creates batches of elements that arrive at specified times. The format is given by the following:

ARRIVALS: *Identifier, Type(Type ID), Time, Batch Size,*
Assignments: repeats;

The operands are described as follows:

Operand	Description	Default
Identifier	Arrival number or name [integer or symbol name]	—
Type(Type ID)	Arrival type: STATION(*Station ID*) QUEUE(*Queue ID*) BLOCK(*Block ID*)	—

Operand	Description	Default
Time	Time for arrival to occur [constant, FIRST, LAST, WARMUP, EVERY (*Interval* [,*Offset*] [,*Max Batches*][*Max Time*])]	Beginning of simulation
Batch Size	Number of entities in batch arrival [expression truncated to an integer]	1
Assignment	Initial values of entity attributes [list of assignments in the form Attribute ID = Value, . . .]	—

ARRIVALS may be referenced in the model via the *Identifier* operand. *Identifier* accepts either an integer or symbol name, but not both. If numbered, ARRIVALS must be in ascending order. Sequential numbering is recommended to conserve space.

The entities arrive as prescribed in the operand *Time*. The options include the following:

FIRST indicates that entities within a batch arrive simultaneously at the location specified at the beginning of each simulation replication.

LAST indicates that entities within a batch arrive simultaneously at the location specified at the end of each simulation replication.

WARMUP indicates that entities within a batch arrive simultaneously at the location specified at the end of the warm-up period specified in the REPLICATE element.

EVERY indicates that entities arrive at the *Offset* time and subsequently at each *Interval* thereafter during the simulation. If *Offset* is not specified, it defaults to *Beginning Time* as specified on the REPLICATE element. Arrivals continue until reaching the value specified in *Max Batches* or *Max Time*. If neither of these is specified, arrivals continue until the end of the simulation. If *Max Batches* is defaulted, *Interval* must be nonzero.

Arriving entities may be given initial attribute values by entering *Assignments* in the form *Attribute ID* = *Value*. *Attribute ID* may be specified as A(*k*), *Attribute Name*, M, NS, or IS, that is, any SIMAN expression.

Examples of the ARRIVALS element are as follows:

ARRIVALS: 1, QUEUE(DrillQ), , 4, A(1)=2.3, A(3)=1.0;

Four entities arrive at *DrillQ* at time 0.0. A(1) and A(3) are given the values indicated. A value for A(2) is not specified by the ARRIVALS element, so it is either 0.0 or some value specified in the ATTRIBUTES element, if any.

```
ARRIVALS: First,STATION(MC1),120.0:
          Second,BLOCK(50),240.0,5,DrillTime=NORM(7.5,1.2):
          Third,BLOCK(Retry),480.0,2,M=7,NS=2,IS=6;
```

One entity arrives at the station named MC1 at time 120.0. Five entities arrive at Block number 50 at time 240.0. The entities have their *DrillTime* attribute set to the value from a sample from a normal distribution with mean 7.5 and standard deviation 1.2. Two entities are scheduled to arrive at block labeled *Retry* at time 480.0. These entities will have M initialized to 7, NS initialized to 2, and IS initialized to 6.

```
ARRIVALS: AtStart,  BLOCK(Options), FIRST:
          AtFinish, BLOCK(DataScan), LAST:
          AtWarm,   BLOCK(30), WARMUP:
          AtDrillQ, QUEUE(DrillQueue), EVERY(15.0, 60.0, 20);
```

One entity arrives at the block labeled *Options* at the beginning of the simulation. One entity arrives at the block labeled *DataScan* at the end of the simulation. (This entity is likely being used to generate a special report with the help of WRITE blocks.) One entity arrives at block 30 after the WARMUP period specified in the REPLICATE element. One entity arrives at the location labeled *DrillQueue* at time 60.0 and every 15.0 time units thereafter until 20 entities have arrived.

Several remarks are in order for appropriate use of the ARRIVALS element as follows: QUEUE-type arrivals must arrive either to a QUEUE block with *Queue ID* operand specified as an integer or symbol name, or to a labeled QUEUE block with the associated QUEUES element definition specifying the block label in operand *Block Label*. BLOCK-type arrivals may not arrive to hold blocks that require a preceding QUEUE block. They must arrive at the QUEUE block itself.

13.3 Entity Manipulations

The blocks in this section have several purposes, which are classified as entity manipulations. Examples are copying entities, cloning them, and removing them from a resource by preemption.

13.3.1 The COPY Block

The COPY block creates a duplicate of an entity residing in a queue, without removing the entity from the queue and sends the duplicate to a labeled block. The format for the COPY block is as follows:

COPY: *Entity Rank, Queue ID, Copy Destination*;

The operands are described as follows:

Operand	Description	Default
Entity Rank	Rank of the entity to be copied [expression truncated to an integer, or the keyword NQ]	—
Queue ID	Queue from which the entity is to be copied [queue symbol name, or expression truncated to an integer queue number]	—
Copy Destination	Label of block to which the created copy is to be sent [block label]	—

The arriving entity proceeds to the next block specified and is processed before the copied entity. The copied entity will have the attributes of the arriving entity. If the last entity is the one to be copied, specify NQ or NQ(*Queue ID*) as the *Rank*.

An example of the COPY block is as follows:

```
COPY:
        NQ, DrillQ, Lathe;
```

This block will make a copy of the last entity in the *DrillQ* and send it to the block with label *Lathe*. The example assumes that there is an entity in the queue *DrillQ*. If there is no entity, a runtime error will occur. Prior to sending an entity to the COPY block, a check using variable *Queue ID* or a SEARCH block should occur. If there is no entity meeting the specifications, do not branch to the COPY block.

13.3.2 The DUPLICATE Block

The DUPLICATE block clones arriving entities and sends them to the block indicated. The format for the block is as follows:

DUPLICATE: *Quantity to Duplicate, Duplicate Destination*:
        repeats;

The operands are described as follows:

Operand	Description	Default
Quantity to Duplicate	Number of duplicate entities to be created [expression truncated to an integer]	0
Duplicate Destination	Label to receive duplicate entities [block label]	Next block

The original entity continues to the following block unless the NEXT modifier is used. If no label is specified, the duplicate entities continue also to the next block. The duplicate entities are clones in that they possess all attributes of the original entity.

A mistake frequently made is to DUPLICATE one more entity than necessary. The DUPLICATE block clones *Quantity to Duplicate* entities, so that the resulting number in the model is *Quantity to Duplicate* + 1.

Examples of the DUPLICATE block are given by the following:

```
DUPLICATE:
      Needed+2;
```

This block duplicates the *Quantity to Duplicate* given by the named attribute or variable *Needed+2* and sends them, along with the original entity, to the next block. In this manner, the DUPLICATE block is performing a role often relegated to a BRANCH block.

```
DUPLICATE:
      3, LocationA:
      2, LocationB;
```

Three copies are sent to the block labeled *LocationA*, and two are sent to the block labeled *LocationB*. The original entity continues to the next block in the model.

```
DUPLICATE:
      1, LocationA:
      NEXT(LocationB);
```

The copy is sent to the block labeled *LocationA*, and the original is sent to the block labeled *LocationB*.

13.3.3 The PREEMPT Block

This is a rather complex block with much capability and many options. We will describe some of the basic features in this text. Additional features can be obtained from the *SIMAN V Reference Guide: Part 1* (see end of chapter reference).

The format for the PREEMPT block is

PREEMPT, *Priority*: *Resource ID, Attribute ID, Destination*;

The operands are described as follows:

Operand	**Description**	**Default**
Priority	Preemption priority [expression]	1.0
Resource ID	Number or name of resource to preempt [expression truncated to an integer, or other options described in *SIMAN V Reference Guide: Part 1*]	—
Attribute ID	Attribute of the preempted entity used to store that entity's remaining delay time [integer attribute number, A(*k*), or attribute symbol name]	Stored internally
Destination	Destination of preempted entity after surrendering the resource [block label or STO(*StorageID*), *Storage ID* is an expression truncated to an integer storage number or a storage symbol name]	Held internally for reallocation

The use of the PREEMPT block can be observed in the examples. Again, consult the *SIMAN V Reference Guide* for additional information concerning the operation of the block.

A few of the basic rules of operation aid in understanding the following three examples. When a resource arrives at a PREEMPT block with its *Resource ID* idle, the entity seizes the resource and continues processing. If the resource is busy, the arriving entity attempts to preempt it from the entity that originally seized it. If preemption occurs, the preempted entity is sent to *Destination*, and the original entity leaves the PREEMPT block to continue processing. If no *Destination* is specified, the interrupted entity is placed in a queue for the resource. After the preemption is completed, the entity is removed from the queue and placed in the block from which it was removed.

The priority operand, *Priority*, controls the preemption process. Entities at PREEMPT blocks always have higher priority than entities at SEIZE blocks, regardless of the SEIZE priority. Entities with lower value PREEMPT priorities receive the resource before entities with higher value PREEMPT priority.

Destination can also be specified as a block label or storage (see Section 13.6). When specified as a block label, the preempted entity transfers to the block associated with the label and continues processing. If *Destination* is specified as STO(*Storage ID*), SIMAN places the preempted entity in the storage until the resource becomes available.

When an entity is preempted, the remaining delay time is retained if *Destination* is defaulted or assigned a storage. However, if the entity is sent to another block, the remaining delay time must be stored in an attribute if it will be needed later.

Consider the following examples:

```
QUEUE,
      MaintQ;
PREEMPT:
      Drill;
```

The entity attempts to preempt the resource *Drill*. If *Drill* is idle, this entity is allocated the resource and continues processing. If *Drill* is busy, and preemption can occur, then the entity that most recently seized *Drill* is removed from its current process and placed in an internal queue. SIMAN stores the remaining delay time for this preempted entity, and it finishes processing as soon as the resource *Drill* becomes available again. If preemption cannot occur, the entity arriving to this PREEMPT block is placed in queue *MaintQ* to wait for the resource to be released.

```
QUEUE,
      M + 4;
PREEMPT:
      Mill(M + 8),
      MillTime,
      AltMill;
```

The entity attempts to preempt the resource *Mill*(*M* + 8). If preemption occurs, the interrupted entity is sent to the block labeled *AltMill* with its remaining delay time stored in attribute *MillTime*.

```
QUEUE,
      FailureQ;
PREEMPT,
      2.0:
      Mill, ,
      STO(FailMill);
```

```
DELAY:
     RepairTime;
RELEASE:
     Mill;
```

The entity will capture the *Mill* unless another entity with a priority less than 2.0 preempts the *Mill*. After *RepairTime*, the *Mill* releases the entity. The preempted entity is placed in the storage named *FailMill* until *Mill* becomes available.

13.4 Input and Output System

The SIMAN input and output (I/O) system is composed of the FILES element, and the READ, WRITE, and CLOSE blocks. The I/O system provides flexibility in initializing a simulation, inputting data for the simulation run, outputting data for analysis, and modeling inventory. This section will briefly describe these four components largely by example. Much more detail and additional examples are available from the *SIMAN V Reference Guide: Part 1*.

13.4.1 The FILES Element

The FILES element must be included whenever external files are accessed using READ and WRITE blocks. The FILES element identifies the system file and defines the access method, formatting, and operational characteristics of the file. The format for the FILES element is as follows:

FILES: *Number, Name, System Name, Access Type, Structure, End of File Action, Comment Character, Initialize Option*: repeats;

Assume that FORTRAN is being used in the examples of this and the next section. (Alternately, C code could be used.)

FILES: 1, ArrivalsA, "April.96", SEQ, "(1x, i4, 5x, 2f7.2)" ;

File 1 is identified as *ArrivalsA*. It has an operating system name *April.96* (possibly a DOS filename) and contains SEQuential records. Thus, record 5 follows record 4 and record 6 is between records 5 and 7, although the records may be of different lengths. The file structure follows the FORTRAN format indicated.

```
FILES:     1, ArrivalsA, "April.96", SEQ, "(1x, i4, 5x, 2f7.2)" :
           2, Trains, "A:\Departs.dat", DIR(10);
```

File 1 was described previously. It can be accessed using the symbol name *ArrivalsA* or file 1. File 2 is identified as *Trains*. It is located on drive A and has the operating system name *Departs.dat*. It is a DIRect or random access file of length 10. This example is used as the FILES element for the remainder of Section 13.4.

13.4.2 The WRITE and READ Blocks

The WRITE block writes data to an output file. The format for the WRITE block is as follows:

WRITE, *File ID, Format, Record Number*: *Variables*;

The READ block reads data from input files and assigns the values to the list of variables. The format for the READ block is as follows:

READ: *File ID, Format, Record Number*: *Variables*;

Numerous examples are now given:

```
WRITE,
      , "(' Enter Drill Time: ',$)";
READ:
      DrillTime;
```

The entity writes the prompt Enter Drill Time to the screen and waits for keyboard input. When a value is entered from the keyboard, it is READ into the variable *DrillTime*. No FILES element is needed for input from the keyboard.

```
READ:
      ArrivalsA, "(6x, 3(f8.0, 6x))":
      PartType, QtyArr, Criticality;
```

The entity reads data from the formatted, sequential file named *ArrivalsA* according to the format indicated. The values read are assigned to the attributes *PartType*, *QtyArr*, and *Criticality*.

```
READ:
      2, ,Record4:
      Cars, ComingFrom, GoingTo;
```

The entity reads three named attributes, *Cars*, *ComingFrom*, and *GoingTo*, from record number *Record4* (named attribute or variable) of a direct access, unformatted file named *TrainsA* or file number 2.

```
WRITE:
      TNOW, NQ(DrillQ);
```

This block writes the values of the current simulation time and the number in the queue named *DrillQ* to the default output device, most likely the screen. A block like this could be helpful in observing the progress of the simulation.

```
WRITE,
      , "(' Time Between Completions ', g10.4)": TAVG(TBC1);
```

This block writes the average value of the tally TBC1 to the screen in the format indicated, preceded by the text Time Between Completions.

```
WRITE,
      ArrivalsA, "(6x, 3(f8.0, 6x))": PartType, QtyArr, Criticality;
```

The entity writes data to the formatted, sequential file named *ArrivalsA* according to the format indicated. The values written are assigned to the attributes *PartType*, *QtyArr*, and *Criticality*.

```
WRITE,
      2, , Record4: Cars, ComingFrom, GoingTo;
```

The entity writes three named attributes, *Cars*, *ComingFrom*, and *GoingTo*, to record number *Record4* (named attribute or variable) of a direct access, unformatted file named *TrainsA* or file number 2.

13.4.3 The CLOSE Block

The CLOSE block closes a file, making it unavailable for input or output operations. When an entity executes a CLOSE block, the file associated with *File ID* becomes inaccessible, unless it is reopened at a READ or WRITE block. The format of the CLOSE block is as follows:

CLOSE, *File ID*;

The first example of this block is

```
CLOSE,
      ArrivalsA;
```

This block closes the file with the symbol *ArrivalsA*. Next, consider the following:

```
CLOSE,
    1;
```

This block accomplishes the same result as in the previous example.

13.5 Process Plans

This section presents the concept of process plans that work collectively with STATIONS, STATICS, RECIPES, and SEQUENCES.

13.5.1 The SEQUENCES Element Revisited

The SEQUENCES element was discussed previously in Chapter 8, where the stations concept was introduced. The format of the SEQUENCES element contains the *Variable* = *Value* operand; a list of attributes, variables, statics, and their values upon entering the station. The *Variable* = *Value* operand can take any of the following forms:

RECIPE = *RecipeID*	Uses the static values defined in the RECIPES element's *RecipeID* for entities arriving to this station with this sequence set (NS) and index (IS).
STEPNAME = *StepNameID*	Defines an alphanumeric symbol name to be associated with a particular index in a particular sequence set. The default *StepNameID* is the value of IS.
NEXT = *StepNameExpr*	Defines the next sequence step to use (the default is to use step number IS+1).
StaticName = *Expression*	Overrides default values and reassigns the value of *StaticName* to the expression evaluated here.
AttOrVarID = *Expression*	Overrides default values and reassigns the value of *Attribute ID* or *Variable ID* to the expression evaluated here.

13.5.2 The STATICS and RECIPES Elements

The STATICS element allows for the definition of location-dependent variables. These variables may then be grouped in the RECIPES element to create a set of instructions that will be used to complete a process. These concepts, as well as the revisited SCHEDULES and STATIONS elements are explained with three examples, the last two of which are from the *SIMAN V Reference Guide*.

```
SEQUENCES:      PartTypeA,  Arrival, &
                            Process,
                                RECIPE = Customer120 &
                            Departure:
                PartTypeB,  Arrival, &
                            Process,
                                RECIPE = Customer144,
                                NewTime = 1.4*RegularTime:
                            Departure;
```

Entities that have their sequence set number equal to 1 first go to the station *Arrival*, then to the station *Process*, and finally to the station *Departure*. When at station *Process* they follow the recipe defined as *Customer120*. Entities that have their sequence set number equal to 2 follow the same sequence as entities that have their sequence set number equal to 1. Their recipe at station *Process* follows that of *Customer144*. Additionally, their *NewTime* is given by 1.4 times *RegularTime*.

Another example is as follows:

```
STATIONS:     Entrance, , Receiving:
              Process:
              Exit, , Shipping;
RECIPES:      Receiving,    OperatorType = Receiver,
                            ProcessTime = TRIA(12.9, 22.3, 26.8):
              Shipping,     OperatorType = Shipper,
                            ProcessTime = UNIF(10.2, 22.4);
STATICS:      OperatorType:
              ProcessTime;
RESOURCES:    Receiver:
              Shipper;
```

Three stations are defined. Entities arriving at station *Entrance* use the recipe called *Receiving* to define their *OperatorType* as *Receiver* and their *ProcessTime* as a random number generated from a triangular distribution. Entities arriving at station *Exit* use the recipe *Shipping* to define their *OperatorType* as *Shipper* and their *ProcessTime* as a random number generated from a uniform distribution.

The next example is more involved than the others in this chapter. Its experiment frame includes the following:

```
STATICS:       MachineType:
               Operator = GeneralLabor:
               UnloadTime = 5.0:
               LoadTime = 5.0:
               ProcessTime = PTime(MachineType, Material);
RECIPES:       Unloading,
                   MachineType = ForkTruck,
                   ProcessTime = UNIF(10.0, 20.0):
               Drilling,
                   MachineType = Drill,
                   Operator = Machinist:
               Milling,
                   MachineType = Mill,
                   Operator = Machinist:
               Shipping,
                   ProcessTime = 22.5;
STATIONS:      Arrival, ,Unloading:
               Process:
               Departure, ,Shipping;
SEQUENCES:     SmallPart,
                   Arrival &
                   Process,
                       RECIPE = Drilling &
                   Departure:
               HeavyPart,
                   Arrival,
                       Operator = LargeCrane &
                   Process,
                       RECIPE = Milling,
                       ProcessTime = 1.2 *
                           PTime(MachineType, Material) &
                   Departure,
                       Operator = LargeCrane:
```

```
                Part527,
                     Arrival &
                     Process,
                          RECIPE = Drilling &
                     Process,
                          RECIPE = Milling &
                     Departure;
```

The model frame for the example includes the following:

```
STATION,
     Arrival-Process;
SEIZE:
     Operator:
     MachineType;
DELAY:
     ProcessTime;
RELEASE:
     Operator:
     MachineType;
ROUTE:
     0, SEQ;
STATION,
     Departure;
SEIZE:
     Operator;
DELAY:
     LoadTime;
RELEASE:
     Operator;
DISPOSE;
```

The STATICS element defines five statics, *MachineType*, *Operator*, *UnloadTime*, *LoadTime*, and *ProcessTime*. Combinations of these statics are grouped on the RECIPES element. Recipe *Milling* contains two statics, *MachineType* and *Operator*. In this recipe, both *MachineType* and *Operator* have their default values overridden. Entities using this recipe will have *MachineType* set to *Mill* and *Operator* set to *Machinist*. All other static values (*UnloadTime*, *LoadTime*, and *ProcessTime*) retain their default values.

Default static values may also be overridden on the SEQUENCES element. The SEQUENCES element in the example illustrates that entities using the *HeavyPart* sequence set and the *Arrival* station will have their *Operator* static set to *LargeCrane*.

The RECIPES element defines four recipes, *Unloading*, *Drilling*, *Milling*, and *Shipping*. Recipe *Unloading* is referenced in the STATIONS element on the *Arrival* station. Entities that enter the *Arrival* station will have a *MachineType* of *ForkTruck* and a *ProcessTime* equal to a value generated from a uniform distribution.

13.5.3 The STATIONS Element Revisited

The STATIONS element was introduced in Chapter 8 and used in Chapters 9, 10, and 11 in conjunction with material handling. The format for the STATIONS element is as follows:

STATIONS: *Number, Name, Intersection ID, Recipe ID*: repeats;

The *Recipe ID* associates the station with a process recipe from the RECIPES element. By specifying *Recipe ID*, entities arriving at the station will have their static variables set to the values defined in *Recipe ID*. The operand is defaulted if the RECIPES element is not being used.

13.6 Storages

Storages are used with the DELAY, MOVE, PREEMPT, and REQUEST blocks to model an entity that experiences a time delay while stationary. The ROUTE, TRANSPORT, and CONVEY blocks simulate time advances while the entity is in motion. Storages are used to obtain statistics on the number of entities in one of the four blocks mentioned previously. Also, storages are used in Cinema as a location to place entity symbols avoiding their disappearance from the animation screen when the entity is stationary.

Using the STORE and UNSTORE blocks allows an entity to enter a storage, undergo other processes, and later leave the storage. The DELAY, REQUEST, and MOVE blocks place the entity in the storage only for the duration of the block. The PREEMPT block places the preempted entity into a storage until its resource is restored.

13.6.1 The STORAGES Element

The STORAGES element specifies the total number of storages and their names. The format for the STORAGES element is as follows:

STORAGES: *Number, Name*: Repeats;

The operands are described as follows:

Operand	Description	Default
Number	Storage number [integer]	Sequential
Name	Storage name [symbol name]	Blank

An example is as follows:

```
STORAGES:
    Assembling:
    Testing:
    Inspecting:
    Packaging;
```

This element defines four storages, with names *Assembling*, *Testing*, *Inspecting*, and *Packaging*, with numbers 1, 2, 3, and 4, respectively.

13.6.2 The STORE Block

The STORE block adds an entity to a storage. When an entity enters the STORE block, the value of NSTO(*Storage ID*) is incremented and the entity immediately moves to the next block. Storages are defined in the STORAGES element. The format for the STORE block is as follows:

STORE: *Storage ID*;

The operand is described as follows:

Operand	Description	Default
Storage ID	Storage to which to add entity [expression]	—

An example is as follows:

```
STORE: Assembling;
```

The entity is added to storage *Assembling* and continues to the next block in the model.

13.6.3 The UNSTORE Block

The UNSTORE block removes the current entity from a storage. When an entity enters the UNSTORE block, the value of NSTO(*Storage ID*) is decremented and the entity immediately moves to the next block. The format for the UNSTORE block is as follows:

UNSTORE: *Storage ID*;

The operand is described as follows:

Operand	Description	Default
Storage ID	Expression that evaluates to a *Storage ID* number	Last storage entered

An example is as follows:

```
STORE:    Assembling;
:
UNSTORE: Assembling;
```

The entity is added to storage *Assembling* and continues to the next block in the model. Then, the entity is removed from storage *Assembling*.

13.7 Miscellaneous

13.7.1 The TABLES Element

The TABLES element defines the dependent values of a variable. The format for the TABLES element is as follows:

TABLES: *Number, Name, Low Value, Fixed Increment,*
Variable Value i, . . . , Variable Value n: repeats;

The operands are described as follows:

Operand	Description	Default
Number	Table number [integer]	Sequential
Name	Table name [symbol name]	Blank
Low Value	Low value for independent variable	0.01
Fixed Increment	Fixed increment between successive values of the independent variable [constant]	1.0
Variable Value *i*	Dependent variable values corresponding to the independent variable [constant]	—

Table data can be accessed with the function TF(*Table ID, X Value*), where *Table ID* is the *Number* or *Name* and *X Value* is the argument. Linear interpolation is performed if the argument is not in the TABLES element. The operation of the TABLES element can be seen in the following example:

```
TABLES:
    ProbFail, 20.0, 10.0, 0.2, 0.4, 0.5, 0.6;
```

This TABLES element indicates the following arguments and responses:

XValue	**TF(*Table ID, XValue*)**
20.0	0.2
30.0	0.4
40.0	0.5
50.0	0.6

Thus, an argument of 40.0 results in a response of 0.5, an argument of 25.0 results in a response of 0.3, an argument of 16.0 results in a response of 0.2, and an argument of 60.0 results in a response of 0.6.

13.7.2 The INCLUDE Statement

The include statement adds the elements contained in a specified file to the current experiment at the location of the INCLUDE statement. The format of the INCLUDE statement is as follows:

INCLUDE: *File Description*;

where *File Description* contains the file to include enclosed in double quotes. An example illustrates the use of the INCLUDE statement.

The EXPERIMENT frame for Example 2.1 is as follows:

```
BEGIN;
!Example 2.1:    The Single Machine Problem
!
PROJECT,         Single Machine, Team;
RESOURCES:       Drill;
QUEUES:          Buffer;
DSTATS:          NQ(Buffer);        Queue Length
EXPRESSIONS:     1, , EXPO(5, 1):   !Exponential interarrivals
                 2, , UNIF(6, 9, 1); Uniform Drill times
COUNTERS:        JobsDone;
REPLICATE,       1, 0, 480;
```

Let us assume that we have a filename Examp21.ctr consisting of the following:

```
RESOURCES:    Drill;
QUEUES:       Buffer;
DSTATS:       NQ(Buffer);        Queue Length
EXPRESSIONS:  1, , EXPO(5, 1):   !Exponential interarrivals
              2, , UNIF(6, 9, 1); Uniform Drill times
COUNTERS:     JobsDone;
```

Now, we can have an equivalent experiment frame as follows:

```
BEGIN;
!Example 2.1:   The Single Machine Problem
!
PROJECT,       Single Machine, Team;
INCLUDE:       "Examp21.ctr"
REPLICATE,     1, 0, 480;
```

13.8 SUMMARY

This chapter describes many features of SIMAN that may be extremely important in some applications. For instance, there are many models that would not be meaningful without the availability of the PREEMPT block. There are other instances of elements and blocks in this chapter that make for more understandable or easier modeling such as with the NICKNAMES element. The information on file transactions is essential when dealing with large input data files that would be encountered in a real problem. The information on process plans is a new feature of SIMAN V.

13.9 REFERENCE

SIMAN V Reference Guide (1994), Systems Modeling Corporation, Sewickley, Penn.

13.10 EXERCISES

E13.1 In Exercise 4.1, rush jobs of Type I arrive according to an exponential distribution with a mean of 24 minutes. A rush job preempts a regular job at the mill, if there is no resource available. The preempted job finishes its operation for the remaining delay after the preempting job is processed at the mill. (Preempted jobs are never ejected from the system because of a lack of room in the buffer in front of the mill.) Compare the time in the system for rush and regular jobs by type of job.

E13.2 In Exercise 4.1, use WRITE and READ blocks that will allow all the operation times to be provided as inputs from the keyboard.

E13.3 In Exercise 4.1, the four operations (Mill, Plane, Drill, and Inspect) are also known as OP10, OP20, OP30, and OP40 with operation times OP10Time, OP20Time, OP30Time, and OP40Time. Using nicknames, let the four operation times be identified as MillTime or OP10Time, and so on. Use both the *Name* and the nickname in the model.

E13.4 Modify Exercise 4.1 so that at the 10th, 20th, and 30th hour, four Type I jobs arrive at the queue before the mill for processing. The delay time for the jobs is 12 minutes for the mill, 8 minutes for the plane, 5 minutes for the drill, and 2 minutes for the inspector.

E13.5 Reconsider Exercise 7.4. Let the emergencies occur as preemptions (at the Loader only) rather than as high priority orders. Compare the flowtimes of emergency and regular jobs.

E13.6 With respect to Exercise 7.4, it has been found that loading time is a function of the hour of the day. From 8:00 A. M. to 10:00 A. M., loading time rises steadily from 65 seconds to 85 seconds. From 10:00 A. M. to 2:00 P. M., loading time drops steadily from 85 to 75 seconds. From 2:00 P. M. to the end of the day, loading takes 75 seconds. Answer the previous questions given this new information.

14

Animation Using Cinema in the ARENA Environment

In earlier versions of SIMAN, a separate program was used to animate models. This program was called Cinema. Using a CAD-like environment, the user could draw static backgrounds, pictures for entities, workstations, and transporters, and status displays. The ARENA environment links SIMAN V and Cinema, making animation more flexible, faster, and simpler. This chapter provides a limited discussion of building models in the ARENA environment. For a detailed description of ARENA, refer to the *ARENA User's Guide* and the *ARENA Template Reference Guide* published by Systems Modeling Corporation (see end-of-chapter references).

14.1 User Interface

ARENA is a menu-driven environment. With the mouse and occasional commands from the keyboard, the user can manipulate SIMAN and Cinema objects, draw pictures, enter data, and control the execution of the model. If the user has worked in other windowing environments, ARENA will seem very familiar. For the inexperienced user, ARENA's menus and dialog boxes work in an intuitive manner. Do not be afraid to experiment with the system.

14.1.1 Windows

The activity of building a model occurs within a window. A window is a rectangular area bounded by a menu bar, scroll bars, and a panel. The activity within one window is separate from another window. For example, the user may be working on two or more models at the same time, but only one window is active at any time. Any commands issued affect only the model in the active window. A window is activated by using the mouse pointer to click within its boundaries. The window will be brought to the top of the stack, and its title bar will be highlighted.

The *menu bar* is a strip along the top of the window that lists the headings of related commands. For example, the View heading groups commands related to how the model appears within the window, whereas the Edit heading groups editing commands like COPY, PASTE, CUT, and SELECT.

Scroll bars appear along the right and bottom edges of a window. By clicking on the arrows at either end of the scroll bar or by moving the slider box, the user may view different portions of the workspace. The actual workspace is many times larger than can be viewed on a monitor. In order to view the entire space, the user must pan the view back and forth, up and down.

The panel is another command area of the window. Like the menu bar, the panel groups related icons. An *icon* is a picture which generally represents an activity or function. The panels are arranged like a stack of file folders, with tabs naming each panel on the side. By clicking on a panel tab, the user may activate a panel and move it to the top of the stack. Up and down arrows at the bottom of a panel indicate that there are more icons on the panel than can be displayed at one time. Small arrows scroll the panel icons one at a time, whereas the large arrows scroll many at a time. From the panel, the user chooses blocks and elements to build the model, issues commands to run the model, and draws pictures to enhance the animation of the model.

All windows may be closed and opened. Many may be resized or moved around the screen.

14.1.2 Menus

A *menu* is a list of related commands. Using the mouse or the directional keys on the keyboard, the user may highlight a command and select it by clicking the mouse button or by pressing the **Return** key.

14.1.3 Dialog Boxes

The *dialog box* is ARENA's method of requesting information from the user. It is essentially a page of headings and empty text boxes. Some headings may contain default values, which can be accepted or changed. Some display a list of acceptable choices or a list of acceptable formats for that particular text box. Some dialog boxes offer buttons that display other pages when selected.

At the bottom of the dialog box are the **Accept** and **Cancel** or **Close** buttons. The **Accept** button should be pressed after the user has entered all required data. If the user has entered the incorrect dialog box or needs to leave the box without saving the data, the **Cancel** or **Close** button should be used. When the **Cancel** or **Close** button is depressed, ARENA does not recognize any changes to the data in the dialog box.

14.2 Building Models with COMMON.TPO

In Chapter 2, building models in ARENA was introduced. There the Blocks and Elements templates were used. In this section, the COMMON.TPO template will be discussed. The modules on this template combine some blocks or elements that are commonly used together. For example, the SERVER module represents the association of the QUEUE, SEIZE, DELAY, and RELEASE blocks. The dialog box for the SERVER requires information such as queue capacity, station name, and delay time.

In the following section, we will revisit Example 3.1 and build the same model in the ARENA environment.

EXAMPLE 14.1 Three-Machine Problem Revisited

Recall Example 3.1, which dealt with three machines in sequence. The time between arrivals was exponentially distributed with a mean of 5 minutes. Other data for the system is shown in Table 14.1. Simulate the system for 40 hours of operation.

At the Arena prompt, type:

```
ARENA>Arena
```

(The diskettes provided with this text do not contain the ARENA environment used in this chapter. See the README.DOC file to obtain more information about obtaining the entire ARENA environment.)

Upon first entering the ARENA environment, there is no active window. With the mouse pointer, click on the Model heading on the menu bar and select **New**. This opens a new window ready for modeling. Notice the panel tabs at the left of the workspace: Run, Draw, Animate,

Table 14.1 Data for Example 14.1

Resource Name	Resource Capacity	Buffer Capacity	Delay Distribution
Drills	2	2	Uniform(6, 9)
Mills	3	3	Triangular(10, 14, 18)
Grinders	2	2	Discrete(.25, 6, .75, 8, 1.0, 12)

and a blank tab. Click on the Menu heading Panel and choose the **Attach** option. Once activated, ARENA will display a pop-up menu with the names of templates. Scroll through the list until COMMON.TPO is shown, click on it to highlight and select it, and then click the **Attach** button to add it to the panel.

The Arrival Station

In Chapter 2, we modeled a system using the BLOCKS and ELEMENTS templates. These templates allowed modeling of the system in nearly the same fashion as using blocks and elements in a text file. The Common template offers modules rather than separate blocks and elements. Each module may combine several blocks or elements into one function.

For example, click on the **Arrive** button on the Common panel. Notice that the pointer arrow has changed to crosshairs. Move the crosshairs into the work space and click again.

"Arrive" will appear enclosed in a box. This graphic represents the Arrive station. Now, double-click on the word Arrive. You have opened a dialog box within the Arrive module.

1. In the Station text box type in the label `Incoming`.
2. In the Time Between text box enter `EXPO(5)`.

This is the interarrival time for the entities. Note: In this dialog box, you can specify an attribute to mark or to make attribute or variable assignments. In SIMAN, this would require the use of several blocks and modifiers; in Arena, however, it is accomplished within a single module.

3. In the Leave Data section, click on the **Connect** button.

When you have finished with this dialog box, click on **Accept** to close it.

Adding Servers

Now back in the workspace, make sure that the Arrive module is highlighted (click on the word Arrive, if it is not). Click on the **Server** button on the Common panel and place a server to the right of the Arrive module. Notice that the Arrive and Server modules have been connected by a black line. The line connects the Arrive module's exit (the black triangle) to the Server module's entrance (the small black square).

ARENA automatically connects a highlighted module to the next one placed in the workspace. This default setting may be turned off by pulling down the Modules menu heading and clicking on **Auto-connect**. You may connect modules manually by clicking on the **Connect** button at the far right of the button bar (it looks like a backwards S) and drawing the connection from a module's exit to another module's entrance. Try manually connecting the Arrive and Server modules by first selecting the connector line with the mouse. Boxes appear at either end of the line indicating the boundary of the item selected. Press the **Delete** key to remove the connector. Now click on the **Connect** button and use the cross hairs to draw a line from the Arrive exit to the Server entrance. Click on the **Connect** button to turn off this function.

Now, double-click on the Server module to open its dialog box. Looking at the information requested in the dialog box, you should notice that the familiar activities of the QUEUE, SEIZE, DELAY, and RELEASE blocks are contained here.

1. In the Station text box, type in `Drilling` and press the Return key.
2. In the Capacity text box, type in `2`.
3. In the Process Time text box, type `UNIF(6,9)`.
4. Click on **Accept** to close the Server dialog box.

Notice that the default resource name and queue name were assigned as Drilling_R and Drilling_R_Q, respectively.

Now, similarly add the Mill and Grinder servers. Use Milling as the Mill Station ID and Grinding as the Grinder Station ID. Remember, if you wish to use the **Auto-connect** feature, then make sure that the function is turned on in the Modules menu and that the preceding module is highlighted before the next module is placed.

Counting and Disposing of Entities

The workspace should now have four modules: the Arrive module and three Server modules. If you cannot see all four modules at once, use the minus key (−) to zoom out. The plus key (+) zooms in the picture.

1. Click on the **Depart** button on the Common panel and place it after the last server.
2. Open the dialog box.
3. In the Station box, type `Shipping`.
4. Click on the **Individual Counter** button.
5. Click on the **Accept** button to close the dialog box.

The **Individual Counter** button will count entities as they depart the model and are disposed. The default name for this counter is Shipping_C.

We must add another Depart module to the model in order to count and dispose of the entities ejected at the queues.

1. Click on the Depart module in the Common panel and place a new Depart module in the workspace.
2. Open the dialog box.
3. In the Label text box, type `Balks`.
4. In the Station text box, type `Ejections`.
5. Click on the **Individual Counter** button.
6. Click on the **Accept** button to close the dialog box.

This module does not need to be connected to any of the other modules because each queue will automatically send its overflowing entities to the module with the label "Balks". Make sure that you have specified this in the Queues dialog box of the Mill and Grinder, as we did in the Drill server.

Describing Queue Information

In the previous section, we described how to count and dispose of entities that overflow the waiting queue before a Server. At this point, we must return to each Server module and describe the queue limits.

1. Click on the Drilling Server module and open its dialog box.
2. Click on the **Queue** button to open a secondary dialog box.
3. Type in `2` for Capacity and press the Return key.
4. Turn off the **Dispose Balked Entities** option by clicking on the button.
5. A new text box will appear asking where to send ejected entities. Type in `Balks` as the label name.
6. Click on the **Accept** button to close the dialog box.
7. Click on **Accept**.

Follow these steps for the Milling and Grinding Server modules using the queue capacities given in Example 3.1.

The Simulate Module

The structure of the model is nearly complete. The definition remaining is that of the simulation length.

1. Click on the **Simulate** button on the Common panel.
2. Place the Simulate module in the lower-right corner.

3. Open the dialog box.
4. Type in `2400` in the Length of Replication text box.
5. You may enter information for "Title," "Analyst," and "Date."
6. Click on the **Accept** button.

The model is complete. Click on the menu heading Model and choose **Save**. Enter `EX14-1` and click on **Save**. It is a good idea to save your model several times during its construction.

Before we run the simulation, you may wish to arrange the layout of the system in the workspace. To move modules, click on the module name, or handle, and drag the graphic to a new position. The connecting lines will stretch to accommodate changes in module location.

14.3 Using the Run Panel

With the Run Panel, it is possible to perform some of the functions of the Interactive Run Controller discussed in Chapter 5. Buttons on this panel are used for beginning, clearing, and ending the simulation.

14.3.1 Running Example 14.1

To run Example 14.1, click on the Run panel tab, then click on the **Go** button. A box will pop up in the middle of the screen displaying the message Validating EX14-1.DOE. During this step, ARENA is generating the .MOD, .M, .EXP, .E, and .P files.

When ARENA finds no errors in the model, it will begin running the simulation with rudimentary animation. By default, the entities are represented by a small red box. When a server is busy, its color changes from blue to red (also a default). As the simulation run progresses, observe the increasing count at both of the Depart stations as good jobs and ejected jobs leave the system. When the simulation run is complete, click on the **End** button.

To observe the Summary Report, click on the ARENA heading in the Main Menu Bar. Then choose the **View Text File** option. In the File Name text box, type

```
*.OUT
```

and press the Return key. A list of all output files will be displayed within the scroll box. Click on the proper name and then click **Load**. The Summary Report will be displayed within a new window that may be resized, shrunk, or closed. This report contains the resource utilizations and the queue information. Observe that some utilizations are greater than one. This occurs because the capacities of the servers were greater than one. We will learn to correct for this factor in the next example.

To check your construction and results of Example 14.1, you may wish to compare it with the model included on the diskettes provided with this text. The file name is "EXMP14-1.DOE".

14.3.2 Starting, Interrupting, and Exiting a Run Session

When you click on the **Go** button, ARENA enters a special mode called the run session. Until the **End** button is pressed, ARENA remains in the run session and all panels except the Run panel are disabled. During a run session, ARENA may be in one of two states: *active* or *interrupt*. In the active state, the mouse pointer is removed from the screen and the animation may be viewed in the workspace. In the interrupt state, entered by pressing the Escape key, the mouse pointer reappears and the user may point to locations on the Run panel to change various options, resize the window, and pan or zoom the model. The run of the simulation has not been terminated, just interrupted.

The **Go** button begins a run session. If the run session is in the interrupted state, it continues the simulation from where it was interrupted.

Clicking on the **Step** button moves an entity one step through the model. It works much the same as the STEP command introduced in Chapter 5. If ARENA is not in a run session when the **Step** button is clicked on, then ARENA begins a run session, initiates the model, moves the first entity through one step, and enters the interrupt state.

The **Clear** button initializes the simulation to its starting conditions. If ARENA is not in a run session, clicking on the **Clear** button starts a run session, initializes the model, and interrupts the run session. From this point, **Go** or **Step** may be used to continue the simulation. If ARENA is in a run session and in the interrupt state, the **Clear** button initializes the simulation to starting conditions but does not terminate the run.

The **End** button terminates the run session, restoring all disabled panels.

14.3.3 Viewing and Correcting Errors

When the **Go** or **Step** button is clicked on to initiate a run session, ARENA performs a validation of the model. If, during the validation, an error is discovered, a special window is displayed giving the error message. There are three types of errors:

> *Type 1* errors occur due to missing information from a required field, unknown symbols, and so on.
>
> *Type 2* errors occur while ARENA is generating the program file to run the simulation.
>
> *Type 3* errors occur during a run session and are due to logical inconsistencies, division by zero, and so on.

If ARENA detects an error, a message is displayed in the Errors/Warnings dialog box. Along the bottom edge of the box are four buttons: **Previous, Next, Show,** and **Done**. The **Next** and **Previous** buttons allow the user to move through the list of errors produced.

If ARENA is able to detect in which module the error occurred, clicking on the **Show** button will cause the relevant dialog box to be displayed. The information may be corrected and the **Accept** button clicked. If ARENA is unable to detect the error's location, the **Show** button is greyed out. Once all errors have been corrected, clicking on the **Done** button will close the Errors/Warnings dialog box.

Type 1 and Type 2 errors may be checked before entering a run session. At the top right of the button bar is a button with a check mark on it. Pressing the button activates validation of the model.

14.3.4 Keyboard Interaction During a Run Session

While the run session is in the active state, the mouse pointer is removed from the screen. However, some control of the simulation is allowed through the keyboard.

Escape Arrows	The Escape key allows the user to interrupt the run session. The Arrow keys may be used to pan up and down, left and right through the workspace.
+, −	The Plus and Minus keys may be used to zoom in (−) or zoom out (+) the view of the workspace.
>, <	The greater than symbol (>) may be used to speed up the simulation, whereas the less than symbol (<) will slow down the simulation. Note that this does not change the length of the simulation run, only how fast it takes place in the animation.

14.4 Drawing

Combined with ARENA's powerful simulation environment is a drawing program, with functions and capabilities similar to other popular drawing and painting programs. By clicking on the panel labeled "Draw," the user is offered several tools to enhance the appearance of the animation.

The Draw Panel

Activate the Draw panel by clicking on its tab. This panel has many standard drawing features that allow changes in color, line type, and line width.

The group of buttons at the top of the panel control the selection of drawing objects. These are the **Line, Polyline, Arc, Box, Polygon, Ellipse,** and **Text** buttons. To activate one of these objects, click on the button with the pointer.

The group below controls the color of lines, fill, and text. To change the color of a line, click on the **Line Color** button and click on a color.

Below the color palette are other groups of buttons that control line type, line width, and fill patterns.

The best way to familiarize yourself with the functions of each of these groups of buttons is through practice. Draw several pictures until you are comfortable with the functions and capabilities of each group of buttons.

EXAMPLE 14.2: Elaboration of Example 14.1

Having run Example 14.1, we know that the model works properly. If we were to show this model to someone who had not participated in its construction, it is unlikely that the blinking and flashing objects on the screen would have much meaning. Through the functions of the Draw panel, we may label the objects on the screen and change their appearance.

To change the appearance of the entity, use the pointer to select the red box just above the Simulate module. When the red box is successfully selected, a larger box made of white dashed lines will enclose it. By double-clicking on this selected area, you may edit the appearance of the entity. A special editing dialog box will appear called Entity Picture Placement. Click on the button that displays the red box. A new window is activated that displays the Draw panel. In this window, you may edit the picture that represents the entity. Try changing its color, or delete the picture and draw one of your own. When you are satisfied with the picture, click on the blue A at the top left of the active window and select **Close**. Then click on the **Accept** button. In the workspace, you will see that the new entity picture has replaced the default red box.

To change the appearance of the servers, you may follow the same procedure as above. ARENA offers, however, a library of pictures suitable for representing servers. Select the server picture in the workspace. Make sure that the blue rectangle is enclosed by the dashed box. Double-click to enter the Resource Picture Placement dialog box. In the lower-right corner of the box is a subsection with a text box for File Name. Click on the upside down triangle to display a list of possible library files. Click on the file called MEN, and then click on **Load**.

Now a series of pictures is displayed in the Current Library box. Scroll through these pictures until you find the picture of a man wearing a red coat standing at a table. Select this picture by clicking on it once. Then select the picture of the default server (colored blue). Then click on the lower **Copy** button. The man in the red coat represents the server in an idle state. Now click on the picture of the default server (colored red). Click on the picture of the man in a red coat standing at a table with his head lowered. Copy it to the other side. You now have an idle and an active representation of the server.

You may do this for the other two servers. To change the appearance of one of these library pictures, double-click on it to bring up the editing window. Change the picture using the drawing functions and then close the window.

The part of the module that takes part in the simulation is really the handle, the name of the module. You may move the graphics freely to suit personal aesthetics without altering the logic.

Using the Draw panel, select the Text object. In the dialog box that opens, type in `Finished Jobs` and click on the **Accept** button. Then place the text in a corner of the screen. Now move the number graphic from the Shipping Depart module next to the text. Do the same for the Ejections Depart module, using the text `Ejected Jobs`. Now the running totals of finished and ejected jobs are labeled and conveniently located.

Try running the model again with these new features. While the new pictures do not change the basic simulation, they certainly do make the animation more interesting and the activities more obvious. The file "EXMP14-2.DOE" contains the changes to Example 14.1 covered in this section.

EXAMPLE 14.3: Example 14.1 Revisited

Using the model of Example 14.1 as a base, we will now increase its complexity. Each drill must be shut down for 10 minutes for cleaning after processing 25 consecutive jobs. The drill is shut down after the 25th job is completed. The mills are subject to breakdowns, which occur according to an exponential distribution with mean 80 minutes. The length of the breakdown follows a normal distribution with a mean of 10 minutes and a standard deviation of 1.2 minutes.

To accomplish this, we will need to take advantage of the SETS, STATESETS, and FAILURES elements discussed in Chapter 7. The COMMON.TPO template contains all the modules needed to create the necessary downtime.

Creating SETS

Go to the Common panel, click on the Sets module, and add it to the workspace of Example 14.1. This module does not take an active part in the animation (like the Simulate module), so it may be placed anywhere. Follow these steps to add the needed sets of resources.

1. Open the dialog box to the Sets module.
2. Click on the **Resources** button. This brings up a secondary button labeled **Resources**. Click on this button to bring up the next dialog box.
3. Click on the **Insert** button.
4. Type in `Drills` as the name of the set and press the Return key.
5. Click on **Insert** in the next dialog box.
6. Type in `Drilling_R` as the resource or choose it from the list.
7. Click on **Accept** to close the Resource window.
8. Click on **Accept** to close the Drill Server window.

Now, add a set called Mills following the same instructions as above. Use the resource called Milling_R as the set member. When completed, click on **Accept** to close the completed dialog boxes. Save Example 14.3 under "EX14-3.DOE".

Now, we must define the nature of the failures within the Drills and Mills sets.

1. Open the Drill server dialog box.
2. Click on the **Resource** button. This opens another dialog box where we may define Statesets, Failures, and Downtime.
3. Type in `DrillStates` in the Statesets text box and press the Return key.
4. Click on **Insert** to add a state. We will add two states, Drilling and Cleaning.

5. Type `Drilling` as the state name. Type `BUSY` as the associated state and click on **Accept**.
6. Click on **Insert** and add a state called `Cleaning`. Type in `DrillClean` as its associated state. Click on **Accept**.
7. Click on **Insert** under the Failures heading.
8. Type `DrillClean` as the name of the Failure. This Failure is based on a count of the entities. Therefore, we will use the default value for the Failure Type.
9. Type `Wait` in the Fail When text box.
10. In the Count text box that appears, type `25`.
11. In the Downtime box, type `10`.
12. Click on **Accept** to close the dialog boxes.

To add the Failure to the Mills set, follow the above steps within the Mill server module. Use the following information.

1. Stateset name is MillStates.
2. Add the following states and associated states: Milling(BUSY), Broken(MillFail), and Waiting(IDLE).
3. Under the Failure heading, insert the following failure:

Failure name:	`MillFail`
Based on:	Time
Fail When:	Preempt
Uptime:	`EXPO(80)`
Downtime:	`NORM(10,1.2)`
Uptime in this...:	`Milling`

Click on **Accept** to clear the windows.

At this point, we have defined failure states for the Drills and Mills. After running this simulation, open the Summary Report output file. The information under the Frequencies heading shows the number of MillFails and Cleanings that the Mills and Drills underwent.

In Example 14.1, the Summary Report showed utilizations greater than 1 for some of the resources. This occurs because the capacity for each resource is also greater than 1. To account for the capacity of a resource, use the Statistics module on the Common panel.

1. Place the Statistics module in the workspace.
2. Open its dialog box.
3. In the Time Persistent section, click on **Insert**. A new box will appear.
4. Under Expression, click on the down arrow to display the list. Choose NR(Grinding_R). Then edit the expression so that it appears like the following: `NR(Grinder_R)/2`

5. Under the Report Label, you may specify a symbol name such as Grinder Utilization.
6. Click on the **Accept** button to clear the dialog box.

Now we have successfully accounted for the capacity of the resource in the calculation of the utilization.

Like examples in the early chapters of this text, the entities travel instantaneously between each server. In the next section, we will add routes between the servers in order to observe the travel of each entity through the system.

14.5 Animating

We have already seen a simple form of the animation produced by ARENA. By using the Draw panel, we have enhanced the look of the servers and the display of relevant information (the counters). The Animate panel offers many functions that allow us to observe other statistics, add transporters and conveyors, and use waiting areas.

Adding Routes

The conversion from moving entities between modules by connector to routing the entities requires a change in each module that handles an entity. These are the Arrive module, the three server modules, and the Shipping Depart module.

1. Open the dialog box to the Arrive module.
2. In the Leave Data section, click on the **Route** button.
3. In the Station text box, type in `Drilling` or select Drilling from the list.
4. Enter `2` for the Route Time.
5. Click on the **Accept** button.

The transformation is similar in the remaining modules, although the placement of the information differs. Make these changes in the other modules, using 2 for the Route Time. Make sure that the Station entry follows the proper sequence: Arrive to Drilling, Drilling to Milling, Milling to Grinding, and Grinding to Shipping. When the changes are accepted to a module, an additional graphic is added, another yellow figure. The two yellow boxes represent the “in” and “out” points for the station. Notice that the black connecting line disappears because now the stations will be joined by a route.

Now, we will add the paths along which the entities will move between stations. First, make sure that you are satisfied with the arrangement of the modules in the workspace. Click on the Animate tab to activate it.

1. Click on the **Route** button.
2. Accept the default choices in the dialog box that appears.
3. Place the crosshairs over the yellow station graphic to the right of the Arrive module and click. Stretch the line to the leftmost station graphic of the Drilling server module and click again. A yellow line should now join the out point for the Arrive station and the in point for the Drilling station. Join the other stations similarly, that is, always join from out point to in point. The Depart module has only an in point. Conversely, the Arrive module has only an out point.

All the stations are now joined with routing paths. Each entity will take 2 minutes of simulation time to travel between stations.

Viewing Status

The Animate panel offers a number of objects that allow for the viewing of the model's status during the simulation. In this section, we will add a clock to view the simulation time, a plot of the number of entities in queue, and an indicator of the number of busy units in a resource.

To add the clock, click on the button labeled CLCK. This is the **Clock** button. The Clock dialog box will appear. The Time Units/Hour check box indicates the number of simulation time units per hour of simulated time. For Example 14.3, there are 60 time units per simulated hour. The Starting Time section allows the specification of a starting time other than 12:00 A.M. You may alter the Time format from A.M./P.M. to 24-hour time and choose from an analog or digital representation of the clock. Choose 24-hour mode with a digital clock. Start the time at 8:00 A.M. and click on the **Accept** button.

Now use the mouse pointer (crosshairs) to indicate the center of the clock. Click on the mouse button and then stretch the box to an appropriate size. Click again. The clock is now an object in the workspace and may be resized or moved like other objects.

The Plot object displays a running plot of some expression. For Example 14.3 we will create a plot of the number in queue at the Milling station.

1. Click on the **Plot** button
2. In the dialog box that appears, click on **Insert**.

3. In the next dialog box, type `NQ(Milling_R_Q)` in the Expression text box. NQ is the SIMAN variable that keeps track of the number in queue. Under Min Y, type `0` and under Max Y, type `3`. Since the capacity of the Mill queue is 3, the size of the plot needs to be only between 0 and 3.
4. Click on the **Accept** button.
5. Use the mouse pointer to locate the upper-left-hand corner of the Plot box and click. Then stretch the box to an appropriate size and click again. Now the plot graphic is displayed. When the simulation begins, the box will clear and begin plotting the number of entities in the Mill queue.

Finally, we will add a level indicator to the workspace.

1. Click on the **LEV** button.
2. In the dialog box that appears, type `NR(Drilling_R)` in the Expression text box.
3. For Min X, type `0` and for Max X, type `2`.
4. Click on the **Accept** button.
5. Use the crosshairs to locate and stretch the level graphic to an appropriate size.

The level will now show the number of busy units of the Drilling station during the simulation. Run the simulation and check your results against the file "EXMP14-3.DOE" on the included diskettes.

EXAMPLE 14.4: Using the TRANSFER.TPO Template

In building Example 14.3, we saw how adding routes between the various stations added more realism to the model. In Example 14.4, we will add a material handling aspect to Example 14.1 in the form of an automated guided vehicle (an AGV) that will carry entities between stations.

Creating a Sequence

To use transporters, entities must follow a sequence rather than being directed by logic from server to server. To add a sequence module to the model use the following steps:

1. Make the Common panel active.
2. Click on the **Seq's** button.

3. Add it to the workspace.
4. Open the dialog box.
5. Click on the **Accept** button.
6. In the Sequence text box, type `AGVSequence` for the name and press the Return key. This brings up another dialog box.
7. Click on **Insert** to add the first step in the sequence.
8. In the text box labeled Station, type `Drilling` and click on **Accept**. Observe that this step has been added to the list of steps in the initial dialog box. Insert the remaining steps, `Milling`, `Grinding`, and `Shipping`.
9. Click on **Accept** to close the Sequences dialog box.

We have created a sequence for the entities to follow through the model.

Adding a Transporter

At this point, we may add a transporter to the model. To do this, click on the Template menu heading and choose **Attach**. Then choose TRANSFER.TPO from the list of templates and click on the **Attach** button. The Transfer panel is now available. This panel has buttons that create transporters, conveyors, links, segments, and distances along with several familiar SIMAN V blocks that allow increased complexity in modeling transporters.

To add the transporter to the model:

1. Click on the **Trnsp** button.
2. Place the Transporter module in the workspace.
3. Open the dialog box.
4. In the Transporter text box, type `AGV` as the transporter's name. We will use the default of 1 transporter unit.
5. Since this is an AGV and not a free path transporter, click on the button next to the words "Guided Path." The name "AGV_Net" will appear as the default Network name.
6. Type in `120` for the velocity.
7. Click on the **Accept** button.

With the Transporter module, the initial position and status of the transporter may be specified, failures may be defined, and special statistics may be included for the summary report. In this example, use the defaults of no failures, no special statistics, initial active status, and initial position at the starting station.

Two important modules have now been added to the model. At this point, the other modules must now be altered in order to use the sequence and transporter definitions.

1. Open the Arrive module's dialog box.
2. Under the Arrival Data section, click on the **Assignments** button. Click on **Insert**. Under Assignment Type, click on the **Other** button and type NS into the Variable text box. NS defines the sequence number that the entities will follow. In the Value text box, type AGVSequence.
3. Click on the **Accept** button.
4. In the Leave Data section, click on the **Tran Out** button. Click on **Request** for Transfer Type. Type in AGV as the Transporter name or choose it from the list. Click on **Accept**.
5. Click on the **Animate** button.
6. In the section labeled Leave for Next Station, click on the **Storage** button. This will add a storage area near the Arrive module where entities will await the AGV. Use the default Storage name (Incoming_S4).
7. Click on the **Unstore** button. This removes an entity's association with a particular storage when it leaves the station. Click on **Accept**.
8. Click on the **Seq** button to indicate that the entities will follow a sequence. Click on **Accept** to close the dialog box.

There is now a dark blue "T" next to the Arrive station. This is the graphic that represents the storage in the animation. It may be resized and moved just like the queue graphic.

1. Open the dialog box to the Drilling Server module.
2. In the Enter Data section, click on the **Tran In** button. Click on the button next to Free Transporter. This will cause the entity to free the transporter when it enters the Drilling station. Click on the **Accept** button to close this box.
3. In the Leave Data section, click on the **Tran Out** button.
4. In the dialog box, click on **Request** for Transfer Type. We will use the default of Transporter under Transporter Name and the CYC default for the Selection Rule.
5. Type in AGV for the Transporter (or choose it from the list).
6. Click on the **Accept** button.
7. Click on the **Seq** button to indicate that the entity follows a sequence.
8. Click on the **Animate** button.

9. In the Leave for Next Station section, click on **Storage** (use the default name) and **Unstore**. Click on the **Accept** button.
10. Click on **Accept** to close the dialog box.

Follow this sequence of steps for both the Milling station and the Grinding station. At the Shipping Depart module, there is no Leave Data section because this is the disposal point for the entity. Therefore, indicate that the entity should free the transporter upon entering the station by clicking the **Tran In** button and choosing the same information as in the other modules. No change is needed in the Depart module, which handles the ejected jobs.

Creating the Network

Because we are using a guided path transporter (an AGV), a network of paths must be created connecting the modules in the system. A network consists of intersections (represented by orange diamonds) and links (represented by orange lines) between intersections. Each intersection and link has a unique name. To build the network follow these steps:

1. Click on the **Links** button in the Transfer panel.
2. Add the link to the workspace. Note: The two intersections connected by a link. The left intersection is the beginning intersection, and the right intersection is the ending intersection.
3. Open the Link dialog box by double-clicking on the L.
4. Type in `IncomingIntx` as the Beginning Intersection.
5. In the Associate with Station text box, type in `Incoming` or choose it from the list.
6. Type in `DrillIntx` as the Ending Intersection.
7. Type in `Drilling` as the associated station (or choose it from the list).
8. The default Link Name appears as `IncomingIntxDrillIntx`. We will accept this default name.
9. In Number of Zones, type the number `6`.
10. In Length of each Zone type the number `10`. In our model, this translates to `6` zones of length 10 units or that the Incoming and Drilling stations are 60 units apart.
11. Click on the down arrow next to the Link Type text box and choose **Bidirectional** from the list.

12. In the Network Name text box, type `AGV_Net` or choose it from the list.
13. Click on **Accept**.
14. Click on the left intersection graphic and place it near the beginning point, the Arrive module. Click on the right intersection graphic and place it near the ending point, the Drilling Server module. The transporter will follow the link between the two intersections.

Similar intersections and links must be created between the other modules. Use the following information to create them.

For this model, use the default Link Name and a Bidirectional Link type. Use 10 units as the length of each zone. The two links that run from Drilling to Grinding and from Incoming to Shipping serve as alternate paths within the network.

It will help the readability of the model if intersections with the same name are overlapped. For example, the ending intersection of the first link (from Incoming to Drilling) is DrillIntx and the beginning intersection of the next link (from Drilling to Milling) is also DrillIntx. The graphics for these identical intersections may be superimposed.

Before running the simulation, arrange the graphics in the workspace so that server, queue, and storage graphics are not overlapping. While the model is being animated, it is important to have a clear view of these constructs as they interact.

Run the simulation for 40 hours and compare your construction and results of the model with the file "EXMP14-4.DOE" on the diskettes provided.

Layouts and Links

In Example 14.4, the links follow straight paths between stations. In future models, you may draw the floor plan of a job shop, for example, where the paths between stations must curve and bend to account for the architecture of the facility. To bend the links between

Table 14.2 Link Information for Example 14.4

Beginning Intx	Associate with Stn	Ending Intx	Associate with Stn	Number of Zones
DrillIntx	Drilling	MillIntx	Milling	4
MillIntx	Milling	GrindIntx	Grinding	4
GrindIntx	Grinding	ShipIntx	Shipping	6
DrillIntx	Drilling	GrindIntx	Grinding	6
IncomingIntx	Incoming	ShipIntx	Shipping	12

two intersections, you may either add a new intersection or use the **NTWK** (Network) button on the Animate panel to create a new link. To accomplish the latter, follow these steps:

1. After you select the link between the two intersections, the name of the network (AGV_Net for Example 14.4) will appear.
2. Delete the link.
3. Click on the Animate tab to activate the Animate panel.
4. Click on the **NTWK** button. The dialog box will open.
5. Click on the **Accept** button.
6. Using the pointer, click on the beginning intersection of the link, and then draw the path to the next intersection. Click the mouse pointer where you need to anchor the line. The line must anchor at the ending intersection of the link.
7. Repeat these steps to create other bending links between intersections.

The file "EXMP14-5.DOE" on the included diskettes shows Example 14.3 within a facility layout.

14.6 SUMMARY

ARENA is a graphical environment that combines the power of SIMAN and Cinema into one format. Rather than writing text files to be compiled, the user develops the model's logic on the screen using modules that represent SIMAN's blocks and elements. Block and element parameters are defined through dialog boxes associated with each module.

The drawing features of ARENA allow the user to develop layouts and graphical representations of entities, resources, queues, and transporters that enhance the aesthetics of the model. ARENA's built-in animation capabilities allow for visual validation of the model's logic and observation of the model's status during the simulation.

In this chapter, three models of increasing complexity were built within the ARENA environment. However, only a few avenues of ARENA's modeling capabilities have been explored.

14.7 REFERENCES

ARENA Template Reference Guide (1994), Systems Modeling Corporation, Sewickley, Penn.

ARENA User's Guide (1994), Systems Modeling Corporation, Sewickley, Penn.

14.8 EXERCISES

14.1 Model Exercise 3.1 in the ARENA environment. Run the simulation for 40 hours, one replication only.

14.2 Adapt Exercise 3.1 for use with an AGV. Use the following information:

AGV velocity: 80 feet per minute
Zone length: 5 feet

Link	Number of Zones
Incoming to Assembly	10
Assembly to Soldering	7
Soldering to Inspection	7
Inspection to Shipping	10
Incoming to Shipping	20

Run the simulation for 40 hours, one replication only.

14.3 Alter Example 14.2 for use with a free-path transporter rather than an AGV. You may need to consult the *ARENA Template Reference Guide* for additional information concerning transporters.

14.4 Alter Example 14.2 for use with a conveyor system. You may need to consult the *ARENA Template Reference Guide* for additional information concerning transporters.

Appendix A

SIMAN Blocks

ACCESS: *Conveyor Name*, *Quantity to Access*;

Operand	Description	Default
Conveyor Name	Name of conveyor to access [Symbol Name]	—
Quantity to Access	Number of consecutive cells to access [expression truncated to an integer]	1

Function: The ACCESS block allocates *Quantity to Access* consecutive cells of the conveyor *Conveyor Name* to an entity.

ACTIVATE: *Transporter Name*;

Operand	Description	Default
Transporter Name	Transporter Name [Symbol Name]	—

Function: The ACTIVATE block sets the operational status of the transporter *Transporter Name* to active (available).

ALLOCATE: *Priority*, *Alternate Path*: *Transporter Name(Unit)*, *Entity Location*;

Operand	Description	Default
Priority	ALLOCATE block priority	1
Alternate Path	Station, link, or intersection to be included in distance calculation when using distance-based transporter selection rule (SDS or LDS) with a guided transporter	Follow system map

Operand	Description	Default
Transporter Name	Transporter to be allocated [symbol name]. Can specify an optional unit in the form shown below	—
Unit	Transporter index [expression truncated to an integer; Transporter Selection Rule; or Transporter Selection Rule, *Attribute ID*]	1
Entity Location	Location of the entity defined by system map [*StationID*, STATION(*StationID*), INTX (*IntxID*), or LINK(*LinkID*, *Zone*)]	Entity M value

Function: The ALLOCATE block assigns a transporter to the entity without actually moving the transporter from its current location.

ALTER: *ResourceID*, *Capacity Change*: repeats;

Operand	Description	Default
ResourceID	Resource to alter [expression truncated to an integer]	—
Capacity Change	The magnitude of the capacity change [expression truncated to an integer]	1

Function: The ALTER block changes the capacity of the resource defined by *ResourceID* by a positive or negative *Capacity Change*.

ASSIGN: *Variable or Attribute* = *Value*: repeats;

Operand	Description	Default
Variable or Attribute	SIMAN variable, attribute, or resource STATE(*ResourceID*)	—
Value	Value to be assigned [expression]	—

Function: The ASSIGN block allows for the assignment of a value to a SIMAN variable, attribute, or resource STATE(*ResourceID*).

BEGIN, *Model Listing*, *Model Name*;

Operand	Description	Default
Model Listing	Option for generating a listing of model statements during model processing [Yes or No]	Yes
Model Name	Model name associated with the statements in this model file [alphanumeric]	Filename

Function: The BEGIN block establishes the model name and determines the listing options for model processing.

BLOCK: *Blockage ID*, *Number of Block Points*;

Operand	Description	Default
Blockage ID	Blockage number or name [integer or symbol name]	—
Number of Block Points	Number of block points to add to Blockage ID [expression truncated to an integer]	1

Function: The BLOCK block adds *Number of Block Points* to *Blockage ID.*

BRANCH, *Max Number of Branches, Random Number Stream*:
Branch Type, Condition or Probability, Destination Label, Primary Entity Indicator;
IF, *Condition*, *Label*, *Primary Entity Indicator*:
WITH, *Probability*, *Label*, *Primary Entity Indicator*:
ELSE, *Label*, *Primary Entity Indicator*:
ALWAYS, *Label*, *Primary Entity Indicator*;

Operand	Description	Default
Max Number of Branches	Maximum number of branches to take [expression truncated to an integer]	Infinite
Random Number Stream	Random number stream to use with the WITH rule [integer]	10 or number specified in SETUP
Branch Type	Type of rule used to branch the entity [IF, WITH, ELSE, or ALWAYS]	—

Operand	Description	Default
Condition	Branch condition [condition of expression] Used only when *Branch Type* is set to IF	—
Probability	Probability of selecting branch [expression, 0 ≤ Probability ≤ 1] Used only when Branch Type is set to WITH	—
Label	Label of block to which copy of entity is routed [block label]	—
Primary Entity Indicator	Indicator that primary (incoming) entity can select this branch [Yes or No]	Yes

Function: The BRANCH block controls the flow of an entity through a set of branches.

CAPTURE: *Quantity to Capture*, *Travel Destination*;

Operand	Description	Default
Quantity to Capture	Number of forward zones or length of guided transporter track to capture [expression truncated to an integer]	—
Travel Destination	Travel destination [*StationID*, STATION(*StationID*), INTX (*IntxID*), or LINK(*LinkID*, *Zone*)]	—

Function: The CAPTURE block seizes *Quantity to Capture* additional zones or length units of a guided transporter track and establishes the additional area (zones or length) required to contain a transporter as it moves through the system.

CLOSE, *File ID*;

Operand	Description	Default
File ID	File to be closed [integer file number or file symbol name]	—

Function: The CLOSE block closes a file, making it unavailable for input or output operations.

COMBINE, *Match Expression*: *Quantity to Match*, *Save Criterion*;

Operand	Description	Default
Match Expression	The expression used to combine matching entities [expression]	No matching expression
Quantity to Match	The number of entities to permanently combine [expression truncated to an integer]	—
Save Criterion	Save criterion [FIRST, LAST, PRODUCT, SUM]	LAST

Function: The COMBINE block forms one permanent representative entity from a number of entities defined by the operand *Quantity to Match.* The attributes of the representative entity are assigned according to the operand *Save Criterion.*

CONVEY: *Conveyor Name*, *Destination Station*;

Operand	Description	Default
Conveyor Name	Name of conveyor [Symbol name]	Conveyor the entity accessed
Destination Station	Destination station [expression truncated to an integer station number, station symbol name, or SEQ]	SEQ

Function: The CONVEY block conveys the arriving entity by way of the conveyor defined by the operand *Conveyor Name* to the *Destination Station.*

COPY: *Entity Rank*, *Queue ID*, *Copy Destination*;

Operand	Description	Default
Entity Rank	Rank of the entity to be copied [expression truncated to an integer, or the keyword NQ]	—

Operand	Description	Default
Queue ID	Queue from which the entity is to be copied [queue symbol name, or expression truncated to an integer queue number]	—
Copy Destination	Label of block to which the created entity copy is to be sent [block label]	—

Function: The COPY block creates a copy of an entity residing in a queue without removing the entity from the queue, and sends the duplicate to a labeled block.

COUNT: *Counter ID*, *Counter Increment*;

Operand	Description	Default
Counter ID	Counter number or name [expression truncated to an integer counter number, or counter symbol name]	—
Counter Increment	Magnitude of counter increment [expression truncated to an integer]	1

Function: The COUNT block increments the counter specified by *Counter ID* by the value of the operand *Counter Increment.*

CREATE, *Batch Size*, *First Creation Time*: *Interval*, *Max Batches*;

Operand	Description	Default
Batch Size	Number of entities in each batch creation [expression truncated to an integer]	1
First Creation Time	Simulation time at which the first batch creation should occur [expression]	Begin time of replication
Interval	Time between batch creations [expression]	Infinite
Max Batches	Maximum number of batches to be created [expression truncated to an integer]	Infinite

Function: The CREATE block generates arriving entities to a process model.

DELAY: *Duration*, *Storage ID*;

Operand	Description	Default
Duration	Length of the delay [expression]	0.0
Storage ID	Storage associated with the DELAY block [expression truncated to an integer storage unit, or storage symbol name]	No storage

Function: The DELAY block delays an entity by *Duration* time units.

DROPOFF, *Rank of Entity*, *Quantity to Dropoff*: *Dropoff Location*, *Attributes*;

Operand	Description	Default
Rank of Entity	Starting rank of entities to drop off [expression truncated to an integer]	1
Quantity to Dropoff	Number of entities to drop off [expression truncated to an integer]	1
Dropoff Location	Label to send entities which are dropped off [block label]	—
Attributes	List of representative entity attributes that are assigned to the entities being dropped off [*, A(*), M, NS, IS, *AttributeID*]	No attribute assignment

Function: The DROPOFF block removes the number of entities defined by the operand *Quantity to Dropoff* from the current entity group and sends them to the block label specified by the operand *Dropoff Location.* Attribute values from the representative entity can be assigned to the entities leaving the group.

DUPLICATE: *Quantity to Duplicate*, *Duplicate Destination*: repeats;

Operand	Description	Default
Quantity to Duplicate	Number of duplicate entities to be created [expression truncated to an integer]	0
Duplicate Destination	Label to send duplicate entities to [block label]	Next block

Function: The DUPLICATE block creates duplicates of the arriving entity and sends them to the *Duplicate Destination* block label. The number of duplicates created is specified by the operand *Quantity to Duplicate.*

ELSE;

There are no operands associated with this block.

Function: The ELSE block is used in conjuction with the IF, ELSEIF, and ENDIF blocks to control the flow of an entity through a group of blocks in the model.

ELSEIF: *Condition* or *Expression*;

Operand	Description	Default
Condition or Expression	Condition or expression to evaluate [logical or mathematical expression]	—

Function: The ELSEIF block is used in conjunction with the IF, ELSE, and ENDIF blocks to control the flow of an entity through a group of blocks in the model.

ENDIF;

There are no operands associated with this block.

Function: The ENDIF block is used in conjunction with the IF, ELSE, and ELSEIF blocks to control the flow of an entity through a group of blocks in the model, and is used to terminate the block combination.

ENDWHILE;

There are no operands associated with this block.

Function: The ENDWHILE block is used in conjunction with the WHILE block to control the flow of an entity through a group of blocks in the model. It terminates the while loop.

EXIT: *Conveyor Name*, *Quantity to Release*;

Operand	Description	Default
Conveyor Name	Name of the conveyor to exit [symbol name]	Conveyor the entity accessed
Quantity to Release	Number of consecutive cells to release [expression truncated to an integer]	All accessed cells

Function: The EXIT block releases cells of the conveyor specified by *Conveyor Name*. The number of cells released is defined by the operand *Quantity to Release*.

FINDJ, *Start of Range*, *End of Range*: *Search Condition*;

Operand	Description	Default
Start of Range	Start limit of index range to be searched [expression truncated to an integer]	—
End of Range	End limit of index range to be searched [expression truncated to an integer]	—
Search Condition	Search condition containing the index J [condition, *Expression*, MIN(*Expression*), or MAX(*Expression*)]	—

Function: The FINDJ block searches from index *Start of Range* to index *End of Range* to find the value of the global variable J that satisfies the specified *Search Condition.* J is set to the value of the first index value that satisfies *Search Condition,* or to zero if no value of J in the specified range satisfies the *Search Condition.*

FREE: *Transporter Name*;

Operand	Description	Default
Transporter Name	Transporter unit to be freed [Transporter Name or Transporter Name(*Index*), *Index* is an expression truncated to an integer]	Transporter that was allocated

Function: The FREE block releases an allocated transporter unit.

GROUP, *Match Expression*: *Quantity to Group*, *Save Criterion*;

Operand	Description	Default
Match Expression	The expression used to group matching entities [expression]	No matching expression
Quantity to Group	Number of entities to temporarily group [expression truncated to an integer]	—
Save Criterion	Save criterion [FIRST, LAST, PRODUCT, SUM]	LAST

Function: The GROUP block forms one temporary representative entity from a number of entities specified by the operand *Quantity to Group.* The attributes of the representative entity are assigned according to the *Save Criterion* operand.

HALT: *Transporter Name*;

Operand	Description	Default
Transporter Name	Transporter unit to deactivate [TrnName or TrnName(*Index*), *Index* is an expression truncated to an integer]	—

Function: The HALT block changes the status of the transporter unit specified by the operand *Transporter Name* to inactive.

IF: *Condition* or *Expression*;

Operand	Description	Default
Condition or expression	Condition or expression to evaluate [logical or mathematical expression]	—

Function: The IF block is used in conjunction with the ELSEIF, ELSE, and ENDIF blocks to control the flow of an entity through a group of blocks in the model.

INCLUDE: *File Name*;

Operand	Description	Default
File Name	File containing blocks to include in the model [system-specific filename enclosed in double quotes]	—

Function: The INCLUDE statement adds the blocks contained in the file specified by the operand *File Name* into the current model at the location of the INCLUDE statement.

INSERT: *Queue Label, Rank to Place Entity*;

Operand	Description	Default
Queue Label	Label corresponding to the QUEUE block where the entity will be inserted [block label]	—
Rank to Place Entity	Queue rank to place the entity [expression truncated to an integer]	Queue ranking rule

Function: The INSERT block places the current entity at the position specified by the operand *Rank to Place Entity* in the queue associated with block label *Queue Label*.

MATCH, *Match Attribute*: *Queue Label*, *Destination Label*: repeats;

Operand	Description	Default
Match Attribute	Attribute [integer attribute number, A(*k*), or attribute symbol name]	No attribute
Queue Label	Label of the detached QUEUE block where entity resides [block label]	—
Destination Label	Destination label for the entity [block label]	Dispose the entity

Function: The MATCH block synchronizes the advance of two or more entities located in different, detached queues. When operand *Match Attribute* is specified, the MATCH block synchronizes the advance of entities with matching values of *Match Attribute.*

MOVE, *Storage ID*, *Alternate Path*: *Transporter Name*, *Destination*, *Velocity*;

Operand	Description	Default
Storage ID	Storage associated with the MOVE block [expression truncated to an integer storage number, or storage symbol name]	No storage
Alternate Path	A station, link, or intersection to be included in transportation path of a guided transporter	Not used
Transporter Name	Transporter unit to move [TrnName or TrnName (*Index*)]	Unit the entity was allocated
Destination	Transporter destination [*StationID*, STATION (*StationID*), INTX(*IntxID*), LINK(*LinkID*), LINK (*LinkID*, *Zone*), or use Keyword FIRSTX]	Entity M value
Velocity	Velocity for MOVE [expression]	Current velocity

Function: The MOVE block moves the transporter specified by *Transporter Name* to the station defined by the operand *Destination.*

PICKQ, *Queue Selection Rule*, *Balk Label*: *Queue Label*: repeats;

Operand	Description	Default
Queue Selection Rule	Rule used to select queue [CYC, ER(*rule*), LNQ, LRC, POR, RAN, SNQ, SRC, UR(*rule number*)]	POR
Balk Label	Label to send balked entities to [block label]	Dispose the entities
Queue Label	The label of each following QUEUE block [block label]	—

Function: The PICKQ block allows an arriving entity to select a particular QUEUE block from a multiple queuing facility. If all queues are full, the entity is sent to the block labeled *Balk Label.*

PICKUP: *Queue ID*, *Starting Rank*, *Quantity to Pickup*;

Operand	Description	Default
Queue ID	Queue from which to pick entities [expression truncated to an integer queue number, or queue symbol name]	—
Starting Rank	Starting rank of the entities in queue *Queue ID* to pick up [expression truncated to an integer]	1
Quantity to Pickup	Number of entities to pick up [expression truncated to an integer]	1

Function: The PICKUP block removes the number of consecutive entities specified by *Quantity to Pickup* from the *Queue ID* starting at position *Starting Rank* and adds them to the group of the entity executing the PICKUP block.

PREEMPT, *Priority*: *Resource ID*, *Attribute ID*, *Destination*;

Operand	Description	Default
Priority	Preemption priority [expression]	1.0

Operand	Description	Default
Resource ID	Number or name of resource to preempt [expression truncated to an integer, or SELECT(*SetName* [,*rule*] [,*AttribID*])]	—
Attribute ID	Attribute of the preempted entity used to store that entity's remaining delay time [integer attribute number, A(*k*), or attribute symbol name]	Stored internally
Destination	Destination of preempted entity after surrendering the resource [block label or *STO*(*StorID*), *StorID* is an expression truncated to an integer storage number, or a storage symbol name]	Held internally for reallocation

Function: The PREEMPT block takes one unit of a resource specified by *Resource ID* away from the entity that originally seized it. The interrupted entity then may be sent back to its queue, sent to a labeled block, or placed in storage. If desired, the preempted entity's remaining delay time can be stored in one of that entity's attributes by specifying the operand *Attribute ID.*

PROCEED, *Priority*: *Blockage ID*: repeats;

Operand	Description	Default
Priority	Proceed priority [expression truncated to an integer]	1
Blockage ID	Blockage identifier [expression truncated to an integer blockage number, or blockage symbol name]	—

Function: The PROCEED block suspends entity movement through the model until all specified *Blockage ID* blockages are clear. If any *Blockage ID* has a block or queue blockages, arriving entities are held in the queue in front of the PROCEED block until all blockages are cleared.

QPICK, *Queue Selection Rule*: *Queue Label*: repeats;

Operand	Description	Default
Queue Selection Rule	Method by which a queue is selected [CYC, ER(*rule*), LNQ, LRC, POR, RAN, SNQ, SRC, UR(*k*)]	POR
Queue Label	The label on each preceding detached QUEUE block [block label]	—

Function: The QPICK block removes a departing entity from a set of parallel QUEUE blocks as the result of a status change in a SELECT block or a hold-type block (ACCESS, ALLOCATE, PREEMPT, REQUEST, or SEIZE).

QUEUE, *Queue ID*, *Capacity*, *Balk Label*: *Blockage Level*, *Blockage ID*: repeats;

Operand	Description	Default
Queue ID	Queue number or name [expression truncated to an integer queue number, or queue symbol name]	—
Capacity	Queue capacity [expression truncated to an integer or FULLWHEN(*expression*)]	Infinite
Balk Label	Label to which balked entity is sent [block label]	Dispose the entity
Blockage Level	Blockage trigger level [expression truncated to an integer or BLOCKWHEN(*expression*)]	—
Blockage ID	Blockage number or name [expression truncated to an integer or symbol name]	—

Function: The QUEUE block is used to model a waiting space before a SELECT or a hold-type block. QUEUE blocks may have a specified *Capacity,* and the arriving entity balks to the block at label *Balk Label* when the queue is full. QUEUE blocks may also specify the *Blockage ID* to increment/decrement if the capacity of the queue surpasses/drops below the specified *Blockage Level.*

READ, *File ID*, *Format*, *Record Number*: *Variables*;

Operand	Description	Default
File ID	File from which to read [expression truncated to an integer file number, file symbol name, or keyword STDIN for standard input]	Standard input
Format	Format to use in interpreting data in FILES [exact FORTRAN or C format enclosed in double quotes, or FREE]	Format specified in FILES element, or FREE for standard input
Record Number	Record number to be read [expression truncated to an integer record number]	Next record
Variables	List of variables to read separated by commas [any user-assignable variable or attribute]	—

Function: The READ block reads data from input files and assigns the values to the list of variables.

RELEASE: *Resource ID*, *Quantity to Release*: repeats;

Operand	Description	Default
Resource ID	Number or name of resource to release [expression truncated to an integer, *resource symbol name*, *SetName*(*Index*), Member(*SetName*, *Index*), or SELECT(*SetName* [, FIRST|LAST][, *AttribID*])]	—
Quantity to Release	Number of resource units to be released [expression truncated to an integer]	1

Function: The RELEASE block releases units of the resource specified by *Resource ID* from the entity. The number of units released is defined by the operand *Quantity to Release*.

RELINQUISH: *Number to Relinquish*;

Operand	Description	Default
Number to Relinquish	Number or length of backward zones to relinquish [expression truncated to an integer]	Current zones occupied by transporter

Function: The RELINQUISH block releases consecutive zones or length units of track occupied by a guided transporter when the entity controlling the transporter executes the RELINQUISH block. The number or length of zones to relinquish is specified by operand *Number to Relinquish.*

REMOVE: *Rank of Entity*, *Queue ID*, *Remove Entity Destination*;

Operand	Description	Default
Rank of Entity	Rank of the entity to be removed [expression truncated to an integer, or the keyword NQ]	—
Queue ID	Queue number from which the entity is to be removed [expression truncated to an integer queue number or queue symbol name]	—
Removed Entity Destination	Label of block to which the removed entity is to be sent [block label]	—

Function: The REMOVE block removes the entity at a position defined by the operand *Rank of Entity* residing in the queue *Queue ID* and sends it to the block label defined by the operand *Removed Entity Destination.*

REQUEST, *Priority*, *Storage ID*, *Alternate Path*: *TransporterName*(*Unit*), *Velocity*, *Entity Location*;

Operand	Description	Default
Priority	REQUEST block priority [expression truncated to an integer]	1
Storage ID	Storage associated with the REQUEST block [expression truncated to an integer storage number, or storage symbol name]	No storage

Operand	Description	Default
Alternate Path	Station, link, or intersection to be included in distance calculation when using distance-based transporter selection rule (SDS or LDS) with a guided transporter	Not used
Transporter Name	Transporter requested [symbol name]. Can specify an optional unit in the form shown below	—
Unit	Transporter index [expression truncated to an integer; Transporter Selection Rule; or Transporter Selection Rule, *Attribute ID*]	1
Velocity	Velocity for REQUEST [expression]	Current velocity
Entity Location	Location of the entity defined by system map [*StationID*, STATION(*StationID*), INTX(*IntxID*), or LINK(*LinkID*, *Zone*)]	Entity M value

Function: The REQUEST block assigns a transporter to the entity and moves the transporter to the entity's current location.

ROUTE: *Duration*, *Destination*;

Operand	Description	Default
Duration	Routing time delay [expression]	0.0
Destination	Destination station [expression truncated to an integer station number, station symbol name, or SEQ]	SEQ

Function: The ROUTE block transfers the entity in *Duration* time units to the station specified by the operand *Destination.*

SCAN: *Condition*;

Operand	Description	Default
Condition	Scan condition [condition or expression]	—

Function: The SCAN block holds an entity in the preceding queue until the scan *Condition* is true.

SEARCH, *Search Item*, *Starting Index*, *Ending Index*: *Search Condition*;

Operand	Description	Default
Search Item	Queue or group to be searched [keyword GROUP, an expression truncated to an integer queue number, or a queue symbol name]	Search current entity's group
Starting Index	Starting index rank for the search [expression truncated to an integer]	1
Ending Index	Ending index rank for the search [expression truncated to an integer]	Last entity
Search Condition	Search condition [condition, MIN(*Expression*), or MAX(*Expression*), *Expression* should contain one or more attributes]	—

Function: The SEARCH block searches a queue or group to find the entity rank that satisfies the *Search Condition.* The value of the global system variable J is set to the rank of the first entity that satisfies the *Search Condition,* or to 0 if the *Search Condition* is not satisfied.

SEIZE, *Priority*: *Resource ID*, *Number of Units*: repeats;

Operand	Description	Default
Priority	Seize priority [expression evaluated as a floating point value]	1.0
Resource ID	Number or name of resource to seize [expression truncated to an integer, resource *symbol name*, *SetName*(*Index*), Member(*SetName*, *Index*), or SELECT(*SetName* [, *rule*] [, *AttribID*])]	—
Number of Units	Number of resource units required [expression truncated to an integer]	1

Function: The SEIZE block allocates units of a resource defined by the operand *Resource ID* to an entity. The number of units seized is specified by the operand *Number of Units.*

SELECT, *Resource Selection Rule*: *Seize Label*: repeats;

Operand	Description	Default
Resource Selection Rule	Method by which resource is selected [CYC, ER(*k*), LNB, LRC, POR, RAN, SNB, SRC, or UR(*k*)]	POR
Seize Label	Label of each following SEIZE block [block label]	—

Function: The SELECT block is used when an entity requires any one of a set of interchangeable resources. The SELECT block uses a *Resource Selection Rule* to specify the logic for selecting among two or more *Seize Label* operands corresponding to SEIZE blocks.

SIGNAL: *Signal Code*, *Release Limit*;

Operand	Description	Default
Signal Code	Signal code [expression truncated to an integer]	—
Release Limit	Total number of entities to release [expression truncated to an integer]	Infinite

Function: The SIGNAL block sends a *Signal Code* to each WAIT block in the model and releases a maximum of *Release Limit* entities.

SPLIT: *Attribute*;

Operand	Description	Default
Attribute	List of representative entity attributes that are assigned to members of the temporary group [*, A(*), M, NS, IS, *AttributeID*]	No attribute assignments

Function: The SPLIT block terminates a temporary representative entity that was formed at a GROUP or PICKUP block, and recovers the original entities that formed the group.

START: *Conveyor*, *Velocity*;

Operand	Description	Default
Conveyor	Conveyor to activate [expression]	—
Velocity	Conveyor velocity [expression]	No change

Function: The START block sets the operational status of the specified *Conveyor* to active with a velocity specified by the *Velocity* operand.

STATION, *Beginning Station ID - Ending Station ID*;

Operand	Description	Default
Beginning Station ID	Lower limit of station range [integer station number, station symbol name, or station set name]	—
Ending Station ID	Upper limit of station range [integer station number, station symbol name, or station set name]	No range

Function: The STATION block represents a point in the model to which entities are transferred. Single stations are identified with *Beginning Station ID* only, and station macro submodels are identified with the range *Beginning Station ID - Ending Station ID.*

STOP: *Conveyor Name*;

Operand	Description	Default
Conveyor Name	Conveyor to deactivate [expression]	—

Function: The STOP block sets the operational status of the specified *Conveyor Name* to inactive.

STORE: *Storage ID*;

Operand	Description	Default
Storage ID	Storage to which to add entity [expression]	—

Function: The STORE block adds an entity to a storage.

TALLY: *Tally ID*, *Value*, *Number of Observations*;

Operand	Description	Default
Tally ID	Tally identifier [expression truncated to an integer tally number, or tally symbol name]	—
Value	Value to be recorded [expression, INTerval(*AttributeID*), BETween(*VariableID*), or BETween]	—

Operand	Description	Default
Number of Observations	Number of observations of value to be recorded [expression truncated to an integer]	1

Function: The TALLY block records the *Number of Observations* of *value* in *tally ID.*

TRACE, *Trace Setting*, *Format*: *Variable or Expression*,...;

Operand	Description	Default
Trace Setting	Integer or expression that evaluates to one of the trace level values [0 = Off, 1 = least detailed, 2 = medium detailed, 3 = most detailed]	—
Format	Format used in interpreting data	Free
Variable or Expression	List of variables or expressions to output	—

Function: The TRACE block is used to output user-defined trace messages to be displayed.

TRANSPORT, *Alternate Path*: *Transporter Name*, *Entity Destination*, *Velocity*, *Guided Trans Dest*;

Operand	Description	Default
Alternate Path	A specified station, intersection, link, or zone to be included in the path of transportation [VIA(*StationID*), VIA(STATION(*StationID*)), VIA(INTX(*IntxID*)), VIA(LINK(*LinkID*)), VIA(LINK(*LinkID*, *Zone*))]	Not used
Transporter Name	Transporter unit [TrnName or TrnName(*Index*), *Index* is an expression truncated to an integer]	Unit the entity was allocated

Operand	Description	Default
Entity Destination	Entity destination station [expression truncated to an integer station number, station symbol name, or SEQ]	SEQ
Velocity	Velocity for transport [expression]	Current velocity
Guided Trans Dest	Guided transport destination [expression]	Intersection associated with *Entity Destination* Station

Function: The TRANSPORT block transports an entity to a station destination using a transporter specified by *Transporter Name.* The entity reappears at station block *Entity Destination* and the transporter's new location will be *Guided Trans Dest.*

UNBLOCK: *Blockage ID*, *Number to Remove*;

Operand	Description	Default
Blockage ID	Blockage Number or Name [integer or symbol name]	—
Number to Remove	Number of block points to remove from Blockage ID [expression truncated to an integer]	1

Function: The UNBLOCK block removes block points from *Blockage ID.* The number of block points to remove is specified by the operand *Number to Remove.*

UNSTORE: *Storage ID*;

Operand	Description	Default
Storage ID	Storage from which to remove entity [expression]	—

Function: The UNSTORE block removes the current entity from a storage.

WAIT: *Signal Code*, *Number to Release*;

Operand	Description	Default
Signal Code	Signal code [expression truncated to an integer signal code]	—
Number to Release	Maximum number of entities to release from this block when the signal code is received [expression truncated to an integer]	Infinite

Function: The WAIT block holds an entity in the preceding QUEUE block until a *Signal Code* is received. When the *Signal Code* is received, the WAIT block will release a maximum of *Number to Release* entities from the preceding QUEUE block.

WHILE: *Condition* or *Expression*;

Operand	Description	Default
Condition or Expression	Condition or expression to evaluate [logical or mathematical expression]	—

Function: The WHILE block is used in conjunction with the ENDWHILE block to control the flow of an entity through a group of blocks in a model.

WRITE, *File ID*, *Format*, *Record Number*: *Variables*;

Operand	Description	Default
File ID	File to be written to [expression truncated to an integer file number, file symbol name, or keywords STDRPT for the summary file or STDOUT for standard output]	Standard output
Format	Format to use in interpreting data [exact FORTRAN or C format enclosed in double quotes, or FREE]	Format specified in FILES element, or FREE for standard output

Operand	Description	Default
Record Number	Record number to be written [expression truncated to an integer record number]	Next record
Variables	List of variables or expressions to write [expression or any SIMAN attribute, variable or experiment element name string]	No value written

Function: The WRITE block writes data to an output file.

Appendix B

SIMAN Elements

ARRIVALS: *Identifier, Type(Type ID), Time, Batch Size, Assignments*: repeats;

Operand	Description	Default
Identifier	Arrival number or name [integer or symbol name]	—
Type	Arrival type [STATION(*TypeID*), QUEUE(*TypeID*), or BLOCK(*TypeID*)]	—
TypeID	Corresponding block identifier for the arrival type. [STATION identifier, QUEUE identifier, or BLOCK identifier]	—
Time	Time for arrival to occur [TIME, FIRST, LAST, WARMUP, EVERY(*Interval* [,*Offset*] [, *Max Batches*] [, *Max Time*])]	Beginning of simulation
Batch Size	Number of entities in batch arrival [expression truncated to an integer]	1
Assignments	Initial values of entity attributes [list of assignments in the form *AttribID* = *Value, . . .*]	—

Function: The ARRIVALS element creates batches of entities that arrive at the system model at specified times.

ATTRIBUTES: *Number, Name(1-D Array Index, 2-D Array Index), Initial Values, . . .* : repeats;

Operand	Description	Default
Number	Attribute number (index into the A(*k*) array) [integer]	Sequential
Name	Attribute name [symbol name]	Blank
1-D Array Index	First dimension index into the named attribute array	No array
2-D Array Index	Second dimension index into the named attribute array	No array
Initial Values	Initial values upon entity creation [constant]	0.0 or last value

Function: The ATTRIBUTES element specifies the total number of general-purpose entity attributes, their names, and attribute initial values.

BEGIN, *Listing, Run Controller*;

Operand	Description	Default
Listing	Option for generating a listing of experiment statements during experiment processing [Yes or No]	Yes
Run Controller	Invoke the Interactive Run Controller [Yes or No]	No

Function: The BEGIN statement specifies the listing option for experiment processing and determines whether or not the SIMAN Interactive Run Controller is activated during the simulation run.

BLOCKAGES: *Number, Name, Initial Blockages, Global Priority(Priority Expression)*: repeats;

Operand	Description	Default
Number	Blockage number [integer]	Sequential
Name	Blockage name [symbol name]	Blank
Initial Blockages	Initial number of blockages [expression truncated to an integer]	—

Operand	Description	Default
Global Priority	Global priority for selecting from among blockages [QTIME, HVF(*Priority Expression*), LVF(*Priority Expression*)]	QTIME
Priority Expression	Expression used to determine global priority. Specified only when Global Priority is set to HVF or LVF.	—

Function: The BLOCKAGES element defines all block and queue blockages used in a model, initializes the number of SIMAN block blockages, and sets the global priority for tie breaking.

CONVEYORS: *Number, Name, Segment Set ID, Velocity, Cell Size, Status, Max Cells per Entity, Type, Accumulation Length*: repeats;

Operand	Description	Default
Number	Conveyor number [integer]	Sequential
Name	Conveyor name [symbol name]	—
Segment Set ID	Segment set name or number [integer or symbol name]	—
Velocity	Conveyor velocity [constant]	1.0
Cell Size	Length of each conveyor cell [integer]	1
Status	Initial status of conveyor [Active or Inactive]	Active
Max Cells per Entity	Maximum number of cells occupied by an entity [integer]	1
Type	Conveyor type [Accumulating or Nonaccumulating]	Nonaccumulating
Accumulation Length	Accumulation length of an entity [constant, attribute name, or attribute number]	Cell size

Function: The CONVEYORS element defines the characteristics of all conveyors used in the model.

COUNTERS: *Number, Name, Limit, Initialize Option, Output File*: repeats;

Operand	Description	Default
Number	Counter number [integer]	Sequential
Name	Counter name and summary report identifier [symbol name]	"Counter *Number*"
Limit	Counter limit [positive integer]	Infinite
Initialize Option	Initialize counter between simulation replications [Yes, No, Replicate]	Replicate
Output File	Output file to which counter observations are written during the simulation run [unique integer number or system-specific filename enclosed in double quotes]	No save

Function: The COUNTERS element specifies parameters for counters that may be used to keep integer count statistics on events occurring in the model.

DISTANCES: *Identifier, Starting Station ID - Ending Station ID - Distance, ...*: repeats;

Operand	Description	Default
Identifier	Distance set number or name [integer or symbol name]	Sequential
Starting Station ID	Beginning station identifier [integer station number or station symbol name]	—
Ending Station ID	Ending station identifier [integer station number or station symbol name]	—
Distance	Distance from *Starting Station ID* to *Ending Station ID* [positive integer]	0

Function: The DISTANCES element is used to define the travel distances between all stations that free-path transporters may visit.

DSTATS: *Number, SIMAN Expression, Name, Output File*: repeats;

Operand	Description	Default
Number	DSTAT number [integer]	Sequential
SIMAN Expression	SIMAN expression on which time-persistent statistics are to be recorded [expression]	—
Name	DSTAT name and label for the SIMAN summary report [symbol name]	*Expression*
Output File	Output file to which DSTATS observations are written during the simulation run [unique integer number or system-specific file name enclosed in double quotes]	No save

Function: The DSTATS element is used to obtain time-persistent statistics in a discrete system.

EXPRESSIONS: *Number, Name (1-D Array Index, 2-D Array Index), Expressions*: repeats;

Operand	Description	Default
Number	Expression number [integer]	Sequential
Name	Expression name [symbol name]	Blank
1-D Array Index	First dimension index into the named distribution array	No array
2-D Array Index	Second dimension index into the named distribution array	No array
Expressions	*SIMAN Expression*	—

Function: The EXPRESSIONS element defines expressions, perhaps associated with their names, or the SIMAN variable ED(*NumberName*).

FAILURES: *Number, Name, Type(Time Between Failures, Duration, State)*: repeats;

Operand	Description	Default
Number	Failure number [integer]	Sequential
Name	Failure name [symbol name]	Blank
Type	COUNT or TIME	—

Operand	Description	Default
Time Between Failures	Time or count between failures [expression or expression truncated to an integer]	—
Duration	Duration of the failure [expression]	—
State	State for which *Between* time accumulates [time-based failures only]	All

Function: The FAILURES element defines the characteristics of failures associated with resources.

FILES: *Number, Name, System Name, Access Type, Structure, End of File Action, Comment Character, Initialize Option*: repeats;

Operand	Description	Default
Number	File number [integer]	Sequential
Name	File name used for identification on READ and WRITE blocks [symbol name]	Blank
System Name	Operating system file name enclosed in double quotes [system-dependent]	—
Access Type	File type [SEQuential; or DIR(*Length*), direct access using record *Length* bytes; or USER]	SEQ
Structure	File structure [UNFormatted; FREE; WKS; or *Format*, the exact FORTRAN or C format, which must be 255 characters or fewer]	UNFormatted
End of File Action	End of file action [ERRor, DISpose, REWind, or IGNore]	ERRor
Comment Character	Character indicating comment record [character enclosed in double quotes, or No]	No

Operand	Description	Default
Initialize Option	Action to be taken at beginning of each simulation replication for sequential files [HOLD, REWind, or CLOSe]	HOLD

Function: The FILES element must be included whenever external files are accessed using the READ and WRITE blocks.

FREQUENCIES: *Number, Type, Expression, Name, Output File, Value Or Range*$_1$*, Category*$_1$*, Category Option*$_1$ *& ... Value Or Range*$_n$*, Category*$_n$*, Category Option*$_n$: repeats;

Operand	Description	Default
Number	Frequency number [integer]	Sequential
Type	Frequency type VALUE or STATE	—
Expression	SIMAN expression on which FREQUENCIES statistics are to be recorded [expression or STATE(*ResourceID*), where *ResourceID* is a valid resource symbol name]	—
Name	Frequency name and identifier for labeling summary report [symbol name]	*Expression*
Output File	Output file to which FREQUENCIES observations are written during the simulation run [unique integer number or system-specific filename enclosed in double quotes]	No save
Value or Range$_i$	Value or range of values of *Expression* to define a particular *Category* [integer or real value, keyword RANGE(*Low*, *High*), valid state names for the *ResourceID* specified in *Expression*, or the autostates IDLE, BUSY, INACTIVE, or FAILED]	—

Operand	Description	Default
$Category_i$	Category name used to label a particular *Value or* $Range_i$ in the summary report [symbol name]	$Value/Range_i$
Category $Option_i$	Option to exclude or include $Category_i$ in the Restricted Column of the summary report [Exclude or Include]	Include

Function: The FREQUENCIES element is used to record the time-persistent occurrence frequency of a SIMAN variable or expression within user-defined limits of the variable.

INCLUDE: *File Description*;

Operand	Description	Default
File Description	File containing elements to include in the model [system-specific file name enclosed in double quotes]	—

Function: The INCLUDE statement adds the elements contained in the file *File Description* into the current experiment at the location of the INCLUDE statement.

INITIALIZE: *Variable* = *Value*: repeats;

Operand	Description	Default
Variable	Variable or integer attribute to be initialized [IS, J, M, NS, or X(k), k is an integer]	—
Value	Value to be assigned to the variable [constant (integer for IS, J, M, NS)]	—

Function: The INITIALIZE element specifies the initial values for SIMAN user-assignable variables and special-purpose attributes.

INTERSECTIONS: *Number, Name, Travel Length, Link Selection Rule(Rule Attribute ID), Velocity Change Factor*: repeats;

Operand	Description	Default
Number	Intersection number [integer]	Sequential
Name	Intersection name [symbol name]	Blank
Travel Length	Travel length through intersection [integer]	0
Link Selection Rule	Link selection rule [FCFS, LCFS, LVF(*Rule Attribute ID*), HVF(*Rule Attribute ID*), CLOSEST, FARTHEST]	FCFS
Rule Attribute ID	Attribute ID used in the *Link Selection Rule.* Used only when the *Link Selection Rule* is LVF or HVF.	—
Velocity Change Factor	Multiplier applied to the current velocity [constant]	1.0

Function: The INTERSECTIONS element defines the characteristics of all intersections in a guided transporter system.

LINKS: *Number, Name, Beginning Intersection ID - Beginning Direction, Ending Intersection ID - Ending Direction, Number of Zones, Length of Zone, Link Type, Velocity Change Factor*: repeats;

Operand	Description	Default
Number	Link number [integer]	Sequential
Name	Link name [symbol name]	Blank
Beginning Intersection ID	Beginning intersection of link [integer intersection number or intersection name]	—
Beginning Direction	Direction of link as it leaves the beginning intersection [integer]	0
Ending Intersection ID	Ending intersection of link [integer intersection number or intersection name]	—

Operand	Description	Default
Ending Direction	Direction of link as it enters the ending intersection [integer]	*Beginning Direction*
Number of Zones	Number of zones in the link [integer]	1
Length of Zone	Length of each zone [integer]	0
Link Type	Link type [Unidirectional, Bidirectional, or Spur]	Unidirectional
Velocity Change Factor	Multiplier applied to the current velocity [constant]	1.0

Function: The LINKS element defines the characteristics of guided transporter track sections between intersection pairs *Beginning Intersection ID* and *Ending Intersection ID.*

NETWORKS: *Number, Name, Starting Link - Ending Link,* . . .:repeats;

Operand	Description	Default
Number	Network number [integer]	Sequential
Name	Network name [symbol name]	Blank
Starting Link	Starting link in range of links to include [integer link number or link symbol name]	—
Ending Link	Ending link in range of links to include [integer link number or link symbol name]	No range

Function: The NETWORKS element lists all links to be included when defining a system map that guided transporters follow.

NICKNAMES: *Nickname, Name* or *Constant*: repeats;

Operand	Description	Default
Nickname	User-defined symbol name	—
Name or Constant	SIMAN variable or attribute name, user-defined symbol, or constant	—

Function: The NICKNAMES element substitutes a user-defined symbol name *Nickname* for a SIMAN attribute or system variable, a previously defined symbol name, or a constant, *Name or Constant.*

OUTPUTS: *Number, SIMAN Expression, Output File, Report Label*: repeats;

Operand	Description	Default
Number	Output element number [integer]	Sequential
SIMAN Expression	SIMAN expression to be recorded	—
Output File	Output file to which observations are to be written at the end of each replication [integer number or system specific file name enclosed in double quotes]	No save
Report Label	Identifier for labeling the output in the output processor in the SIMAN summary report [alphanumeric]	*SIMAN Expression*

Function: The OUTPUTS element defines SIMAN expressions whose values are to be reported in the SIMAN Summary Report and optionally recorded in output files at the end of each replication of a simulation.

PARAMETERS: *Number, Name, Value*$_1$, ..., *Value*$_n$: repeats;

Operand	Description	Default
Number	Parameter set number [integer]	Sequential
Name	Parameter set name [symbol name]	Blank
Value$_i$	Parameters values [constant]	—

Function: The PARAMETERS element defines parameters associated with SIMAN random variables.

PROJECT, *Title, Analyst Name, Date, Summary Report*;

Operand	Description	Default
Title	Project title [alphanumeric]	Blank
Analyst Name	Analyst name [alphanumeric]	Blank
Date	Date in form or *month / day / year*	System specific
Summary Report	Create a summary report at the end of each replication [**Yes** or **No**]	Yes

Function: The PROJECT element is used to label the SIMAN summary report.

QUEUES: *Number, Name, Ranking Criterion(Rule Expression), Block Label*: repeats;

Operand	Description	Default
Number	Queue number [integer]	Sequential
Name	Queue name [symbol name]	Blank
Ranking Criterion	Ranking criterion [FIFO, LIFO, HVF(*Rule Expression*), or LVF(*Rule Expression*)]	FIFO
Rule Expression	Expression used in *Ranking Criterion*. Specified only when *Ranking Criterion* is set to HVF or LVF.	—
Block Label	Associated QUEUE block [block label or SHARED keyword]	Determined during run

Function: The QUEUES element specifies the total number of queues along with their name, ranking criterion, and associated model queue block.

RANKINGS: *Identifier, Ranking Criterion(Rule Expression)*: repeats;

Operand	Description	Default
Identifier	Queue identifier or lower limit of a queue range for which the ranking criterion applies [integer queue number, queue symbol name, or queue range]	—
Ranking Criterion	Ranking criterion [FIFO, LIFO, HVF(*Rule Expression*), or LVF(*Rule Expression*)]	—
Rule Expression	Expression used in *Ranking Criterion*. Specified only when *Ranking Criterion* is set to HVF or LVF.	—

Function: The RANKINGS element defines the order in which entities reside in queues.

RECIPES: *Name, Static Name = Value*: repeats;

Operand	Description	Default
Name	Recipe name [symbol name]	—

Operand	Description	Default
Static Name	Static name [symbol name]	—
Value	Value to assign to *Static Name* [expression]	—

Function: The RECIPES element allows grouping of statics variables (which have been defined on the STATICS element) together to form a procedure or set of instructions necessary to accomplish a process.

REDIRECTS: *Number, Name, Network ID, Beginning Intersection ID - Ending Intersection ID - Next Intersection ID, . . .*: repeats;

Operand	Description	Default
Number	Redirect number [integer]	Sequential
Name	Redirect name [symbol name]	Blank
Network ID	Network identifier [integer network number or network name]	—
Beginning Intersection ID	Beginning intersection ("from") [integer intersection number or intersection symbol name]	—
Ending Intersection ID	Ending intersection ("to") [integer intersection number or intersection symbol name]	—
Next Intersection ID	Intersection to go to next [integer intersection number or intersection symbol name]	—

Function: The REDIRECTS element overwrites the individual shortest-distance entry of the System Map from intersection *Beginning Intersection ID* to intersection *Ending Intersection ID* in network *Network ID* by forcing the transporter to move through intersection *Next Intersection ID* on its way to *Ending Intersection ID.*

REPLICATE, *Number of Replications, Beginning Time, Replication Length, Initialize System, Initialize Statistics, Warm Up Period*;

Operand	Description	Default
Number of Replications	Number of simulation replications to execute [integer]	1

Operand	Description	Default
Beginning Time	Beginning time for the first replication [constant]	0.0
Replication Length	Maximum length of each replication [constant]	Infinite
Initialize System	Initialize system status between replications [Yes or No]	Yes
Initialize Statistics	Discard previous observations between replications [Yes or No]	Yes
Warm Up Period	Warm-up time for system to reach steady state conditions [constant]	0.0

Function: The REPLICATE element specifies the number of simulation replications, the beginning time of the first replication, the maximum length of each replication, the type of initialization to be performed between replications, and the time needed to reach steady-state conditions.

RESOURCES: *Number, Name, Capacity Keyword (Capacity expression, Capacity Entity Rule), StateSetID–Initial State, FAILURE(Failure ID, Failure Entity Rule)*:. . .: repeats;

Operand	Description	Default
Number	Resource number [integer]	Sequential
Name	Resource name [unique symbol name]	—
Capacity Keyword	Keyword defining the resource capacity type [CAPACITY or SCHEDULE]	Capacity
Capacity Expression	Initial capacity [integer] if type is CAPACITY or *Schedule ID* [integer or schedule symbol name] if type is SCHEDULE	1
Capacity Entity Rule	Rule identifier [PREEMPT, IGNORE or WAIT]	Ignore
StateSetID	StateSet number or name that defines the states for this resource	No associated StateSet

Operand	Description	Default
Initial State	Initial state of resource [integer or state name]	Inactive or idle AutoState
Failure ID	Failure identifier [integer or failure symbol name]	No associated failures
Failure Entity Rule	Rule Identifier [PREEMPT, IGNORE or WAIT]	Ignore

Function: The RESOURCES element defines the characteristics of the resources, including the resource names, initial capacities, and related states and failures.

RULES: *Identifier, Selection Rule*: repeats;

Operand	Description	Default
Identifier	Rule number or name [integer or symbol name, or range]	—
Selection Rule	Queue, Resource, or Transporter selection rule	—

Function: The RULES element specifies the decision rules to be used by ALLOCATE, PICKQ, QPICK, REQUEST, and SELECT blocks.

SCHEDULES: *Identifier, Resource Capacity * Capacity Duration,* . . .: repeats;

Operand	Description	Default
Identifier	Schedule number or name [integer or symbol name]	—
Resource Capacity	Resource capacity [integer]	—
Capacity Duration	Length of time at capacity [expression]	Infinite

Function: The SCHEDULES element defines a time-dependent schedule of resource capacity changes.

SEEDS: *Identifier, Seed Value, Initialize Option*: repeats;

Operand	Description	Default
Identifier	Number or name of stream to initialize [integer or symbol name]	—

Operand	Description	Default
Seed Value	Initial seed value [integer]	Machine dependent
Initialize Option	Reinitialize stream between replications [Yes, No, Common, Antithetic]	No

Function: The SEEDS element defines seed values for random number streams and determines how the stream should be reinitialized between replications.

SEGMENTS: *Identifier, Beginning Station, Next Station - Length, ...*: repeats;

Operand	Description	Default
Identifier	Segment set number or name [integer or symbol name]	Sequential
Beginning Station	Beginning station identifier [integer station number or station symbol name]	—
Next Station	Next station identifier [integer station number or station symbol name]	—
Length	Distance from the previous station [integer]	—

Function: The SEGMENTS element defines the segments that make up the path of the conveyor.

SEQUENCES: *Number, Name, Station ID, Variable = Value & ...*: repeats;

Operand	Description	Default
Number	Sequence set number [integer]	Sequential
Name	Sequence set name [symbol name]	Blank
Station ID	Next station in visitation sequence [integer station number or station symbol name]	—

Operand	Description	Default
Variable	Attribute, variable or static to be assigned. *Value* upon entering the station.	—
Value	Value assigned to *Variable* upon entering the station.	—

Function: The SEQUENCES element defines an ordered set of stations and entity visits, as well as attribute and/or static assignments to be made at each of the stations in the sequence.

SETS: *Number, Name, Member, . . .*: repeats;

Operand	Description	Default
Number	Set number [integer]	Sequential
Name	Set name [symbol name]	—
Member	Resource to include as a member [resource symbol name or LowIndex..HighIndex]	—

Function: The SETS element defines a group of resources that may be referenced by way of a common name and a unique index.

STATESETS: *Number, Name, State Name(Stateset Type), . . .*: repeats;

Operand	Description	Default
Number	Stateset number [integer]	Sequential
Name	Stateset name [symbol name]	Blank
State Name	Name of state [symbol name]	Blank
Stateset Type	AutoState or *Failure ID*	—

Function: The STATESETS element defines statesets and their associated individual states.

STATICS: *Name = Default Value*: repeats;

Operand	Description	Default
Name	Static name [symbol name]	—
Default Value	Value to assign to static Name [expression]	0

Function: The STATICS element defines a series of location-dependent variables.

STATIONS: *Number, Name, Intersection ID, Recipe ID*: repeats;

Operand	Description	Default
Number	Station number [integer]	Sequential
Name	Station name [symbol name]	Blank
Intersection ID	Associated intersection [integer intersection number or intersection symbol name]	No intersection
Recipe ID	Associated recipe [integer recipe number or recipe symbol name]	No recipe

Function: The STATIONS element specifies the total number of stations, their names, and their associated intersection and/or recipes.

STORAGES: *Number, Name*: repeats;

Operand	Description	Default
Number	Storage number [integer]	Sequential
Name	Storage name [symbol name]	Blank

Function: The STORAGES element specifies the total number of storages and their names.

TABLES: *Number, Name, Low Value, Fixed Increment, Variable Value*$_1$, . . ., *Variable Value*$_n$: repeats;

Operand	Description	Default
Number	Table number [integer]	Sequential
Name	Table name [symbol name]	Blank
Low Value	Low value for the independent variable [constant]	0.01
Fixed Increment	Fixed increment between successive values of the independent variable [constant]	1.0
Variable Value$_i$	Dependent variable values corresponding to the independent variable [constants]	—

Function: The TABLES element defines the dependent values of a variable specified by the low value of the independent variable (*Low Value*), the fixed increment between successive values of the independent variable (*Fixed Increment*), and a list of the dependent values

(*Variable* $Value_i$) corresponding to each of the successive independent variable values.

TALLIES: *Number, Name, Output File*: repeats;

Operand	Description	Default
Number	Tally number [integer]	Sequential
Name	Tally name and identifier for labeling summary report [symbol name]	"Tally *Number*"
Output File	Output unit or filename for saving observations [integer unit number, or system-specific file name enclosed in double quotes]	No save

Function: The TALLIES element defines all information needed for TALLY blocks in a SIMAN model.

TRACE, *Beginning Time, Ending Time, Condition, Expression, ...*;

Operand	Description	Default
Beginning Time	Time after the beginning time of the simulation at which the trace is to begin [constant]	0.0
Ending Time	Time after the beginning of the simulation at which the trace is to end [constant]	Infinite
Condition	Trace condition [logical condition]	Trace ALL
Expression	SIMAN expression to display during trace	No values

Function: The TRACE element is used to generate a detailed trace report of entity processing.

TRANSPORTERS: *Number, Name, Number of Units, System Map Type(Map ID) - Type of Control, Velocity - Acceleration - Deceleration - Turning Velocity, Initial Position(Position ID, Zone) - Initial Status - Vehicle Size(Size Integer)*: repeats;

Operand	Description	Default
Number	Transporter name [integer]	Sequential
Name	Transporter name [symbol name]	—

Operand	Description	Default
Number of Units	Number of available units [integer]	1
System Map Type	Keyword defining the map type used for the transporter [DISTANCE or NETWORK]	DISTANCE
Map ID	Map ID defining the Distance Set ID or the Network Set ID	None
Control	Type of zone control [START, END, or *k*]	Start
Velocity	Transporter velocity [constant]	1.0
Acceleration	Transporter acceleration [constant]	0.0
Deceleration	Transporter acceleration [constant]	0.0
Turning Velocity	Turning velocity factor [constant]	1.0
Initial Position	Initial position of transporter Type [INTERSECTION, LINK, or STATION]	—
Position ID	Station, intersection, or link identifier corresponding to the initial position type [construct ID]	—
Zone	Zone number if initial position is on LINK [integer]	—
Initial Status	Initial status of transporter [Active or Inactive]	Active
Vehicle Size	Measure type of vehicle size [LENGTH or ZONE]	Zone
Size Integer	Value for initial vehicle size: Number of length units or zones [integer]	1

Function: The TRANSPORTERS element establishes characteristics of SIMAN transporters that can be of two general types, free-path or guided.

VARIABLES: *Number, Name(1-D Array Index, 2-D Array Index), Values, . . .*: repeats;

Operand	Description	Default
Number	Rate variable number (index into the V(*k*) array) [integer]	Sequential

Operand	Description	Default
Name	Rate variable name [symbol name]	Blank
1-D Array Index	First index into the named variable array	No array
2-D Array Index	Second index into the named variable array	No array
Value	Initial values for level variable at beginning of run [constant]	0.0 or last value

Function: The VARIABLES element specifies the total number of entity variables, their names, and, if desired, initial values.

Appendix C

Attributes, Variables, and Mathematical Functions

Attributes

*A(*Attribute Number [, Entity Number])*	Returns the value of attribute *Attribute Number* for entity *Entity Number*. Returns the value of attribute *Attribute Number* of the active entity when *Entity Number* is defaulted.
IS([Entity Number])*	Returns the sequence index number for entity *Entity Number*. Returns the sequence index number of the active entity when defaulted.
M([Entity Number])*	Returns the current station number for entity *Entity Number*. Returns the current station number of the active entity when defaulted.
NS([Entity Number])*	Returns the sequence set number for entity *Entity Number*. Returns the sequence set number of the active entity when defaulted.

Calendar Variables

FIRSTINCAL	Returns first *Entity Number* from the event calendar. May be 0.
NEXTINCAL *(Entity Number)*	Returns the next entity number following *Entity Number*. Nonassignable integer values are returned.

*Note: User assignable

Conveyor Variables

CLA*(Conveyor ID)*	Returns length of accumulated entities on conveyor *Conveyor ID*. CLA is based on entity size since accumulated entities do not occupy cells.
ICS*(Conveyor ID)*	Returns status of conveyor *Conveyor ID*. (0 = Idle; 1 = Moving; 2 = Blocked, nonaccumulating conveyors, only; 3 = Inactive).
LC*(Conveyor ID)*	Returns number of cells conveying on conveyor *Conveyor ID*.
LEC*(Conveyor ID)*	Returns length of entities on conveyor *Conveyor ID*. LEC is based on cell size.
MLC*(Conveyor ID)*	Returns length of conveyor *Conveyor ID*.
NEA*(Conveyor ID)*	Returns number of accumulated entities on conveyor *Conveyor ID*.
NEC*(Conveyor ID)*	Returns number of entities on conveyor *Conveyor ID*.
VC(Conveyor ID)*	Returns velocity of conveyor *Conveyor ID* (user assignable).

Defined Constructs (Number of)

MXARR	Arrivals
MXASM	Attribute symbols (named attributes)
MXATT	Attributes
MXBKG	Blockages
MXBLK	Blocks
MXCNT	Counters
MXCNV	Conveyors
MXDSB	Distributions
MXDST	DSTATS
MXENT	Entities
MXEXP	Expressions
MXFAL	Failures
MXFIL	Files

Defined Constructs (Number of)

MXFRQ	Frequencies
MXINX	Intersections
MXLNK	Links
MXNET	Networks
MXPAR	Parameter sets
MXQUE	Queues
MXREC	Recipes
MXRUL	Rules
MXSCH	Schedules
MXSEE	Seeds
MXSEQ	Sequences
MXSET	Sets
MXSTA	Stations
MXSTR	Storages
MXSTT	Statics
MXTAB	Tables
MXTAL	Tallies
MXTRN	Transporters
MXVAR	Variables
MXVSM	Variable symbols (named variables)
NMPAR	Values in parameter set

Frequency Variables

FAVG(*Frequency Number, Category)*	Returns average time for frequency *Frequency Number* in category *Category*. (FAVG = FRQTIM/FCOUNT)
FCATS(*Frequency Number)*	Returns number of categories for frequency *Frequency Number*.
FCOUNT(*Frequency Number)*	Returns number of observations for frequency *Frequency Number* in category *Category*.
*FHILIM(*Frequency Number, Category)*	Returns upper limit for frequency *Frequency Number* in category *Category*.

Frequency Variables

FLOLIM(Frequency Number, Category)*	Returns lower limit for frequency *Frequency Number* in category *Category*.
FRQTIM*(Frequency Number, Category)*	Returns time for frequency Frequency Number in category *Category*.
FTOT*(Frequency Number)*	Returns total frequency time for frequency *Frequency Number*.
FTOTR*(Frequency Number)*	Returns restricted frequency time for frequency *Frequency Number*.

Functions

ED*(Expression Number)*	Returns a value from expression *Expression Number* as defined in the EXPRESSIONS element.

Group Variables (see Comments)

AG*(Rank, Attribute Number)*	Returns the value of general purpose attribute *Attribute Number* for the entity at rank *Rank* of the group associated with the current entity.
ENTINGROUP*(Rank, [,Entity Number])*	Returns the entity number (i.e., IDENT value) for entity at rank *Rank* in group of representative *Entity Number*. If *Entity Number* is defaulted, the active entity's group is utilized.
ISG*(Rank)*	Returns the value of sequence index attribute IS for the entity at rank *Rank* of the group associated with the active entity's group.
MG*(Rank)*	Returns the value of station attribute M for entity at rank *Rank* of the group associated with the active entity.
NG*[(Entity Number)]*	Returns the number of entities in the group associated with *Entity Number*. Returns the size of the active entity's group if *Entity Number* is defaulted.

Group Variables (see Comments)

NSG*(Rank)*	Returns the value of sequence set attribute NS for the entity at *Rank* rank.
SAG*(Attribute Number)*	Returns the sum of values of the specified attribute *Attribute Number* of all members of the active entity's group.

Miscellaneous

ATTR(Attribute Number, [Index 1, Index 2])*	Returns the value of attribute given by *Attribute Number* with associated indices *Index 1* and *Index 2*. Returns value of attribute given by *Attribute Number* if indices are defaulted.
EXPR*(Expression Number, [Index 1, Index 2])*	Returns the value of expression given by *Expression Number* with associated indices *Index 1* and *Index 2*. Returns the value of expression given by *Expression Number* if indices *Index 1* and *Index 2* are defaulted.
IDENT	The entity's identifier as it appears in the trace. These numbers are reused as entities are disposed.
*J	Returns general purpose integer variable.
NB*(Blockage Number)*	Returns the value of blockage given by *Blockage Number*.
NSYM*(Symbol Name)*	Returns the number associated with the symbol name *Attribute Name*. See comments.
NUMBLK*(Block Label)*	Returns the block number given by *Block Label*.
NUMENT	Returns the current number of entities active in the model.
P(Parameter Set ID, Parameter Number)*	Returns the value of parameter *Parameter Number* in set *Parameter Set ID*.
V(Variable Number)*	Returns the current value of the specified *Variable Number*.

Miscellaneous

*VAR(*Variable Number, [Index 1, Index 2]*)	Returns the value of a variable given by *Variable Number* with associated indices *Index 1* and *Index 2*. Returns the value of a variable given by *Variable Number* if indices *Index 1* and *Index 2* are defaulted.
*X(*Index*)	Returns general purpose global variable *Index*.

Queue Variables (see Comments)

AQUE(*Queue ID, Rank, Attribute Number*)	Returns the value of attribute *Attribute Number* of the entity ranked at rank *Rank* in queue *Queue ID*.
ENTATRANK(*Rank, Queue ID*)	Returns the entity number (i.e., IDENT value) of entity in queue *Queue ID* at rank *Rank*.
FIRSTINQUE(*Queue ID*)	Returns the entity number (i.e., IDENT value) of first entity in queue *Queue ID*.
ISQUE(*Queue ID, Rank*)	Returns the value of attribute IS of the entity at rank *Rank* in queue *Queue ID*.
LASTINQUE(*Queue ID*)	Returns the entity number (i.e., IDENT value) of last entity in queue *Queue ID*.
MQUE(*Queue ID, Rank*)	Returns the value of attribute M of the entity at rank *Rank* in queue *Queue ID*.
NQ(*Queue ID*)	Number of entities in queue *Queue ID*.
PREDECESSOR(*Entity Number*)	Returns the entity number of entity preceding queued entity (i.e., IDENT value). Returns 0 if no predecessor.
SAQUE(*Queue ID, Attribute Number*)	Returns the sum of attributes given by *Attribute Number* of all entities in queue *Queue ID*.
SUCCESSOR(*Entity Number*)	Returns entity number for entity following queued entity (i.e., IDENT value). Returns 0 if entity is last in queue.

Resource Variables

BUSY_RES	Used in an expression to check if a resource is currently in the busy state.
FAILED_RES	Used in an expression to check if a resource is currently in the failed state.
IDLE_RES	Used in a expression to check if a resource is currently in the idle state.
INACTIVE_RES	Used in an expression to check if a resource is currently in the inactive state.
IRF(*Resource ID*)	Returns a number greater than 0 if *Resource ID* is currently failed and returns 0 if not currently failed.
MR(Resource ID)*	Returns capacity of resource *Resource ID*.
NR*(Resource ID)*	Number of busy units of resource *Resource ID*.
STATE(Resource ID)*	Returns current state of resource *Resource ID* as defined in the STATESETS element.

Set Information

MEMBER*(Set ID, Index)*	Returns construct (e.g., resource) number for *Index* (order) in set *Set ID*. Set member.
MEMIDX*(Set ID, Member ID)*	Returns *Index* (order) of construct *Member ID* in set *Set ID*.
NUMMEM*(Set ID)*	Returns number of constructs in set *Set ID*.

Simulation Replication Variables

MREP	Returns the *Number of Replications* value in the REPLICATE element.
NREP	Returns the current replication number.
*TFIN	Returns the ending time scheduled for the current replication.
TNOW	Returns the current simulation time.

Station Variables

MSQ*(Sequence ID, Sequence Index)*	Returns the station to be visited in sequence set number *Sequence ID* with index number *Sequence Index.*
NE*(Station ID)*	Returns the number of entities in transit to station given by *Station ID.*

Statistics Collection Variables

DAVG*(DSTAT ID)*	Returns the current discrete time weighted average value of *DSTAT ID.*
DMAX*(DSTAT ID)*	Returns the current maximum value of DSTAT *DSTAT ID.*
DMIN*(DSTAT ID)*	Returns the current minimum value of DSTAT *DSTAT ID.*
DSTD*(DSTAT ID)*	Returns the current standard deviation of DSTAT *DSTAT ID.*
DTPD*(DSTAT ID)*	Returns the current time period of DSTAT *DSTAT ID.*
MC(Counter ID)*	Returns the maximum limit for counter *Counter ID.*
NC*(Counter ID)*	Returns the current count of counter *Counter ID.*
TAVG*(Tally ID)*	Returns the current average value of tally *Tally ID.*
TMAX*(Tally ID)*	Returns the current maximum value of tally *Tally ID.*
TMIN*(Tally ID)*	Returns the current minimum value of tally *Tally ID.*
TNUM*(Tally ID)*	Returns the current number of observations of tally *Tally ID.*
TATD*(Tally ID)*	Returns the current standard deviation of tally *Tally ID.*

Storage Variable

NSTO*(Storage ID)*	Returns the number of entities in storage *Storage ID.*

Table Look-up

TF*(Table ID, X Value)*	Returns the value of the dependent variable from table *Table ID* given the independent variable *X Value.*

Transporter Variables (General Status)

IT(Transporter ID, Unit Number)*	Returns the status of transporter unit *Unit Number* of set *Transporter ID*. (0 = idle, 1 = busy, 2 = inactive). IT is user assignable for values 0 and 2, only.
MT*(Transporter ID)*	Returns the number of active units of transporter set *Transporter ID*.
NT*(Transporter ID)*	Number of busy transporter units of set *Transporter ID*.
VT(Transporter ID)*	Velocity of transporter set *Transporter ID*.
VTU(Transporter ID, Unit Number)*	Velocity of transporter unit *Unit Number* of transporter set *Transporter ID*.

Transporter Variables (Free-Path)

ID*(Transporter ID, Station ID)*	Returns distance from active entity's location M to destination *Station ID* in the distance set followed by the specified *Transporter ID*.
IDIST*(Distance Set ID, Beginning Station ID, Ending Station ID)*	Returns distance from distance set *Distance Set ID* between stations *Beginning Station ID* and *Ending Station ID*.
LT*(Transporter ID, Unit Number)*	Returns current location of transporter unit *Unit Number* of set *Transporter ID*.

Transporter Variables (Guided)

ACC*(Transporter ID)*	Returns acceleration of transporters in set *Transporter ID*.
DEC*(Transporter ID)*	Returns deceleration of transporters in set *Transporter ID*.
ISZT*(Transporter ID, Unit Number)*	Returns size type of transporter unit *Unit Number* of set *Transporter ID*, (1 = ZONE, 2 = LENGTH). See NSZT, also.
LDL*(Transporter ID, Unit Number)*	Returns current destination link of transporter unit *Unit Number* of set *Transporter ID*. LDL returns 0 if the current destination is an intersection.

Transporter Variables (Guided)

LDX*(Transporter ID, Unit Number)*	Returns current destination intersection of transporter unit *Unit Number* of set *Transporter ID*. LDX returns 0 if the current destination is a link.
LDZ*(Transporter ID, Unit Number)*	Returns current destination zone of transporter unit *Unit Number* of set *Transporter ID*.
LT*(Transporter ID, Unit Number)*	Returns current intersection of transporter unit *Unit Number* of set *Transporter ID*.
LTL*(Transporter ID, Unit Number)*	Returns current link location of transporter unit *Unit Number* of set *Transporter ID*. If the transporter is in an intersection, LTL returns 0.
LTZ*(Transporter ID, Unit Number)*	Returns current zone number of transporter unit *Unit Number* of set *Transporter ID*. If the transporter is in an intersection, LTZ returns 0.
NSZT*(Transporter ID, Unit Number)*	Returns current size value of transporter unit *Unit Number* of set *Transporter ID*. See ISZT, also.
TAZ*(Transporter ID, Unit Number)*	Returns the simulation time when transporter unit *Unit Number* of set *Transporter ID* arrived at its current location.
TVF*(Transporter ID)*	Returns turning velocity factor of transporter set *Transporter ID*.
TWZ*(Transporter ID, Unit Number)*	Returns the accumulated time that transporter unit *Unit Number* of set *Transporter ID* has been waiting for access to zones while traveling.

Transporter Variables (Guided Network)

IDSNET*(Network ID, Beginning Intersection ID, Ending Intersection ID)*	Returns the length of travel between intersections *Beginning Intersection ID* and *Ending Intersection ID* in network *Network ID*.
LENZ*(Link ID)*	Returns zone length of zones in link *Link ID*.

Transporter Variables (Guided Network)

LNKNUM*(Beginning Intersection ID, Ending Intersection ID)*	Returns the link number connecting intersections *Beginning Intersection ID* and *Ending Intersection ID* (0 if the links are not connected).
LTYP*(Link ID)*	Returns link type *Link ID*. (1 = unidirectional, 2 = bidirectional, 3 = spur).
LX*(Intersection ID)*	Returns length of intersection *Intersection ID*.
MZ*(Link ID)*	Returns number of zones in link *Link ID*.
NDX*(Link ID)*	Returns destination intersection of vehicles on link *Link ID*.
NEXTX*(Network ID, Current Intersection ID, Destination Intersection ID)*	Returns the next intersection of travel between the intersection *Current Intersection ID* and intersection *Destination Intersection ID* in network *Network ID*.
NL*(Link ID)*	Returns number of occupied zones in link *Link ID*.
NX*(Intersection ID)*	Returns status of intersection *Intersection ID*. (0 = empty, 1 = occupied)
NXB*(Link ID)*	Returns *Beginning Intersection ID* of link *Link ID*.
NXE*(Link ID)*	Returns *Ending Intersection ID* of link *Link ID*.
NZ*(Link ID, Zone ID)*	Status of zone *Zone ID* of link *Link ID*. (1 = occupied, 0 = empty, 1 = captured)
VL*(Link ID)*	Link velocity factor of link *Link ID*.
VX*(Link ID)*	Velocity factor of intersection *Link ID*.

Comments

All operands except *Symbol Name* on NSYM*(Symbol Name)* accept expressions. An expression can be an integer, a real number, a mathematical equation, or a symbol name. When used with the MARK modifier and the INTERVAL keyword from the TALLY block, named attributes and variables return their value when entered in an ex-

Comments
pression. All other symbol names return their number. To obtain an attribute or number, use the NSYM*(Symbol Name)* function. For example, if the *ATTRIBUTES* element is *ATTRIBUTES: Job Type, 15;,* the symbol name *JobType* returns its initial value of 15 until reassigned, but NSYM*(JobType)* always returns the *Symbol Number 1*. Care should be exercised when entering expressions containing symbol names for the *Attribute Number* operand of GROUP and QUEUE variables. *Attribute Number* represents the attribute number of the entity in the group or queue that the current entity wishes to access. Use NSYM*(Symbol Name)* for a named attribute's *Attribute Number*.

Mathematical Functions	Description
ABS*(Argument)*	Absolute value
ACOS*(Argument)*	Arccosine
AINT*(Argument)*	Truncate
AMOD*(Argument 1, Argument 2)*	Real remainder, returns $(A_1 - (AINT(A_1/A_2)*A_2))$ A_1 = $Argument_1$, A_2 = $Argument_2$
ANINT*(Argument)*	Round to nearest integer
ASIN*(Argument)*	Arcsine
ATAN*(Argument)*	Arctangent
COS*(Argument)*	Cosine
EP*(Argument)*	Exponentiate
HCOS*(Argument)*	Hyperbolic cosine
HSIN*(Argument)*	Hyperbolic sine
HTAN*(Argument)*	Hyperbolic tangent
MN*($Argument_1$, $Argument_2$, ...)*	Minimum value
MOD*($Argument_1$, $Argument_2$, ...)*	Integer remainder, same as AMOD except that the arguments are truncated to integer values first
MX*($Argument_1$, $Argument_2$, ...)*	Maximum value
LN*(Argument)*	Natural logarithm
LOG*(Argument)*	Common logarithm
SIN*(Argument)*	Sine
SQRT*(Argument)*	Square root
TAN*(Argument)*	Tangent

Appendix D

Distribution Definitions

SIMAN provides 13 random variable distributions and three special purpose functions, RA, CO, and ED. These 13 random variable distributions can be accessed two different ways from a SIMAN model. The four-letter function requires the parameter list enclosed in parentheses following the distribution name. Four-letter functions may also be specified in the model or experiment using the full distribution name. The two-letter function has its parameter list stored in the PARAMETERS element, and the operand *IP* specifies which parameter set to use.

The four-letter functions provide a convenient, clear method of entering distributions in the model and do not require the PARAMETERS element. Four-letter functions are useful when using the distribution in a limited number of blocks. The two-letter functions reference the PARAMETERS element. The use of the EXPRESSIONS element, as explained shortly, often makes the two-letter functions and PARAMETERS element unnecessary.

The three special purpose functions, RA, CO, and ED only use two-letter abbreviations. The RA(*IST*) function returns a random number between 0 and 1 using random number stream IST, ED(k) returns a value from expression number k as defined in the EXPRESSIONS element, and CO(*IP*) returns the Constant value stored in parameter set *IP*. The EXPRESSIONS element provides a convenient method for defining expressions and distributions in the experiment. With the EXPRESSIONS element, any expression may be defined including the four-letter random functions, which allows defining the distribution and its parameters in the experiment without constructing a PARAMETERS element.

The operands P_k and V_k represent values entered with four-letter functions or values stored in the PARAMETERS element for two-letter functions. These values define the characteristics of the distributions.

Operand *IST*, the random number stream, can be specified as a number between 1 and 10. IF *IST* is omitted, SIMAN uses stream 10.

Four-Letter Function	Two-Letter Function	Parameters
BETA(P_1, P_2, IST) (Beta)	BE(IP, IST)	P_1 = shape parameter, α_1 P_2 = shape parameter, α_2
Constant value	CO(IP)	P_1 = constant value
CONT(P_1, V_1, P_2, $V_2, \ldots$, IST) (Empirical Continuous Probability)	CP (IP, IST)	P_k, k = 1, 2, . . . denote associated cumulative probabilities V_k, k = 1, 2, . . . denote values of the random variable
DISC(P_1, V_1, P_2, V_2, . . . , IST) (Empirical Discrete Probability)	DP(IP, IST)	P_k, k = 1, 2, . . . denote associated cumulative probabilities V_k, k = 1, 2, . . . denote values of the random variable
ERLA(P_1, P_2, IST) (Erlang)	ER(IP, IST)	P_1 = exponential mean, $1/\lambda$ P_2 = Erlang order k
EXPO(P_1, IST) (Exponential)	EX(IP, IST)	P_1 = mean, $1/\lambda$
Expression Definition	ED(k)	k = index into EXPRESSIONS
GAMM(P_1, P_2, IST) (Gamma)	GA(IP, IST)	P_1 = scale parameter, β P_2 = shape parameter, α
JOHN(P_1, P_2, P_3, P_4, IST) (Johnson)	JO(IP, IST)	P_1 = shape parameter 1, γ P_2 = shape parameter 2, $\delta*$ P_3 = scale parameter, λ P_4 = location parameter, ξ
LOGN(P_1, P_2, IST) (Lognormal)	RL(IP, IST)	P_1 = lognormal mean P_2 = lognormal standard deviation
NORM(P_1, P_2, IST) (Normal)	RN(IP, IST)	P_1 = mean, μ P_2 = standard deviation, σ
POIS(P_1, IST) (Poisson)	PO(IP, IST)	P_1 = mean, α

Four-Letter Function	Two-Letter Function	Parameters
Random number between 0 and 1	RA(IST)	
TRIA(P_1, P_2, P_3, IST) (Triangular)	TR(IP, IST)	P_1 = minimum P_2 = mode P_3 = maximum
UNIF(P_1, P_2, IST) (Uniform)	UN(IP, IST)	P_1 = minimum P_2 = maximum
WEIB(P_1, P_2, IST) (Weibull)	WE(IP, IST)	P_1 = scale parameter, β P_2 = shape parameter, α

*Note: P_2 is a positive number. To specify the unbounded Johnson distribution, specify $-P_2$. For example, JOHN(4, 2, 6, 15) specifies a bounded Johnson distribution, but JOHN(4, −2, 6, 15) specifies an unbounded Johnson distribution.

Index